INTRODUCTION TO SIMULATION AND RISK ANALYSIS

SECOND EDITION

INTRODUCTION TO SIMULATION AND RISK ANALYSIS

James R. Evans

University of Cincinnati

David L. Olson

University of Nebraska

Prentice
Hall

Upper Saddle River, New Jersey 07458

Library of Congress Cataloging-in-Publication Data

Evans, James R. (James Robert)
    Introduction to simulation and risk analysis / James R. Evans, David L. Olson.— 2nd ed.
      p. cm.
    Includes bibliographical references and index.
    ISBN 0-13-032928-2
    1. Decision making—Computer simulation.   2. Decision support systems.
   3. Business—Computer simulation.   4. Electronic spreadsheets.   5. Crystal ball
(Computer file) I. Olson, David Louis. II. Title.
    HD30.213 .E93 2002
    658.4′03′002855369—dc21   2001021777

Executive Editor: Tom Tucker
Editor-in-Chief: P. J. Boardman
Assistant Editor: Nancy Welcher
Editorial Assistant: Virginia Sheridan
Media Project Manager: Cindy Harford
Senior Marketing Manager: Debbie Clare
Marketing Editor (Production): John Roberts
Production Editor: Maureen Wilson
Permissions Coordinator: Suzanne Grappi
Associate Director, Manufacturing: Vincent Scelta
Manufacturing Buyer: Diane Peirano
Cover Design: Bruce Kenselaar
Full-Service Project Management: BookMasters, Inc.
Full-Service Project Manager: Sharon Anderson
Printer/Binder: Hamilton Printing Co.

Credits and acknowledgments borrowed from other sources and reproduced, with permission, in this textbook appear on appropriate page within text

Microsoft Excel, Solver, and Windows are registered trademarks of Microsoft Corporation in the U.S.A. and other countries. Screen shots and icons reprinted with permission from the Microsoft Corporation. This book is not sponsored or endorsed by or affiliated with Microsoft Corporation.

10 9 8 7 6 5 4 3 2 1
ISBN 0-13-032928-2

Brief Contents

Contents

Preface

The purpose of this book is to provide an introduction to the concepts, methodologies, and applications of simulation in business. One of the difficulties in teaching and learning simulation is the choice of a software platform. There exists a wide variety of excellent simulation software packages and languages, most of which require significant start-up on the part of the student (and in many cases, the instructor as well). A course can easily deteriorate into a language course, and students can easily lose sight of the basic concepts and principles. Moreover, most packages are generally oriented toward engineering applications and are not suitable for applications of risk analysis, a subject that is becoming increasingly used in practical business applications. We avoid these problems by using spreadsheets as the principal means to illustrate simulation modeling concepts, computational issues, and analysis of results to provide a foundation for learning more powerful simulation software.

Spreadsheets provide the ideal environment with which to introduce simulation to business students. First, spreadsheets are nearly as common as calculators and provide a way to convey quantitative methodologies in a language that business students can most easily understand. Second, spreadsheets allow one to address the elementary concepts of both risk analysis and systems simulation approaches in a common framework. For these reasons, spreadsheets are used in this book as the foundation for conveying basic principles about simulation models and allowing students considerable hands-on experience with minimal frustration. With such a foundation, the advanced student can more easily learn to use commercial simulation software. However, beyond the basics, we use two commercial packages, Crystal Ball for risk analysis and ProcessModel—new to the second edition—for systems simulation to illustrate more complex and robust applications.

This book is aimed at upper-level undergraduate and beginning graduate students in business administration and related disciplines. Microsoft Excel is used exclusively throughout the book, although most models can easily be translated into other spreadsheet formats. The book is logically divided into four parts. Part 1 consists of three chapters that provide the basic concepts of simulation. Chapter 1 describes the nature of simulation models, provides examples of pure Monte-Carlo (repeated sampling) approaches, and introduces the concept of systems (time/event driven) simulation. The simulation process and benefits and limitations of simulation are also discussed. Chapter 2 describes how to implement simple simulation models on Excel spreadsheets. Methods for generating probabilistic outcomes and performing simple Monte-Carlo simulations are also introduced. The appendix to Chapter 2 presents optional material about random number generation techniques. Chapter 3 focuses on probability and statistics in simulation. It provides a comprehensive review of statistical concepts and methods important in simulation analysis, probability distributions commonly used in simulation, issues related to modeling probabilistic inputs, random variates and their generation, and statistical issues of analyzing the output from Monte-Carlo simulations. We assume that students will have had at least a basic course in business statistics.

Part 2 consists of two chapters that focus exclusively on risk analysis. The Excel add-in, Crystal Ball, is introduced as a practical method for Monte-Carlo simulation. It is used throughout the book, and a time-limited version of the full software is included with the book. Chapter 4 provides a comprehensive overview of Crystal Ball as well as an original application developed by Cinergy Corporation, a major Midwest gas and electric utility. Chapter 5 presents a variety of applications in operations management, finance, and marketing. These examples show the variety of uses of Monte-Carlo simulation as well as the flexibility of Crystal Ball in addressing risk.

Part 3 consists of four chapters that deal with systems simulation. In Chapter 6, we describe the fundamentals of simulating inventory and queueing systems. This chapter includes a review of essential analytical models; simulation model development from a process view, activity-scanning view, and event-driven view; spreadsheet implementation; and continuous simulation modeling. Chapter 7 discusses output analysis and experimentation in systems simulation, including issues of transient behavior, statistical methods for comparing different systems, and experimental design in simulation. Chapter 8 provides an introduction to ProcessModel, an easy-to-use, yet powerful software package for systems simulation. Chapter 9 presents additional applications using ProcessModel in operations scheduling, information systems, and medicine.

Part 4 consists of the concluding chapter of the book. Chapter 10 discusses simulation in forecasting and optimization, using companion products to Crystal Ball—CB Predictor and OptQuest—to illustrate the approaches. The chapter also includes elementary introductions to time series forecasting and optimization modeling.

Several features have been designed into this book to improve pedagogy. First, cell formulas and detailed explanations are presented for most spreadsheet models. Second, each chapter has at least one or more "Simulation in Practice" feature that describes real applications of simulation in various businesses. Finally, each chapter has numerous questions and problems that provide a means of review of important concepts and allow students to work with and extend models in the chapter or apply the concepts to new situations.

New in the Second Edition

We have taken considerable time in revising this edition to provide more in-depth coverage of key topics and to make the book appeal to a broader audience. Specific changes include:

- Expanded presentation of basic Excel skills needed to understand and implement the examples in the text (Chapter 2)
- "SkillBuilder Exercises" throughout the chapters, designed to provide hands-on exercises for the student to practice and develop spreadsheet and software application skills
- Full (time restricted) student version of Crystal Ball, a powerful Excel add-in for Monte-Carlo simulation
- Full (time restricted) student version of ProcessModel simulation software for systems simulation
- New "Simulation in Practice" cases and revised problems and exercises throughout the book
- Many new examples and illustrations of simulation models, including casino games to convey elementary simulation modeling and analysis principles and more sophisticated financial models and applications

- Expanded coverage of basic statistical methods important in simulation (Chapter 3)
- New and expanded discussion of risk analysis concepts (Chapter 4)
- New chapter on simulation with ProcessModel (Chapter 8)
- New chapter on simulation in forecasting and optimization (Chapter 10)

Supplemental Material

This edition includes an Instructor's Solutions Manual.

Acknowledgments

We express our appreciation to the reviewers of the first edition manuscript: Vaidyanathan Jayaraman, University of Southern Mississippi; Ralph Badinelli, Virginia Tech; Arnold Buss, U.S. Naval Academy, Postgraduate School; and Linda Friedman, Baruch College, CUNY. In addition, we would like to thank the following individuals for many insightful comments and suggestions that have guided us in this revision:

William C. Giauque, Brigham Young University
William V. Harper, Otterbein College
Armann Ingolfsson, University of Alberta
Kellie Keeling, Virginia Tech
Scott Malcolm, University of Delaware
Susan Palocsay, James Madison University

and to Professor Xixi Hong of Xiamen University, China, whose careful reading of the first edition revealed several errors and inconsistencies that we have corrected.

A special note of thanks goes to Eric Wainwright, Mike Nagel, and Terry Hardy of Decisioneering, Inc., for their cooperation in providing the student version of Crystal Ball, comments on the manuscript, and permission to use Decisioneering's models and material from user manuals; and also to Mathew Greenfield and Tony Aust of Process-Model, Inc., for their support in providing the student version of ProcessModel and assistance in developing the manuscript and examples. Last, but certainly not least, we wish to thank our editor, Tom Tucker, the editorial staff at Prentice Hall, and Book-Masters, Inc., for their outstanding support and assistance.

James R. Evans, University of Cincinnati
David L. Olson, University of Nebraska

Introduction to Simulation

Chapter Outline

Management scientists use a wide variety of tools and techniques to model, analyze, and solve complex decision problems. These tools—many of which you may be familiar with at a basic level—include statistics, linear programming, decision analysis, queueing (waiting line) theory, forecasting, and simulation. Many of these tools often require the analyst to make some highly simplifying model assumptions. Linear programming, for example, applies to well-structured situations that can be modeled with a linear objective function and linear constraints, and do not include any probabilistic elements. Furthermore, we typically assume that all data are known with certainty. Unfortunately, this is seldom true in practice. For instance, market uncertainties relative to competition may make predicting unit profit very difficult; the rate at which resources are consumed may vary; availability of resources from suppliers may not be assured; and clearly, demand is almost always uncertain.

Another example is queueing theory. Basic analytical queueing models assume that the number of arrivals within a period of time follows a Poisson probability distribution, that service times are exponentially distributed, and that these distributions do not change over time. These assumptions lead to simple and elegant mathematical solutions for the long-term expected waiting time or the expected number in the queue. However, they provide little information about the actual short-term *dynamic behavior* of the systems they model; that is, the changes that we would observe in the system over time. For instance, if we were to view bank security videotapes, we would probably see little activity when the bank opens. The first few customers will get served immediately and not have to wait. However, as more customers arrive, we would probably observe some waiting lines. Analytical models generally cannot answer such questions as: How long will it take until five customers are waiting? or How will the system react if the rate of arrivals increases during lunchtime? We are by no means implying that linear programming or queueing models are not useful; indeed, they have proven to be very effective techniques for many business problems. However, they may not always be able to capture important elements of real situations that may have a significant impact on the model results or the decisions made from them.

For situations in which a problem does not meet the assumptions required by standard analytical modeling approaches, simulation can be a valuable approach to modeling and solving the problem. **Simulation** is the process of building a mathematical or logical model of a system or a decision problem, and experimenting with the model (usually with a computer) to obtain insight into the system's behavior or to assist in solving the decision problem. The two key elements of this definition are *model* and *experiment*. The principal advantage of simulation lies in its ability to model any appropriate assumptions about a problem or system, making it the most flexible management science tool available. Of course, it may take considerable time and effort to develop the model, particularly for very complex models or systems. Throughout this book, we will see many examples of different types of simulation models. However, a model is worthless unless it provides some insight to the user. Thus, a major focus of simulation is conducting experiments with the model and analyzing the results. This requires some basic knowledge of statistics, which we review in Chapter 3 and use throughout the book.

Simulation is particularly useful when problems exhibit significant uncertainty, which generally is quite difficult to deal with analytically. One situation that most people think about is whether or not they will have enough money for retirement (trust us—you will be thinking about this sooner than you think!).[1] Most "financial calculators" available on many Web sites make assumptions that the average annual rate of return is constant and base their calculations accordingly. For example, if you have $100,000 to invest for 20 years and assume a 9 percent rate of return, simple calculations will show that you will end up with $560,441. What they do not tell you, however, is the *likelihood* of achieving that amount, given uncertain fluctuations in the stock market, interest rates, inflation, and so on. What if you were told that there was a 40 percent chance that you would accumulate less than $230,000? Might that change your investment decisions or savings habits? Traditional financial models do not incorporate variability in their assumptions or provide statistical output information. Simulation models, however, can do this easily. Investment firms like T. Rowe Price are now using such simulation models as part of their Retirement Income Manager[SM] analysis to advise clients.[2]

Surveys of management science practitioners showed that simulation and statistics have the highest rate of application over all other tools by over a 2-to-1 margin.[3] Simulation usage was even higher than statistics. We do warn you, however, that simulation should not be used indiscriminately in place of sound analytical models. Many situations exist when approaches such as linear programming or queueing theory are more appropriate. The task of the modeler is to understand the pros and cons of different approaches and use them appropriately.

In this chapter, we

1. Discuss the nature of simulation models and their role in management science.
2. Illustrate basic concepts of simulation by presenting some simple examples.
3. Discuss some benefits and limitations of simulation as a problem-solving tool.

---

[1] Jeff D. Opdyke, "Will My Nest Egg Last?" *The Wall Street Journal Interactive Edition,* June 5, 2000.
[2] Ed McCarthy, "Monte-Carlo Simulation: Still Stuck in Low Gear," *Journal of Financial Planning,* Vol. 13, no. 1, January 2000, pp. 54–60.
[3] Linda Leon, Zbigniew Przasnyski, and Kala Chand Seal, "Spreadsheets and OR/MS Models: An End-User Perspective," *Interfaces,* Vol. 26, no. 2, March–April 1996, pp. 92–104.

The Nature of Simulation

Throughout history, simulation has been used for analyzing systems and decision problems. The Prussian army used to simulate wars by holding field exercises, making soldiers march through the woods of central Europe in all forms of wind and weather at the whims of their general staff (a tradition maintained by armies throughout the world today). Troop leaders would maneuver their units over the countryside to simulate war, with officers serving as umpires to interpret outcomes.

We often see examples of simulation all around us. Simulation is used to forecast the weather and develop the graphical (and now 3-dimensional) weather maps we see on television. Airplane pilots use simulators that show how an airplane would react in specific conditions to pilot actions. This helps the pilot learn how to cope with emergency situations without the need for traumatic experience. The movie *Apollo 13* illustrated the use of simulation to train astronauts as well as to solve the problem of finding the best "power up" sequence within the electrical system limitations that the astronauts faced before their critical reentry. (In fact, the launch sequence in the movie was in itself a simulation.) NASA also uses simulations to predict the rocket and satellite trajectories. Simulation can even be observed in popular board games. For example, "Monopoly" simulates the old Atlantic City real estate market, using dice as a means to identify random events. Several companies even sell simulations of major league baseball, which is one of the most statistically analyzed social environments in existence.

Manual simulations, such as moving troops through the field, operating a space shuttle simulator, or playing a board game, are very time-consuming. People tire easily, and one or two simulated outcomes provide very little information on which to base a decision. However, if the simulation can be implemented on a computer, thousands of simulated outcomes can be processed in a matter of seconds, providing a wealth of information. With the increasing availability of faster and more powerful computers, and better understanding of quantitative modeling because of extensive spreadsheet usage, simulation has become a very popular approach in recent years among practicing managers for the analysis of business problems.

Today, simulation is widely accepted in the world of business to predict, to explain, to train, and to help identify optimal solutions. Simulation is used extensively in manufacturing to model production and assembly operations; develop realistic production schedules; study inventory policies; analyze reliability, quality, and equipment replacement problems; and design material handling and logistics systems. It is used in designing and evaluating computer and communication networks and scheduling resources in complex projects. Simulation also finds extensive application in both profit-seeking service firms such as financial and retail organizations and in nonprofit service organizations such as healthcare, government, and education. These applications might involve studying customer waiting-line behavior, evaluating surgical schedules, designing efficient work flows in offices, determining scheduling policies in the criminal justice system, or evaluating deployment of police or fire personnel. For instance, simulation models can be used by a bank to help identify the number of tellers required to maintain a specified level of customer service as measured by waiting time or line length. Rather recently, due to increased availability and power of personal computers, simulation capabilities have been linked with spreadsheets to allow managers to evaluate risks of financial investment, retirement plans, marketing, real estate, and other common types of business decisions.

MODELS AND SIMULATION

Simulation, like all management science approaches, revolves around models. A **model** is an abstraction or representation of a real system, idea, or object. Models in management science take many different forms. Some models are *prescriptive;* that is, they determine an optimal policy. Linear programming models are prescriptive because the solution to a linear program suggests the best course of action that a decision maker should take. Other models are *descriptive;* they simply describe relationships and provide information for evaluation. Queueing models, which provide measures of system performance such as the average number in the queue and the average waiting time, are an example of descriptive models. Descriptive models are used to explain the behavior of systems, to predict future events as inputs to planning processes, and to assist decision makers in choosing the best solution or systems design. For example, a model of a factory operation can help production managers understand why bottlenecks occur. It can also be used to predict factory output as resources are added or changed in the system. By experimenting with different factory configurations, descriptive models can assist managers in selecting the best design to meet system objectives.

Models may also be *deterministic* or *probabilistic.* In a deterministic model, all data are known, or assumed to be known, with certainty. For example, the rate of return used by "financial calculators" for retirement planning that we discussed earlier is assumed to be constant each year. In a probabilistic model, some data (such as the rate of return in a retirement planning model) are described by probability distributions. Using this classification, linear programming models are deterministic; queueing models are probabilistic.

Finally, models may be *discrete* or *continuous.* In mathematical programming, this dichotomy refers to the types of variables in the model. For example, linear programming models are continuous; integer programming models are discrete. It may also refer to how model variables change over time. For instance, in multiperiod production planning models, we often assume that demand takes place at discrete points in time, such as the beginning of every week; this would classify the model as discrete. In modeling a chemical process, variables such as temperature and pressure would change continuously over time, and it would be important to capture this in the model.

How does simulation fall within these model classification schemes? Simulation models are *descriptive;* they simply estimate measures of performance or evaluate the behavior of a system for a specific set of inputs, as shown in Figure 1-1. Model inputs include the controllable (decision) variables specified by the user and uncontrollable variables or constants that capture the problem's environment. The simulation model itself is literally a set of assumptions that define the system or the problem. Probably the closest analogy with which you are familiar is a spreadsheet. A spreadsheet is a descriptive model in which the assumptions are the formulas entered in the cells. For any set of inputs, the spreadsheet calculates some output measures of interest.

Figure 1-2 shows an example of a financial spreadsheet for evaluating the cash flow of a proposed retail store in a shopping mall, not unlike what you have probably seen in accounting and finance classes. The model assumptions are given in rows 4 through 18. The key inputs are the first-year sales, annual growth rates, various cost factors, associated inflation assumptions, and so on. Formulas for computing the model outputs are entered in the lower section of the spreadsheet. Calculating the outputs basically consists of "stepping through" the formulas. Thus, a spreadsheet is essentially a deter-

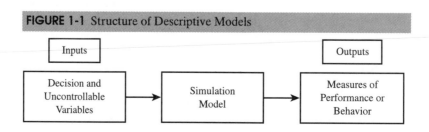

FIGURE 1-1 Structure of Descriptive Models

	A	B	C	D	E	F	G
1	New Store Financial Analysis						
2							
3	**Model Assumptions**		Year 1	Year 2	Year 3	Year 4	Year 5
4	Annual Growth Rate			20%	12%	9%	5%
5	Sales Revenue		$ 800,000				
6							
7	Cost of Merchandise (% of sales)	30%					
8	Operating Expenses						
9	Labor Cost	$ 200,000					
10	Rent Per Square Foot	$ 28					
11	Other Expenses	$ 325,000					
12							
13	Inflation Rate	2%					
14	Store Size (square feet)	$ 5,000					
15	Total Fixed Assets	$ 300,000					
16	Depreciation period (straight line)	5					
17	Discount Rate	10%					
18	Tax Rate	34%					
19							
20	**Model Outputs**		Year 1	Year 2	Year 3	Year 4	Year 5
21	Sales Revenue		$ 800,000	$ 960,000	$ 1,075,200	$ 1,171,968	$ 1,230,566
22	Cost of Merchandise		$ 240,000	$ 288,000	$ 322,560	$ 351,590	$ 369,170
23	Operating Expenses						
24	Labor Cost		$ 200,000	$ 204,000	$ 208,080	$ 212,242	$ 216,486
25	Rent Per Square Foot		$ 140,000	$ 142,800	$ 145,656	$ 148,569	$ 151,541
26	Other Expenses		$ 325,000	$ 331,500	$ 338,130	$ 344,893	$ 351,790
27	Net Operating Income		$ (105,000)	$ (6,300)	$ 60,774	$ 114,674	$ 141,579
28	Depreciation Expense		$ 60,000	$ 60,000	$ 60,000	$ 60,000	$ 60,000
29	Net Income Before Tax		$ (165,000)	$ (66,300)	$ 774	$ 54,674	$ 81,579
30	Income Tax		$ (56,100)	$ (22,542)	$ 263	$ 18,589	$ 27,737
31	Net After Tax Income		$ (108,900)	$ (43,758)	$ 511	$ 36,085	$ 53,842
32	Plus Depreciation Expense		$ 60,000	$ 60,000	$ 60,000	$ 60,000	$ 60,000
33	Annual Cash Flow		$ (48,900)	$ 16,242	$ 60,511	$ 96,085	$ 113,842
34	Discounted Cash Flow		(44,454.55)	13,423.14	50,008.96	79,409.11	94,084.46
35	Cumulative Discounted Cash Flow		(44,454.55)	(31,031.40)	18,977.55	98,386.67	192,471.13

FIGURE 1-2 Financial Spreadsheet Model

ministic simulation. The user may experiment with the model by using different assumptions, for example, changing inflation factors or baseline values to answer a variety of "what if?" questions. We will introduce the use of Excel for modeling and simulation in the next chapter.

Simulation models may be either deterministic or probabilistic. The spreadsheet in Figure 1-2 is essentially a deterministic simulation: The inputs are fixed and generate unique outputs. A probabilistic simulation model generally includes one or more probabilistic elements that reflect uncertainty. For example, in Figure 1-2, we might assume that unit sales are normally distributed or that the unit cost of goods sold is uniformly distributed. Under such assumptions, the model outputs will not have a unique value, but rather will be characterized by a probability distribution. Knowing the probability distribution of outputs provides insights into risks involved in making decisions. In later chapters, we will demonstrate how to incorporate probabilistic elements into spreadsheets and perform simulations. The types of simulation models that we discuss in this book are probabilistic because they account for randomness in the data or the systems they represent. Some common examples of data that exhibit random behavior and are modeled probabilistically in simulation models are consumer sales, machine operating times until failure, customer service times, and project activity completion times.

Finally, simulation models can be either discrete or continuous. For most systems, events occur at discrete points in time. Some examples are customer arrivals, start times of production jobs, departures of ships from ports, and so on. However, in simulating operations like an oil refinery, we must incorporate variables such as temperature, pressure, and material flow rates. These occur continuously over time, and they need to be reflected in the simulation model. Many problems exhibit a combination of discrete and continuous behavior.

Types of Simulation Models

In this book, we will study two distinct types of simulation models: **Monte-Carlo simulation** models and **systems simulation** models. Monte-Carlo simulation is basically a sampling experiment whose purpose is to estimate the distribution of an outcome variable that depends on several probabilistic input variables. For example, we might be interested in the distribution of the cumulative discounted cash flow for the fifth year (cell G35) for the financial model in Figure 1-2 when first-year sales, growth rate, operating expenses, and inflation factor assumptions are uncertain. We could input many different values for these factors into the spreadsheet model and record the value of cell G35 for each combination of inputs. If we use many different combinations of inputs, we will have created a distribution of possible values of the cumulative discounted cash flow that provides an indication of the likelihood of what we might expect. The term *Monte-Carlo simulation* was first used during the development of the atom bomb as a code name for computer simulations of nuclear fission. Researchers coined this term because of the similarity to random sampling in games of chance such as roulette in the famous casino in Monte Carlo (see Figure 1-3).

Monte-Carlo simulation is often used to evaluate the expected impact of policy changes and risk involved in decision making. **Risk** is often defined as the probability of occurrence of an undesirable outcome. Thus, we might be interested in the probability that 3-year profit will be less than a required amount. Monte-Carlo simulation is the principal focus of Chapters 2 through 5.

Systems simulation, on the other hand, explicitly models sequences of events that occur over time. Thus, inventory, queueing, manufacturing, and business process analysis problems are among the types of situations addressed with systems simulation. Systems simulation models are discussed in detail in Chapters 6 through 9. Some authors use the terms *static* and *dynamic* to distinguish between Monte-Carlo and systems simulation models, respectively. A static model is independent of time; a dynamic model is time-dependent.

We will illustrate a Monte-Carlo simulation first, using the following example.

AN EXAMPLE OF MONTE-CARLO SIMULATION

Dave's Candies is a small family-owned business that offers gourmet chocolates and ice cream fountain service. For special occasions, such as Valentine's Day, the store must place orders for special packaging several weeks in advance from their supplier. One product, Valentine's Day Chocolate Massacre, is bought for $7.50 a box and sells for $12.00. Any boxes that are not sold by February 14 are discounted by 50 percent and can always be sold easily. Historically, Dave's Candies has sold between 40 and 90 boxes each year with no apparent trend (either increasing or decreasing). Dave's dilemma is deciding how many boxes to order for the Valentine's Day customers. If demand exceeds the purchase quantity, then Dave loses profit opportunity. On the other hand, if too many boxes are purchased, he will lose money by discounting them below cost.

FIGURE 1-3 The Casino Monte Carlo in Monaco

We can easily develop an expression for Dave's profit if Q boxes are purchased and sales demand is D:

$$\text{Profit} = 12D - 7.50Q + 6(Q - D) \quad \text{if } D \leq Q \tag{1.1}$$
$$= 12Q - 7.5Q \quad\quad\quad\quad\quad\quad \text{if } D > Q \tag{1.2}$$

In the first case, if demand is less than the amount ordered, Dave receives full revenue from the sales of D boxes, must pay for the Q boxes purchased, and receives half revenue for the surplus. In the second case, if demand exceeds the amount ordered, Dave can only sell Q boxes and makes a net profit of $\$12.00 - \$7.50 = \$4.50$ per box.

The inputs to a simulation model of this situation would be

1. the order quantity, Q (the decision variable),
2. the various revenue and cost factors (constants), and
3. the demand, D (uncontrollable and probabilistic).

The model output we seek is the net profit.

If we know the demand, then we can use equation (1.1) or (1.2) to compute the profit. Because demand is probabilistic, we need to be able to "sample" a value from the probability distribution of demand. For now, we will simplify this problem by assuming that demand will be either 40, 50, 60, 70, 80, or 90 boxes with equal probability (1/6). This will allow us to generate samples by rolling a die. (In a later chapter, we will see how to do this quite easily on a spreadsheet.) The following table associates the value of the roll of a die with one of the demand outcomes.

Roll of Die	Demand
1	40
2	50
3	60
4	70
5	80
6	90

We will perform a Monte-Carlo simulation for an order quantity $Q = 60$. The simulation proceeds as follows:

1. Roll a die.
2. Determine the demand, D, from the preceding table.
3. Using $Q = 60$, compute the profit using equation (1.1) or (1.2).
4. Record the profit.

For example, suppose that the first roll of the die is 4. This corresponds to a demand of 70. Since $D = 70 > Q = 60$, we use equation (1.2) to compute the profit:

$$\text{Profit} = 12(60) - 7.5(60) = \$270$$

However, one sample does not give a very good picture of what might happen or what the risks are of ordering 60. By repeating the simulation, we can develop a distribution of profit and assess risk. Table 1-1 summarizes the results for 10 trials of this experiment.

From Table 1-1, the average profit that Dave might expect using $Q = 60$ is $246. We may also construct a relative frequency distribution of profits. From the table, we see that one trial had a profit of $150, two trials yielded a profit of $210, and seven trials resulted in a profit of $270. The frequency distribution is

Profit	Probability
$150	.10
$210	.20
$270	.70

This frequency distribution of profit provides an assessment of the risk involved in making the decision to order 60 boxes. For example, there is a 30 percent chance that the profit will be $210 or less. If Dave needs a profit of, say, $225 to meet his other business expenses, then he might want to choose a different order quantity to reduce this risk. We will discuss issues of using Monte-Carlo simulation to assess risk more fully in a later chapter.

We might observe that if we repeated the simulation, we would expect to roll different values of the die and will probably obtain a different value for the average profit as well as a different frequency distribution (Try it!). This is an important insight into the nature of simulation: It is a sampling experiment that is itself uncertain. Therefore, we need to be able to quantify the uncertainty in our simulated results. Later in this book, we shall see how to do this using basic statistical principles.

We might also observe that 10 trials will provide only limited results. For a larger number of trials, we would expect roughly an equal number of rolls of each value of the die. In this small experiment, we rolled a 2, 3, and 5 twice as often as a 1, 4, or 6. Thus,

TABLE 1-1	Ten Trials of Dave's Candies Simulation Using $Q = 60$		
Trial	*Roll of Die*	*Demand*	*Profit*
1	5	80	$270
2	3	60	$270
3	2	50	$210
4	4	70	$270
5	1	40	$150
6	3	60	$270
7	5	80	$270
8	6	90	$270
9	2	50	$210
10	3	60	$270
		Average	*$246*

we might expect that our conclusions about the average profit and risk are somewhat biased and that the frequency distribution we obtained does not represent the true distribution of profit. We repeated this simulation for 100 trials and obtained the following frequency distribution:

Profit	*Frequency*
$150	20
$210	22
$270	58

with an average profit of $232.80. This distribution probably will be closer to the true expected values than using only 10 trials; however, the average will generally not be the same as the true expected value because of statistical sampling error. Therefore, to obtain valid results with Monte-Carlo simulation, we need to use a sufficiently large number of trials. Again, in a later chapter, we will address this issue statistically.

Finally, the results in Table 1-1 are only descriptive; they do not tell us whether the order quantity $Q = 60$ is best. To find the best decision, we would have to experiment with different order quantities. Using a spreadsheet (which we will describe in Chapter 2), we repeated the simulation for 100 trials, using order quantities of 40, 50, 60, 70, 80, and 90. The summary results are shown in Table 1-2. We see that the order quantity that maximizes the average profit is $Q = 70$, yielding an average profit of $264.00. You might observe that the average profit for $Q = 60$ is slightly different in Table 1-2 than the value found earlier ($234.00 versus $232.80). Again, this is to be expected because of sampling error.

For a simple problem as this, we can easily determine the optimal order quantity analytically (see problem 8 at the end of the chapter). In fact, you will find that the analytical solution shows that the expected profit for $Q = 70$ is $255.00, whereas the (optimal) expected profit for $Q = 80$ is $260.00. However, our simulation results suggested a

TABLE 1-2	Summary Results of 100 Trials					
Order Quantity:	40	50	60	70	80	90
Average Profit:	$180.00	$215.40	$234.00	$264.00	$251.40	$258.60

different solution. What happened? Again, we point to sampling error as the culprit. This again emphasizes the necessity of evaluating simulation results statistically in order to draw appropriate conclusions. Simulation experiments must be planned carefully and the results analyzed thoroughly! This also suggests that simulation should not be used when appropriate analytical models are available. A good analyst should explore options for solving any decision problem and not use simulation as a crutch.

This example showed the nature of Monte-Carlo simulation—repeated sampling from probability distributions to develop the distribution of an output variable. The next example illustrates the nature of a systems simulation model—one that depends on the sequence of prior events and the passage of time.

AN EXAMPLE OF SYSTEMS SIMULATION

Mantel Manufacturing supplies various automotive components to major automobile assembly divisions on a just-in-time basis. The company has received a new contract for water pumps. Planned production capacity for water pumps is 100 units per shift. Because of fluctuations in customers' assembly operations, demand fluctuates and is historically between 80 and 130 units per day. To maintain sufficient inventory to meet its just-in-time commitments, Mantel's management is considering a policy to run a second shift if inventory falls to 50 or below. For the annual budget planning process, managers need to know how many additional shifts will be needed.

For this situation, an analytical model is not obvious, particularly because we need to know exactly when inventory, which depends on the prior history of demand, falls to 50 or below. Thus, we must keep track of the passage of time in a systems (dynamic) simulation model in order to know when to run additional shifts. The fundamental equation that governs this process each day is

$$\text{Ending inventory} = \text{Beginning inventory} + \text{Production} - \text{Demand} \qquad \textbf{(1.3)}$$

Suppose we begin with an inventory of 100 units. As in the previous example, we will simplify the problem by assuming that demand occurs in increments of 10 so that we may use the roll of a die to randomly generate the demand each day. Thus, we associate the demand with the roll of a die as follows:

Roll of Die	Demand
1	80
2	90
3	100
4	110
5	120
6	130

The simulation process can be depicted by the flowchart in Figure 1-4. We would proceed as follows:

1. Initialize the inventory at the start of the simulation. Begin new simulated day.
2. Set beginning inventory equal to the ending inventory from the previous day.
3. Determine the demand by rolling the die.
4. If beginning inventory is 50 or less, then the day's production is 200 units; otherwise, production is 100 units.
5. Use equation (1.3) to compute the ending inventory.
6. Stop if enough days have been simulated; otherwise, return to step 1.

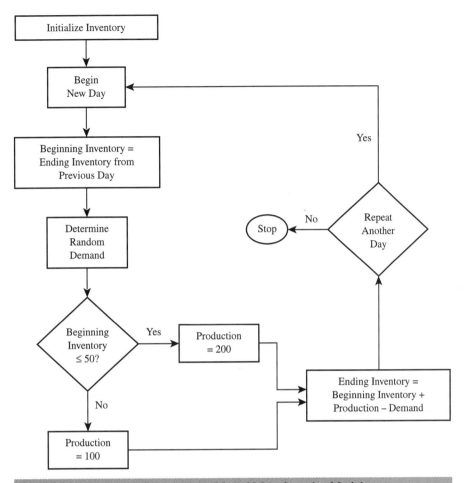

FIGURE 1-4 Simulation Flowchart for Mantel Manufacturing Model

Table 1-3 shows 5 simulated days. We see that the inventory falls below 50 on day 4, so a second shift is run, increasing the day's production to 200. Of course, 5 simulated days provides little meaningful information. Figure 1-5 shows the result of simulating this process for 100 days. The graph shows that six additional shifts were needed over the 100 days to maintain the desired inventory level. Extrapolating this to a 250-day working year, the company should expect to need about $2.5(6) = 15$ additional shifts. As we noted in the previous example, we should expect some variability in this result if we repeat the simulation or run it for a longer time period. Statistical methods will help us to quantify this variation. Management may also wish to experiment with different overtime policies to weigh the risks of running out of stock versus the costs of additional shifts.

TABLE 1-3 Results of 5 Simulated Days for Mantel Manufacturing Example

Day	Beginning Inventory	Roll of Die	Demand	Production	Ending Inventory
1	100	5	120	100	80
2	80	4	110	100	70
3	70	6	130	100	40
4	40	6	130	200	110
5	110	1	80	100	130

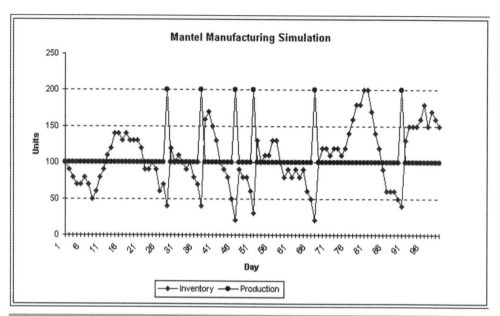

FIGURE 1-5 100-Day Production and Inventory Simulation

Most systems simulation models are far more complex than this example. For instance, consider a simple waiting-line situation (like a drive-through window at a quick service restaurant). Customers arrive at random times and either proceed directly to the window or wait in line behind other customers. Service times, which would depend on the order size and type, are random. If we are interested in knowing such statistics as the average waiting time, average number in the line, and proportion of time the window is being used, we would have to record quite a bit of information, including the actual arrival times for each customer, time service starts, time service is completed, whether any customers are waiting when a customer leaves, and so on.

Now think of a situation in a factory in which jobs move from one machine to another, and waiting lines build up at *each* machine. Add to that any logic for selecting which job to process next and where a completed job goes. You can see the complexity of even moderate-size systems simulation models. In practice, systems simulation is facilitated by commercial software that is generally quite easy to use.

The Simulation Process

Using simulation effectively requires careful attention to the modeling and implementation process. The simulation process consists of five essential steps:

1. *Develop a conceptual model of the system or problem under study.* This step begins with understanding and defining the problem, identifying the goals and objectives of the study, determining the important input variables, and defining output measures. It might also include a detailed logical description of the system that is being studied. Simulation models should be made as simple as possible to focus on critical factors that make a difference in the decision. The cardinal rule of modeling is to build simple models first, then embellish and enrich them as necessary.

2. *Build the simulation model.* This includes developing appropriate formulas or equations, collecting any necessary data, determining the probability dis-

tributions of uncertain variables, and constructing a format for recording the results. This might entail designing a spreadsheet, developing a computer program, or formulating the model according to the syntax of a special computer simulation language (which we discuss further in Chapters 6 and 8).

3. *Verify and validate the model.* **Verification** refers to the process of ensuring that the model is free from logical errors; that is, that it does what it is intended to do. **Validation** ensures that it is a reasonable representation of the actual system or problem. These are important steps to lend credibility to simulation models and gain acceptance from managers and other users. These approaches are described further in the next section.

4. *Design experiments using the model.* This step entails determining the values of the controllable variables to be studied or the questions to be answered in order to address the decision-maker's objectives.

5. *Perform the experiments and analyze the results.* Run the appropriate simulations to obtain the information required to make an informed decision.

As with any modeling effort, this approach is not necessarily serial. Often, you must return to previous steps as new information arises or as results suggest modifications to the model. Therefore, simulation is an evolutionary process that must involve not only analysts and model developers but also the users of the results.

VERIFICATION AND VALIDATION

An old adage states, "If it walks like a duck, sounds like a duck, and looks like a duck, then it must be a duck." Therefore, if we are simulating a duck (business system), it had better walk, sound, and look like a duck (business system) or else the hunters (managers) will easily recognize it as a decoy (useless). A simulation model should be good enough for a manager to make decisions using the simulation results similar to those he or she would make with the same types of data from the real system. Managers must be confident that they are dealing with useful models.

The first step is verification—ensuring that the model is free from logical errors. Verification methods include many standard techniques from software engineering, such as building and programming the model in small modules, debugging each module before putting them together, having several experts review the model, testing the model using simplified assumptions so that the output can be compared with analytical solutions, using realistic input data sets and checking for corresponding realistic outputs, and tracing through detailed logic as the simulation runs. Many simulation software packages now include animation capabilities, whereby an animated graphical representation of a model can be viewed as the simulation runs. For instance, a model of a factory would show the flow of parts from one machine to another, whether machines were running or idle, and breakdowns and repairs. Animation provides a means to observe how the model behaves and can help identify logical problems that you would normally not expect to see. These are often symptoms of programming errors.

Validation—ensuring that the model is a good representation of reality—can be viewed from many different perspectives. *Face validity* refers to asking experts about the simulation model or results to determine whether the model and/or the results are reasonable. This might include comparing the structure of the simulation model to the actual system, focusing attention on linkages among smaller parts of the model. Computer animation can greatly assist in this effort. A useful approach, sometimes called *historical data validation,* is to compare the output of the model to historical data from the real system when the same data inputs are used. *Data validity* includes ensuring that all input data and probability distributions are truly representative of the system being modeled. This might entail statistical tests of goodness-of-fit or assessments of the sensitivity of outputs to variations in the model inputs. Many of the Simulation in Practice cases at the end of each chapter illustrate some practical approaches to validation.

The difficulty of verification and validation depends on the real system's complexity and whether it even exists. No specific procedure exists to select different verification and validation techniques; however, it is extremely important that some methods of both verification and validation be used.

Benefits and Limitations of Simulation

Simulation has many benefits. First, it allows managers and analysts to evaluate proposed systems or decisions without building or actually implementing them, or to experiment with existing systems without disturbing them. This "what if?" capability is a significant advantage. Second, simulation models are generally easier to understand than many analytical approaches. For example, it would probably be difficult to explain expected value decisions to the owner of Dave's Candies. However, the simulation process of selecting an order quantity, determining the demand, and computing the profit corresponds to the actual situation. Thus, the simulation approach delivers a measure of confidence in the solution. Third, the ability to model any assumption, particularly when analytical models are inappropriate or do not exist, sets simulation distinctly apart from other management science approaches.

However, simulation is not without disadvantages. A significant amount of time often is required to obtain the necessary input, develop the simulation model and computer program, and interpret the results. As Hewlett-Packard, for instance, was installing a system for manufacturing ink-jet printers in Vancouver, Washington, the ink-jet printer market was exploding, and it realized that the system would not be fast enough or reliable enough to meet its production goals. The company undertook a simulation project to develop recommendations for design changes to improve the system performance but concluded that the project would take too long to be useful. Instead, the company sought analytical modeling assistance from MIT.[4] In addition to time issues, analysts and programmers are expensive commodities, and simulation model development and testing can be expensive. Although computer time is relatively inexpensive today and computer speed continually improves, simulations can be time-consuming. If a decision is needed quickly, for instance, to reschedule jobs on a factory floor, a simulation model that requires an hour to run on a PC is impractical.

A second major limitation of simulation is the lack of precise answers. We saw through the examples in this chapter that simulation is prone to sampling errors. Therefore, it is necessary to design simulation experiments carefully to characterize the distribution of outcomes, to identify any extreme outcomes that may occur and draw useful statistical conclusions. Analytical models, on the other hand, usually do not have such "fuzziness."

These limitations can be overcome if simulation is used correctly and intelligently. The many advantages of simulation make it an important tool for contemporary decision making.

Simulation in Practice

At the end of every chapter, we present some examples of real applications of simulation. These examples provide further insight into the nature of simulation, the types of questions that simulation studies address, and how the results of simulations are ana-

---

[4]Mitchell Burman, Stanley B. Gershwin, and Curtis Suyematsu, "Hewlett-Packard Uses Operations Research to Improve the Design of a Printer Production Line," *Interfaces,* Vol. 28, no. 1, January–February 1998, pp. 24–36.

lyzed to make good decisions. In this chapter, we illustrate how simulation may be used for education and training and analyzing environmental decisions.

BUSINESS SIMULATION FOR DENTAL PRACTICE MANAGEMENT[5]

Business simulations are games in which the participants make management decisions about a business entity based on past performance and the strategic direction they wish to take. A computer program analyzes the decisions based on the actual competitive and simulated economic environments. The analysis then determines the players' outcomes based on a set of previously determined algorithms, which are then used for a new set of decisions for the next simulated time period. Outcomes may consist of profit, market share, customer satisfaction, shareholder equity, or any other measure on which learning is focused. Business simulation games have been developed for manufacturing, marketing, airlines, global strategy, and many other applications.

Although dental students use simulations to practice clinical techniques, diagnosis, and patient management, the reality of modern dentistry involves managing a dental practice from a business perspective. Few students have any business experience or have worked in a business environment. As a result, the University of Louisville School of Dentistry developed a simulation of the business decisions required of dental practitioners. The specific behavioral objectives the game intends to accomplish are:

1. Manage the information needs of a dental practice.
   a. Read and interpret common financial statements.
   b. Calculate and interpret common financial ratios.
   c. Cull dental practice data for important management information, and react and adapt to changing signals.
   d. Make practice management decisions in the face of ambiguity and incomplete information.
2. Demonstrate proper office management technique, including planning, budgeting, forecasting, and control.
   a. Describe the relationships among operational, staffing, marketing, and financial decisions.
   b. Apply common personal and business financial control techniques.
   c. Apply analytic methods to the management of a dental practice.
3. Develop a strategic view of dental practice management.
   a. Establish a mission, goals, and objectives for practice performance.
   b. Develop a longitudinal perspective in managing a dental practice.
4. Identify external and internal environmental factors that affect the operation of a dental practice.
   a. Recognize the implications of general business prosperity for the dental practitioner.
   b. Define social responsibility and explain its implication.
   c. Describe the relationship between a dental practitioner and the community.
   d. Analyze the interrelationship among business, ethical, and social decisions in the dental practice.

The simulation creates a practice scenario in which participants have borrowed capital to purchase the ongoing practice of a retiring dentist. The first task is to develop a practice philosophy, strategy, and plans for achieving success. Participants then face

[5]David O. Willis, Jerald R. Smith, and Peggy Golden, "A Computerized Business Simulation for Dental Practice Management," *Journal of Dental Education,* October 1997, pp. 821–828.

decisions concerning practice operations, staffing levels, marketing efforts, and financial management that help them to carry out their strategy. These decisions are summarized in Table 1-4. Participants make these decisions on a simulated quarterly basis, based on past practice performance, the strategic direction they have chosen for their practice, and the prevailing economic conditions. They then give those decisions to the game administrator, who enters the information into the computer program outside the classroom. The program analyzes the decisions and produces an income statement and other business information, such as the number of patient visits, resulting from the decisions made. An example is shown in Figure 1-6. The process is repeated for 3 simulated years.

Students found the simulation to be quite valuable, particularly in raising their general understanding of business and financial management. The simulation helped them learn how to make management decisions and appreciate the complexity of dental practice management.

SIMULATION MODELING FOR HAZARDOUS WASTE REMEDIATION ALTERNATIVES[6]

The U.S. Department of Energy (DOE) has over 100 nuclear processing sites requiring remediation to contain and treat hazardous nuclear waste. The waste management program has been responsible for identifying and reducing health and safety risks, and

TABLE 1-4   Decision Variables in Dental Practice Business Simulation Game

Operational Decisions

Choose hours of operation.
Decide whether or not to add operatories.
Decide a credit and collection policy.
Determine participation in managed-care programs.

Staffing Decisions

Hire or terminate staff members—
    receptionist, chairside assistant, hygienist.
Determine staff raises—percentage increase or decrease.
Establish staff benefits.
Decide continuing education participation.

Marketing Decisions

Choose from several practice styles.
Decide a fee to charge for services.
Determine an advertising budget.

Financial Decisions

Decide the amount of draw to take.
Determine how much additional loan or extra repayment.
Decide how much to invest in a 91-day CD.

Other Decisions

Decide whether to purchase certain marketing information.
Respond to an incident that occurs in the office.

---

[6]T. P. White, R. Toland, J. A. Jackson, Jr., and J. M. Kloeber, Jr., "Simulation and Optimization of a New Waste Remediation Process," *Omega*, Vol. 24, no. 6, 1996, pp. 705–714; and Ronald J. Toland, Jack M. Kloeber, Jr., and Jack A. Jackson, "A Comparative Analysis of Hazardous Waste Remediation Alternatives," *Interfaces*, Vol. 28, no. 5, September–October 1998, pp. 70–85.

*** DENTAL PRACTICE MANAGEMENT SIMULATION ***
River City Dental Clinic

Quarter # 5   **Practice # A / 1**

** PRODUCTION INFORMATION **

Regular Production	53,171	
PPO Production	0	
Medicaid Production	9,237	
Capitation Production	0	
Gross Production		62,408

** INCOME AND EXPENSE STATEMENT **

INCOME:		
Current Production		62,408
+Pymts on Previous A/R		20,678
Uncollectible =	1,541	
+Interest Income		0
Billed to Accts Rec		22,467
Adjustments		3,318
Net Cash Revenue		57,301
EXPENSES:		
Variable Costs		
Clinical Supplies	3,761	
Office Supplies	1,605	
Dental Lab	7,033	
Fixed Costs:		
Staff Salaries	14,288	
Payroll Tax/Ins	1,857	
Optional Benefits	1,175	
Hiring/Tng Cost	0	
Utilities	1,442	
Office Rent/Equip	3,500	
Continuing Ed.	1,000	
Prof Serv/Mkt Res	300	
Insurance	1,215	
Depreciation	3,850	
Advertising	1,800	
Loan Interest	4,317	
Misc. Exp	2,225	
Business Taxes	750	
Other Expenses	0	
Total Costs		50,118

PROFIT/LOSS THIS QUARTER	7,182

** CASH MANAGEMENT **

Ending Cash Last Qtr	6,826
+ Profit This Qtr	7,182
+ Loan Addition/Xtra Pymt	0
+ CD Redeemed	0
+ Depreciation Exp	3,850
− Personal Estimated Tax Payment	1,795
− Draw This Qtr	5,000
− Loan Principal Payment	6,263
Total Cash Available	4,800
− CD Purchased	0
Ending Cash This Qtr	4,800

FIGURE 1-6 Example Result from Dental Simulation Business Game

Source: Copyright 1997 © Jerald Smith, David Willis, and Peggy Golden

managing waste at 136 sites in 34 states and territories. Tight fiscal constraints force DOE to look for alternatives to reduce the cost of hazardous waste remediation. Although new technological advancements in waste remediation, such as vitrification using minimum additive waste stabilization (MAWS), have shown promise in effectively treating waste at a lower cost in small-scale tests, a great deal of uncertainty remained about the true cost of the technology. Vitrification, which encapsulates waste material within glass beads for long-term storage, was expected to have high initial costs but would reduce waste volume by one-third, and long-run costs seemed likely to be attractive due to reduced storage and monitoring costs. However, other alternatives also existed, including a cementation process whereby waste is mixed with cement and poured into a metal container, and dry removal, in which chemicals are added to a mixture of water and toxic waste to produce an insoluble compound, which is then formed into a gel and compressed to leave a dry substance that is easily handled and stored.

The DOE needed a life-cycle cost (LCC) model to help evaluate and select appropriate waste remediation technologies. Although deterministic cost-estimating tools are available for net present value analysis for the waste remediation alternatives, they do not allow for probabilistic inputs or produce distributions of potential results. Air Force Institute of Technology analysts developed a generic LCC model that used Monte-Carlo simulation to model the cost and time uncertainties for all three alternatives. The Monte-Carlo approach generated the total cost-risk profile through repeated sampling from various distributions of cost estimates. Because of past difficulties with innovative technologies and cost estimates, DOE required the model to estimate the 95 percent value that remediates a broad range of sites across the DOE complex. This figure would represent the cost that would be exceeded with a probability of only 0.05.

The LCC model features included the following:

- An unlimited number of variables and cost elements
- A maximum project life of 200 years
- Inflated and deflated cost discounting
- Automation of multiple simulations
- User-specified cost correlations
- The ability to handle any type of cost formula (constants, random variables, functions, or a combination)
- LCC probability and cumulative distribution functions
- LCC sensitivity to cost distributions
- Data for break-even analysis between alternatives

In addition to the Monte-Carlo simulation approach, the analysts developed systems simulation models of the vitrification alternative glass production process. The model incorporated linear programming to select the process design that would minimize system costs. Further statistical analyses were conducted on the simulation results to estimate system performance and associated costs for a broad range of waste volumes. The simulation was verified by using hand calculations to check the expected glass volume and process time for various sets of inputs with the simulation outputs. In addition, the characterization process for batches of waste was analyzed to determine if the mean and variation produced by the simulation matched the desired parameters. Finally, the reliability of components modeled by the simulation was compared to the expected availability provided by manufacturers and contractors. To validate the results, the analysts worked closely with process engineers throughout the development of the simulation model.

To make the actual decision, multicriteria decision analysis was applied to include social and political considerations in addition to cost. This allowed the decision maker to easily conduct sensitivity analyses, as the DOE had requested. Using this approach, the analysts demonstrated that the MAWS technique could save DOE $3 billion over 8 years at a typical contaminated site.

Questions and Problems

1. Define the term *simulation* and explain the differences between simulation and analytical modeling.
2. Provide some examples of simulation that you have seen in the media or in your daily life that are different from the examples in this chapter.
3. Describe some types of decision problems for which simulation might be more appropriate than analytical models.
4. Explain the different classifications of models in management science. How does simulation fall into these classifications?
5. Explain why a spreadsheet is an example of a deterministic simulation. Can you provide other examples of deterministic simulations?
6. Explain the difference between Monte-Carlo simulation and systems simulation.
7. Using a die, simulate 25 replications of demand for the Dave's Candies example. Using equations (1.1) and (1.2) and your results, compute the average profit for order quantities of 40, 50, 60, 70, 80, and 90. How do your results compare with those in Table 1-2?
8. Develop an analytical model for the Dave's Candies problem and show that the optimal order quantity is indeed $Q = 80$. (Hint: Use the profit equations (1.1) and (1.2) to find the expected value of a specific order quantity. The expected value is equal to the probability of an outcome times that outcome's payoff, summed over all possible outcomes.)
9. Describe how a deck of cards might be used to simulate random outcomes. Show how a deck of cards can be used to simulate the demands (40, 50, 60, 70, 80, and 90) for the Dave's Candies problem with probabilities 0.1, 0.2, 0.3, 0.2, 0.1, and 0.1, respectively. Use your approach to find the best order quantity for this problem.
10. A neighborhood grocery store orders a weekly entertainment magazine. Demand varies each week, but historical records show the following distribution:

Number of Magazines	Probability
20	1/36
21	4/36
22	17/36
23	12/36
24	2/36

Each magazine costs $1.50 and sells for $2.50. Any magazines remaining at the end of the week are donated to a local retirement home. Develop a simulation model to determine the best order quantity. (Hint: Use dice to simulate the weekly demand.)

11. An electrical system consists of two components in parallel. The probability that each individual component will fail is 0.25. The system fails if both components fail. Estimate the probability of system failure of each component based on 10, 20, and 40 tests. Compare results with the analytical solution.
12. Replicate the Mantel Manufacturing simulation example 30 times using a die to determine the daily demand. Do your results generally compare with Figure 1-5? How many additional shifts would you estimate to have in a 250-day year from your simulation?
13. Suppose in the Mantel Manufacturing example that, in addition to uncertain demand, production output varies each day because of labor availability, production

defects, material delays, and so on. Assume that, on average, in 4 of every 10 days, production is only 90; and that in 1 of every 10 days, production is 110. The rest of the time, production is 100. Develop a physical experiment to sample production from this distribution. How would Figure 1-4 change? Simulate 30 days and compare your results to problem 12.

14. The recent performance of a new stock after an initial public offering has the changes given.

Change in Price	Probability
−1/8	2/36
no change	5/36
+1/8	15/36
+1/4	8/36
+1/2	5/36
+1	1/36

Simulate the stock performance over the next 20 days, assuming that the initial value is $100 per share. (Hint: Use two dice to simulate the price change.)

15. Describe the five steps of the simulation process. How might this process have been used in building the dental practice simulation game described in the "Simulation in Practice" section of this chapter?

16. Define the terms *validation* and *verification*. Describe approaches that can be used to verify and validate simulation models.

17. Explain the concepts of *face validity* and *data validity*.

18. What benefits do simulation analyses have over other methods of analysis? What are some limitations of simulation analysis?

19. Suppose that you were asked to develop a simulation model to improve the performance at a local fast-food restaurant. How would you approach this problem? What types of data would you need? What types of simulation outputs would be useful? What information would you include in the model? What policy alternatives might you consider? Draw a simple flowchart of the current system's operation.

20. Recall your experiences in buying groceries. Develop a flowchart that describes the process that you encounter. If you were to simulate this system, what types of data would you need? What outputs might you compute? What policy alternatives might you consider?

References

Bratley, P., B. L. Fox, and L. E. Schrage. *A Guide to Simulation,* 2d ed. New York: Springer-Verlag, 1987.

Carroll, J. M. *Simulation Using Personal Computers.* Englewood Cliffs, NJ: Reston, 1987.

Pritsker, A. A. B. *Introduction to Simulation and SLAM II,* 3d ed., New York: Halsted Press, 1986.

Pritsker, A. A. B., C. Elliott Sigal, and R. D. Jack Hammesfahr. *Slam II Network Models for Decision Support.* New York: Prentice Hall, 1989.

Sargent, Robert G. "A Tutorial on Validation and Verification of Simulation Models," *Proceedings of the 1988 Winter Simulation Conference,* M. Abrams, P. Haigh, and J. Comfort, eds.

CHAPTER

2

Simulation Using Excel

Chapter Outline

Spreadsheet software such as Microsoft Excel has dramatically changed the way that management science is applied in business. In the "old days," management science techniques would have to be run on large mainframe computers staffed by information systems professionals. Turnaround time was at best a few hours or at worst several days, making it quite difficult to run many scenarios or freely experiment with models. Today, spreadsheets provide all business professionals with the means to model and analyze complex business problems easily by using powerful management science tools.

The flexibility of spreadsheets and their statistical capabilities make them a natural framework for simulation modeling, especially for Monte-Carlo simulation. In this chapter, we present some basic concepts and approaches for implementing simulation models with Microsoft Excel. Although we will use Microsoft Excel exclusively in this book, most of the principles can be applied with other spreadsheet software packages such as Lotus 1-2-3® or Quattro Pro®. We do warn you, however, that some significant differences exist between Excel and other software, particularly with respect to data analysis and statistical functions. Thus, everything we describe may not be transferable exactly to other spreadsheet software. All spreadsheet models used in this book are available on the supplementary CD-rom provided with the book and are noted in the text by their file names (e.g., *model_name.xls*).

Basic Excel Skills

To fully understand the models we develop in this book, it is necessary for you to know many of the basic capabilities of Excel. We will assume that you are familiar with the most elementary spreadsheet concepts and procedures:

- Opening, saving, and printing files
- Moving around a spreadsheet
- Selecting ranges
- Inserting/deleting rows and columns
- Entering and editing text, numerical data, and formulas
- Formatting data (number, currency, decimal places, etc.)
- Working with text strings
- Performing basic arithmetic calculations
- Formatting data and text
- Modifying the appearance of the spreadsheet

Excel has extensive online help, and many good manuals and training guides are available both in print and online. We urge you to take advantage of these. However, to facilitate your understanding of the simulation models we discuss and your ability to develop them yourself, we will review some of the more important topics in Excel that you may not have used. We will also describe other features of Excel as necessary throughout the text.

COPYING FORMULAS AND CELL REFERENCES

Excel provides several ways of copying formulas to different cells. This is extremely useful in spreadsheet simulation, because many models require replication of formulas for different trials. One way is to select a cell, choose *Edit . . . Copy* from the menu bar (or click on the *Copy* icon or simply press *Ctrl-C* on your keyboard), click on the cell you wish to copy to, and then choose *Edit . . . Paste* (or click on the *Paste* icon or press *Ctrl-V*). To copy a formula from a single cell or range of cells down a column or across a row, select the cell or range, click and hold the mouse on the small square in the lower right-hand corner of the cell (the "drag handle"), and drag the formula to the "target" cells you wish to copy to. (See Figures 2-1 and 2-2 for an example of copying the formulas for years 3 through 5 in the financial spreadsheet model introduced in Figure 1-2.) You may enter a formula directly in a range of cells without copying and pasting by selecting the range, typing in the formula, and then pressing *Ctrl-Enter*.

Year 1	Year 2	Year 3	Year 4	Year 5
$ 800,000	$ 960,000			
$ 240,000	$ 288,000			
$ 200,000	$ 204,000			
$ 140,000	$ 142,800			
$ 325,000	$ 331,500			
$ (105,000)	$ (6,300)			
$ 60,000	$ 60,000			
$ (165,000)	$ (66,300)	Click here and		
$ (56,100)	$ (22,542)			
$ (108,900)	$ (43,758)	drag to right		
$ 60,000	$ 60,000			
$ (48,900)	$ 16,242			
(44,454.55)	13,423.14			
(44,454.55)	(31,031.40) ←			

FIGURE 2-1 Highlighting a Range of Formulas to Copy

In any of these procedures, the structure of the formula is the same as in the original cell, but the cell references have been changed to reflect the relative addresses of the formula in the new cells. That is, the new cell references have the same relative relationship to the new formula cell(s) as they did in the original formula cell. Thus, if a formula is copied (or moved) one cell to the right, the relative cell addresses will have their column label increased by one; if we copy or move the formula two cells down, the row number is increased by two. Figure 2-3 shows the formulas for the financial spreadsheet model. For example, note that the formulas in row 21 for years 3, 4, and 5 are the same as for year 2, except for the column reference.

Sometimes, however, you do not want to change the relative addresses because you would like all the copied formulas to point to a certain cell. We do this by using a $ before the column or row address of the cell. This is called an *absolute address*. For example, in Figure 2-3, we want to use the same inflation factor for all years; therefore, we define the reference to the inflation factor in cell B13 as $B$13. Then, if we copy this formula in rows 24 through 26 in column D across to columns E, F, and G, the reference will still point to cell B13. If we had not used an absolute address and copied the formulas, then the reference to cell B13 in column E, for instance, would have been changed to C13,

FIGURE 2-2 Result of Copying a Range of Formulas

Year 1	Year 2	Year 3	Year 4	Year 5	
$ 800,000	$ 960,000	$ 1,075,200	$ 1,171,968	$ 1,230,566	
$ 240,000	$ 288,000	$ 322,560	$ 351,590	$ 369,170	
$ 200,000	$ 204,000	$ 208,080	$ 212,242	$ 216,486	
$ 140,000	$ 142,800	$ 145,656	$ 148,569	$ 151,541	
$ 325,000	$ 331,500	$ 338,130	$ 344,893	$ 351,790	
$ (105,000)	$ (6,300)	$ 60,774	$ 114,674	$ 141,579	
$ 60,000	$ 60,000	$ 60,000	$ 60,000	$ 60,000	
$ (165,000)	$ (66,300)	$ 774	$ 54,674	$ 81,579	
$ (56,100)	$ (22,542)	$ 263	$ 18,589	$ 27,737	
$ (108,900)	$ (43,758)	$ 511	$ 36,085	$ 53,842	
$ 60,000	$ 60,000	$ 60,000	$ 60,000	$ 60,000	
$ (48,900)	$ 16,242	$ 60,511	$ 96,085	$ 113,842	
(44,454.55)	13,423.14	50,008.96	79,409.11	94,084.46	
(44,454.55)	(31,031.40)	18,977.55	98,386.67	192,471.13	

	C	D	E	F	G
19					
20	Year 1	Year 2	Year 3	Year 4	Year 5
21	=C5	=C21*(1+D4)	=D21*(1+E4)	=E21*(1+F4)	=F21*(1+G4)
22	=$B$7*C21	=$B$7*D21	=$B$7*E21	=$B$7*F21	=$B$7*G21
23					
24	=B9	=C24*(1+$B$13)	=D24*(1+$B$13)	=E24*(1+$B$13)	=F24*(1+$B$13)
25	=$B$10*$B$14	=C25*(1+$B$13)	=D25*(1+$B$13)	=E25*(1+$B$13)	=F25*(1+$B$13)
26	=B11	=C26*(1+$B$13)	=D26*(1+$B$13)	=E26*(1+$B$13)	=F26*(1+$B$13)
27	=C21-C22-C24-C25-C26	=D21-D22-D24-D25-D26	=E21-E22-E24-E25-E26	=F21-F22-F24-F25-F26	=G21-G22-G24-G25-G26
28	=$B$15/$B$16	=$B$15/$B$16	=$B$15/$B$16	=$B$15/$B$16	=$B$15/$B$16
29	=C27-C28	=D27-D28	=E27-E28	=F27-F28	=G27-G28
30	=C29*$B$18	=D29*$B$18	=E29*$B$18	=F29*$B$18	=G29*$B$18
31	=C29-C30	=D29-D30	=E29-E30	=F29-F30	=G29-G30
32	=C28	=D28	=E28	=F28	=G28
33	=C31+C32	=D31+D32	=E31+E32	=F31+F32	=G31+G32
34	=C33/(1+$B$17)^1	=D33/(1+$B$17)^2	=E33/(1+$B$17)^2	=F33/(1+$B$17)^2	=G33/(1+$B$17)^2
35	=C34	=C35+D34	=D35+E34	=E35+F34	=F35+G34

FIGURE 2-3 Financial Spreadsheet Model Formulas

resulting in an incorrect formula. You should be very careful to use relative and absolute addressing appropriately in your models. An easy way to make a cell reference absolute or partially absolute is to press the F4 key after entering the cell reference in a formula. For instance, if you enter =A1 in a cell and press F4 repeatedly, the formula changes to =$A$1, then A$1, then $A1, and then back to A1.

FUNCTIONS

Functions are used to perform special calculations in cells. Some of the more common functions that we will use in simulation models are as follows:

MIN(*range*)—finds the smallest value in a range of cells
MAX(*range*)—finds the largest value in a range of cells
SUM(*range*)—finds the sum of values in a range of cells
AVERAGE(*range*)—finds the average of the values in a range of cells
STDEV(*range*)—finds the standard deviation for a sample in a range of cells
AND(*condition 1, condition 2, . . .*)—a logical function that returns TRUE if all conditions are true, and FALSE if not
OR(*condition 1, condition 2, . . .*)—a logical function that returns TRUE if any condition is true, and FALSE if not
IF(*condition, value if true, value if false*)—a logical function that returns one value if the condition is true and another if the condition is false
VLOOKUP(*value, table range, column number*)—looks up a value in a table

Excel has many other functions for statistical, financial and other applications, many of which we will use throughout the text. The easiest way to locate a particular function is to select a cell and then click on the *Paste Function* button "f_x" on the toolbar. This is particularly useful if you know what function to use, but you are not sure of what arguments to enter. Figure 2-4 shows the dialog box from which you may select the function you wish to use, in this case, the NORMINV function. Once this is selected, the dialog box in Figure 2-5 appears. When you click in an input cell, a description of the argument is shown. Thus, if you were not sure what to enter for the argument *Probability,* the explanation in Figure 2-5 will help you. For further information, you could click on the help button in the lower left-hand corner.

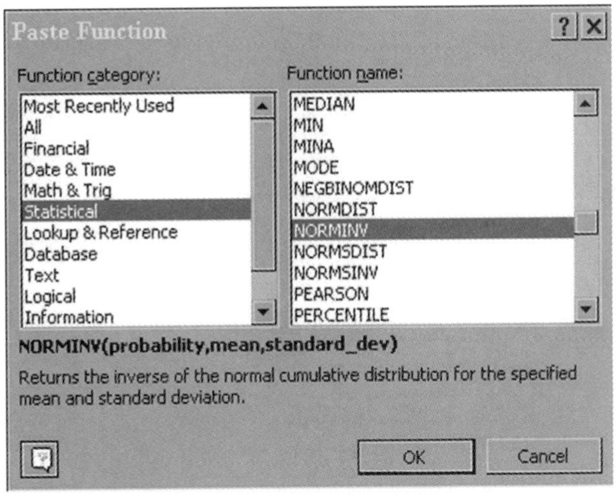

FIGURE 2-4 *Paste* Function Dialog Box

Functions may be linked with other standard arithmetic operations and can also be combined with one another. For instance, we may wish to compute a standard normal value using the formula (where μ = mean, and σ = standard deviation):

$$z = \frac{x - \mu}{\sigma}$$

Suppose that 50 values that define the distribution are stored in the range A1:A50 and that the value of *x* is in cell B1. If we wish to place the value of *z* in cell C2, we would write this formula in cell C2:

=(B1-AVERAGE(A1:A50))/STDEV(A1:A50)

One function that we will use quite often is *IF(condition, value if true, value if false)*. This function allows you to choose one of two values to enter into a cell. If the specified condition is true, *value if true* will be put in the cell. If the condition is false, *value if false* will be entered. For example, if cell C2 contains the function

=IF(A8=2,7,12)

this states that if the value in cell A8 is 2, the number 7 will be assigned to cell C2; if the value in cell A8 is not 2, the number 12 will be assigned to cell C2. "Conditions" may include

= equal to
> greater than

FIGURE 2-5 NORMINV *Paste* Function Input Dialog Box

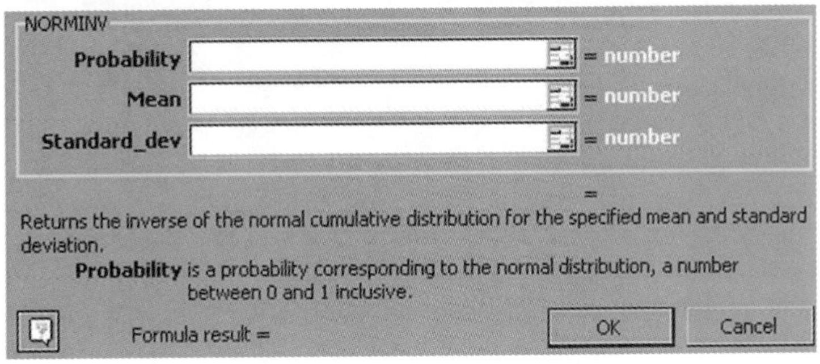

< less than
>= greater than or equal to
<= less than or equal to
<> not equal to

You may "nest" *IF* functions by replacing *value if true* or *value if false* in an *IF* function with another *IF* function. For example:

=IF(A8=2, IF(B3=5,"YES"," "),15)

This says that if the content of cell A8 is 2, then check the content of cell B3. If cell B3 contains 5, then the value of the function is the text string YES; if not, it is a blank space (a text string that is blank). However, if the content of cell A8 is not 2, then the value of the function is 15, no matter what cell B3 is. You may use *AND* and *OR* functions as the condition within an *IF* function, for example: IF(AND(B1 = 3,C1 = 5),12,22). Here, if cell B1 = 3 and cell C1 = 5, then the value of the function is 12; otherwise it is 22.

CHARTS AND GRAPHS

The Excel Chart Wizard is accessed from either the *Insert . . . Chart . . .* menu selection or by clicking on the *Chart Wizard* icon (the colored bar chart on the menu bar). The Chart Wizard guides you through four dialog boxes; the first is shown in Figure 2-6. The following steps outline the process of creating a chart:

1. Select the chart type from the list (e.g., *Bar*) and then click on the specific chart subtype option. Click *Next* or press *Enter* to continue.
2. The second dialog box asks you to define the data to plot. You may enter the data range directly or highlight it in your spreadsheet with your mouse. You also need to define whether the data are stored by rows or columns.

FIGURE 2-6 Excel Chart Wizard

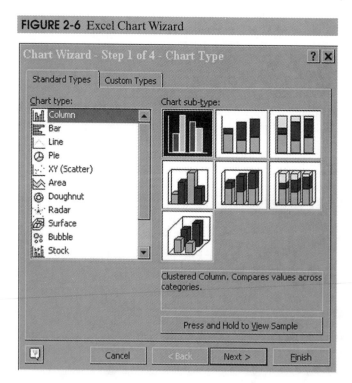

(Note: If the data you wish to plot are not stored in contiguous columns, hold down the *Ctrl* key while selecting each block of data; then start the Chart Wizard.) The *Series* tab allows you to check and modify the names and values of the data series in your chart.

3. The third dialog box allows you to specify details to customize the chart and make it easy to read and understand. You may specify titles for the chart and each axis, axis labels, style of gridlines, placement of the legend to describe the data series, data labels, and even a data table of values from which the chart is derived.

4. Finally, the last dialog box allows you to specify whether to place the chart as an object in an existing worksheet or as a new sheet in the workbook.

OTHER USEFUL EXCEL TIPS

- *Split Screen.* You may split the worksheet horizontally and/or vertically to view different parts of the worksheet at the same time. The vertical splitter bar is just to the right of the bottom scroll bar, and the horizontal splitter bar is just above the right-hand scroll bar. Position your cursor over one of these until it changes shape and then click and drag the splitter bar to the left or down.

- *Paste Special.* When you normally copy (one or more) cells and paste them in a worksheet, Excel places an exact copy of the formulas or data in the cells (except for relative addressing). Often you simply want the result of formulas so that the data will remain constant even if other parameters used in the formulas change. To do this, use the *Edit . . . Paste Special* option shown in Figure 2-7 instead of the normal *Paste* command. Checking the box for *Values* will paste the result of the formulas from which the data were calculated.

- *Column and Row Widths.* Many times, a cell contains a number that is too large to display properly because the column width is too small. You may change the column width from the menu option *Format . . . Column . . . Width.* However, an easier way to change a column width to fit the largest value or text string anywhere in the column is to position the cursor to the right of the column label so that it changes to a cross with horizontal arrows, and then double click. You may also move the arrow to the left or right to manually change the column width. You may change the row heights in a

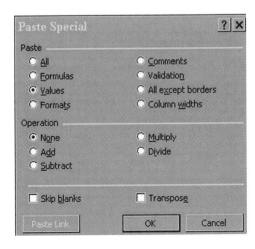

FIGURE 2-7 *Edit . . . Paste Special* Dialog Box

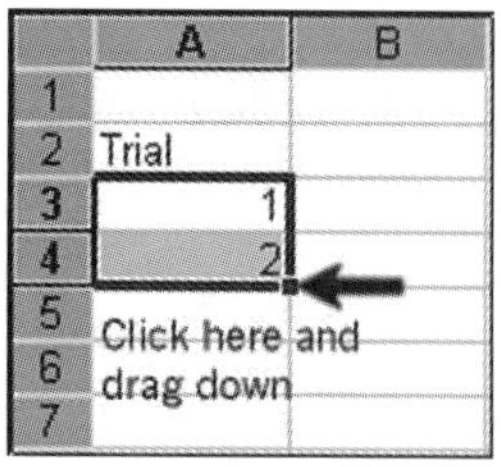

FIGURE 2-8 Filling a Range
with a Series

similar fashion by moving the cursor below the row number label. This can
be especially useful if you have a very long formula to display. To break a
formula within a cell, position the cursor at the break point in the formula
bar and press *Alt* and *Enter* simultaneously.

- *Displaying Formulas in Worksheets.* Choose *Tools . . . Options* from the
  menu bar and click on the *View* tab. Check the box for *Formulas.* You will
  probably need to change the column width to display the formulas properly.
- *Displaying Grid Lines and Row and Column Headers for Printing.* Choose
  File . . . Page Setup from the menu and click on the *Sheet* tab. Check the
  boxes for *Gridlines* and *Row and Column Headings.*
- *Filling a Range with a Series of Numbers.* Suppose you want to build a worksheet for simulating 100 trials in a Monte-Carlo simulation similar to Table 1-1.
  It would be tedious to have to enter the number of each trial from 1 to 100.
  Filling in a column of numbers in a series can be done using the *Edit . . .
  Fill . . . Series* command. Or, simply fill in the first few values in the series as
  shown in Figure 2-8 and highlight them. Now click and hold the mouse, and
  drag the small square in the lower right-hand corner down until you have
  filled in the column to 100; then release the mouse. Excel will show a small
  pop-up window that tells you the last value in the range.
- *Comment Boxes.* You may add "hidden" comment text boxes by clicking on
  some cell (that is perhaps labeled "Comment" or "Formulas," for example),
  and then selecting *Insert . . . Comment* from the menu. This allows you to include descriptive comments about the spreadsheet to assist other users without taking up valuable space on the worksheet itself. Comment cells are
  identified by a small red triangle in the upper right-hand corner of the cell.
  By positioning the cursor over the cell, the comment box is displayed.

Building Simulation Models Using Excel

Any spreadsheet should be built using principles of good design. A good layout is essential to user understanding. It should include a descriptive title, an input data section
area that is separate from the output and any working space, and a separate output section that provides the model results. Input data should be referenced only with cell references or range names so that any changes need only be made to the input data section
and can be reflected throughout the spreadsheet. Formats, such as currency or comma
formats, should be used appropriately. Complex calculations should be divided into several cells to minimize the chances for error and enhance understanding. Comments
should be placed next to formula cells or in comment boxes for explanation, if appropriate. We suggest that you consult one of the references by Thommes or Smith at the
end of this chapter for further discussion of good spreadsheet design.

A SIMULATION MODEL FOR DAVE'S CANDIES

Figure 2-9 shows a spreadsheet for modeling and simulating the order quantity problem for Dave's Candies that we introduced in Chapter 1. The input data are given in columns A and B. A table format for the simulation results is shown in columns D through F. We used the *IF* function to select between equations (1.1) and (1.2) given in Chapter 1, depending on whether demand is greater than, less than, or equal to the order quantity. Note that the formulas for profit in column F do not contain any specific numbers; they reference the prices, cost, and order quantity in column B. Thus, if we wish to change any of these inputs, we need not modify any formulas in the results table; they will be updated automatically.

Each row in the results table represents one trial of the simulation. To this point, we have not described how to generate demands from the probability distribution; the values in column E were generated by rolling a die and entered into the worksheet to verify the formulas for profit in column F. Later in this chapter, we will describe how to generate the demands randomly and will modify this spreadsheet accordingly.

Spreadsheet models should also include calculations of performance measures and key outputs for evaluating the model. For example, we would probably be interested in

FIGURE 2-9 Spreadsheet Model and Selected Cell Formulas for Dave's Candies

	A	B	C	D	E	F
1	**Dave's Candies Simulation**				**Simulation Results**	
2				*Trial*	*Demand*	*Profit*
3	*Selling price*	$ 12.00		1	50	$ 195.00
4	*Cost*	$ 7.50		2	40	$ 135.00
5	*Discount price*	$ 6.00		3	60	$ 255.00
6				4	90	$ 315.00
7	*Demand*	*Probability*		5	70	$ 315.00
8	40	1/6		6	50	$ 195.00
9	50	1/6		7	70	$ 315.00
10	60	1/6		8	70	$ 315.00
11	70	1/6		9	60	$ 255.00
12	80	1/6		10	50	$ 195.00
13	90	1/6				
14				Average Profit		$ 249.00
15	*Order Quantity*	70		Standard Deviation		$ 66.03

(a)

	C	D	E	F
1				
2		*Trial*	*Demand*	*Profit*
3		1	50	=IF(E3<=$B$15,$B$3*E3-$B$4*$B$15+$B$5*($B$15-E3),$B$3*$B$15-$B$4*B$15)
4		2	40	=IF(E4<=$B$15,$B$3*E4-$B$4*$B$15+$B$5*($B$15-E4),$B$3*$B$15-$B$4*B$15)
5		3	60	=IF(E5<=$B$15,$B$3*E5-$B$4*$B$15+$B$5*($B$15-E5),$B$3*$B$15-$B$4*B$15)
6		4	90	=IF(E6<=$B$15,$B$3*E6-$B$4*$B$15+$B$5*($B$15-E6),$B$3*$B$15-$B$4*B$15)
7		5	70	=IF(E7<=$B$15,$B$3*E7-$B$4*$B$15+$B$5*($B$15-E7),$B$3*$B$15-$B$4*B$15)
8		6	50	=IF(E8<=$B$15,$B$3*E8-$B$4*$B$15+$B$5*($B$15-E8),$B$3*$B$15-$B$4*B$15)
9		7	70	=IF(E9<=$B$15,$B$3*E9-$B$4*$B$15+$B$5*($B$15-E9),$B$3*$B$15-$B$4*B$15)
10		8	70	=IF(E10<=$B$15,$B$3*E10-$B$4*$B$15+$B$5*($B$15-E10),$B$3*$B$15-$B$4*B$15)
11		9	60	=IF(E11<=$B$15,$B$3*E11-$B$4*$B$15+$B$5*($B$15-E11),$B$3*$B$15-$B$4*B$15)
12		10	50	=IF(E12<=$B$15,$B$3*E12-$B$4*$B$15+$B$5*($B$15-E12),$B$3*$B$15-$B$4*B$15)
13				
14		Average Profit		=AVERAGE(F3:F12)
15		Standard Deviation		=STDEV(F3:F12)

(b)

some basic descriptive statistical measures, such as the average profit and standard deviation as shown in cells F14 and F15.

Open the file *Daves Candies.xls.* Change the order quantity in cell B15 from 40 to 90 in increments of 10 and observe how the simulation results change. Record the average profit and standard deviation for each order quantity. What is the best order quantity? How does the standard deviation change as the order quantity increases? What does this mean?

A SIMULATION MODEL FOR MANTEL MANUFACTURING

Figure 2-10 shows a spreadsheet model for the Mantel Manufacturing example in Chapter 1. The demand distribution and initial inventory are given in columns A and B. The simu-

FIGURE 2-10 Spreadsheet Model and Selected Cell Formulas for Mantel Manufacturing

	A	B	C	D	E	F	G	H
1	Mantel Manufacturing				Simulation Results			
2								
3	Demand	Probability			Average ending inventory		84.00	
4	80	1/6			Number of additional shifts		1	
5	90	1/6						
6	100	1/6			Beginning			Ending
7	110	1/6		Day	Inventory	Demand	Production	Inventory
8	120	1/6		1	100	130	100	70
9	130	1/6		2	70	80	100	90
10				3	90	110	100	80
11	Initial Inventory			4	80	90	100	90
12	100			5	90	110	100	80
13				6	80	130	100	50
14				7	50	120	100	30
15				8	30	110	200	120
16				9	120	110	100	110
17				10	110	90	100	120

(a)

	D	E	F	G	H
1			Simulation Results		
2					
3		Average ending inventory		=AVERAGE(H8:H17)	
4		Number of additional shifts		=COUNTIF(G8:G17,"=200")	
5					
6		Beginning			Ending
7	Day	Inventory	Demand	Production	Inventory
8	1	=A12	130	100	=E8+G8-F8
9	2	=H8	80	=100+IF(E9<50,100)	=E9+G9-F9
10	3	=H9	110	=100+IF(E10<50,100)	=E10+G10-F10
11	4	=H10	90	=100+IF(E11<50,100)	=E11+G11-F11
12	5	=H11	110	=100+IF(E12<50,100)	=E12+G12-F12
13	6	=H12	130	=100+IF(E13<50,100)	=E13+G13-F13
14	7	=H13	120	=100+IF(E14<50,100)	=E14+G14-F14
15	8	=H14	110	=100+IF(E15<50,100)	=E15+G15-F15
16	9	=H15	110	=100+IF(E16<50,100)	=E16+G16-F16
17	10	=H16	90	=100+IF(E17<50,100)	=E17+G17-F17

(b)

lation results are given in columns D through H. For day 1, the beginning inventory is copied from cell A12; for all other days, it is simply the ending inventory from the previous day. Again, we have generated demands in column F simply by rolling a die for illustrative purposes. Note that an *IF* function is used to test whether the day's beginning inventory is less than 50. If it is, then an additional 100 units is added to the normal production run of 100 units. The ending inventory is computed using equation (1.3) from Chapter 1. Output results that we might be interested in are the average ending inventory and the number of additional shifts. The number of additional shifts is computed using the Excel function =COUNTIF (G8:G17,"=200"), which counts the number of values in column G that equal 200.

SKILLBUILDER EXERCISE

Open the file *Mantel Manufacturing.xls*. What happens to the simulation results if the initial inventory is changed to 30? Why doesn't the average inventory change? Modify the spreadsheet to compute the average demand and average production.

CONSTRUCTING FREQUENCY DISTRIBUTIONS AND HISTOGRAMS

Simulation models generate quite a bit of output data that need to be summarized. For instance, in the Dave's Candies simulation model in Figure 2-9, we would be interested in not only the average profit over the 10 trials, but also the distribution of profit in order to gain some insight about the likelihood of realizing different values of profit. We do this using frequency distributions and histograms.

In Excel, frequency distributions and histograms may be constructed in two principal ways. One approach is to use the Histogram tool by clicking on *Tools . . . Data Analysis* and selecting *Histogram* from the list. In the dialog box (Figure 2-11), specify the *Input Range*. Check the *Labels* box if the range contains a label. If you do not specify a *Bin Range* ("bin" refers to the upper limit of cell intervals used to group continuous data or the cell value for discrete data), Excel will automatically determine cell

FIGURE 2-11 Excel Histogram Tool Dialog Box

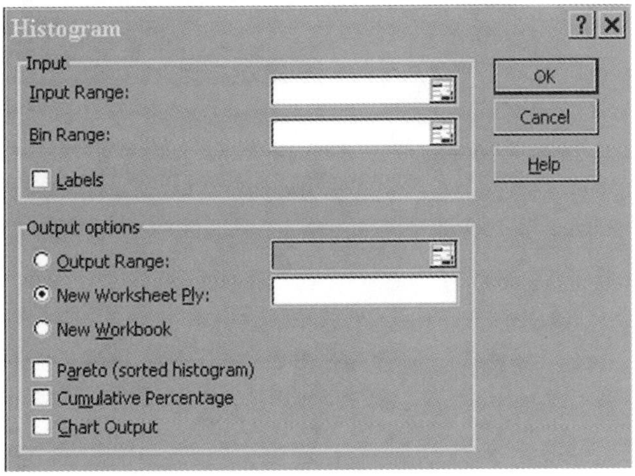

ranges for the frequency distribution and histogram. It is best to define the cell ranges yourself by specifying the upper limits for cell intervals or discrete cell values in a column in your worksheet. (If you check the *Data Labels* box, be sure you include a column label such as "Upper Cell Limit" or "Value.") Check the *Chart Output* box to create a histogram in addition to the frequency distribution.

To illustrate this for the Dave's Candies example, note that the profits are discrete values, which, for an order quantity of 70, can assume only the values $135, $195, $255, and $315. We define these as the bin range in cells A19:A22. Applying the Histogram tool to the profit data in column F results in the frequency distribution and histogram shown in Figure 2-12. The tool always adds an overflow cell for any values larger than the upper limit of the last bin (that is, Excel assumes the data are continuous within the intervals defined by the bin range).

A serious limitation of the Excel Histogram tool is that the frequency distribution and histogram are not linked to the data; thus, if you change any of the data, you must repeat the entire procedure to construct a new frequency distribution and histogram. An alternative is to use Excel's *FREQUENCY* function and the Chart Wizard. First, define the bins as in the first approach. Select the range of cells adjacent to the bin range for a discrete distribution. If your data are continuous, add one additional empty cell below it. (This provides an overflow cell.) Then enter the formula =FREQUENCY(*range of data, range of bins*), and press *Ctrl-Shift-Enter* simultaneously. This is necessary because *FREQUENCY* is an array function in Excel. This will create the frequency distribution. You may then construct a histogram using the Chart Wizard for a column chart, customizing it as appropriate. For the Dave's Candies example, we input the range of data as F3:F12 and the range of bins as B19:B22 in the *FREQUENCY* function, resulting in the output shown in Figure 2-13. The Chart Wizard was used to construct the histogram, which is the same as in Figure 2-12. Now, if the data are changed, the frequency distribution and histogram will be updated automatically.

SKILLBUILDER EXERCISE

Open the file *Mantel Manufacturing.xls*. Apply both the Excel Histogram tool and the *FREQUENCY* function approach to construct frequency distribu-tions and histograms for both the demand and ending inventory values.

FIGURE 2-12 Histogram Tool Results

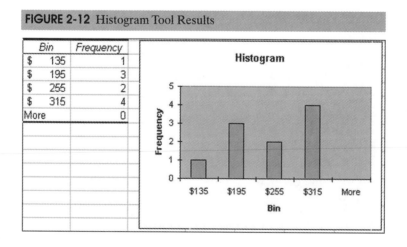

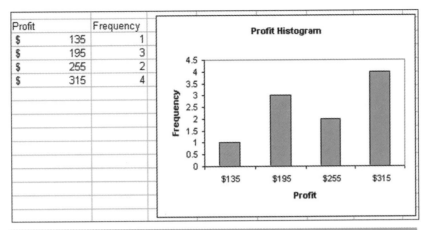

FIGURE 2-13 Using Excel's *FREQUENCY* Function

Random Variates

In both examples, we need to simulate the process of rolling a die on a spreadsheet in order to generate the outcomes that drive the simulation. Rolling a die is a physical experiment that randomly generates a number between 1 and 6, each with equal probability. Many other physical processes can be used to generate random outcomes. These include spinning a roulette wheel, drawing from a deck of shuffled cards, or selecting numbered balls drawn from a cage as is done for state lotteries. Clearly, these outcomes are limited in scope. For practical simulation problems, we will need to generate outcomes efficiently from many different types of probability distributions, such as an arbitrary discrete distribution or a normal, exponential, or Poisson distribution, to name just a few. An outcome generated from a probability distribution is called a **random variate.** Fortunately, a combination of computing technology and a bit of mathematical theory allow us to generate random variates from any distribution we desire.

RANDOM NUMBERS

The basis for generating random variates is the concept of random numbers. In simulation terminology, a **random number** is one that is uniformly distributed between 0 and 1. Recall from statistics that a uniform probability distribution characterizes a random variable for which all outcomes between a minimum value a and a maximum value b are equally likely. (In the next chapter, we will formally review this and other useful probability distributions in simulation.) Thus, all real values between zero and one are equally likely. You might think of a random number as the outcome resulting from the spinner shown in Figure 2-14. Virtually all computer programming languages have the capability of generating a stream of independent random numbers.

Technically speaking, computers cannot generate numbers that are truly random because they must use a deterministic algorithm; that is, a clearly defined sequence of steps. However, random number algorithms are designed to generate a stream of numbers that *appear* to be random—you would not be able to predict the next number simply by looking at the previous numbers that have been generated. Because they are not truly random, computer-generated streams of uniformly distributed numbers between 0 and 1 are often called **pseudorandom numbers.** Because this book is focused solely on computer simulation, we will refer to pseudorandom numbers simply as *random numbers.* The appendix to this chapter provides some technical information on how random numbers are generated on a computer.

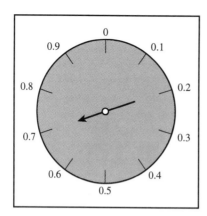

FIGURE 2-14 Spinner Analogy for Generating Random Numbers

In Excel, we may generate a random number within any cell by using the function =RAND( ). This function has no arguments; therefore, nothing should be placed within the parentheses (but they must be included!). Figure 2-15 shows a table of 100 random numbers generated in Excel. You should be aware that unless the automatic recalculation feature is suppressed, whenever any cell in the spreadsheet is modified, the values in any cell containing the *RAND( )* function will change. Automatic recalculation can be changed to manual in the *Tools/Options/Calculation* menu. Under manual recalculation mode, the worksheet is recalculated only whenever the F9 key is pressed.

Figure 2-16 shows a frequency distribution of the 100 random numbers in Figure 2-15. If the random numbers are truly uniformly distributed between 0 and 1, we would expect an equal amount (10) in each cell. However, we see that some variation exists, al-

FIGURE 2-15 100 Random Numbers Generated in Excel

	A	B	C	D	E	F	G	H	I	J
1	**Random Numbers**									
2										
3	0.2014	0.0600	0.2597	0.9992	0.4529	0.8057	0.1539	0.3389	0.0593	0.1766
4	0.4699	0.7001	0.8196	0.0402	0.5299	0.7954	0.4316	0.1440	0.8935	0.1687
5	0.3586	0.6013	0.3320	0.7444	0.4777	0.8264	0.3347	0.1681	0.7261	0.1372
6	0.6043	0.8375	0.5392	0.3784	0.2122	0.5367	0.3429	0.2743	0.3593	0.0952
7	0.1440	0.9043	0.0265	0.8876	0.8719	0.8429	0.1008	0.4919	0.1679	0.7084
8	0.8986	0.3382	0.6960	0.2134	0.6857	0.8938	0.3105	0.9710	0.2757	0.4975
9	0.8412	0.1241	0.5489	0.2772	0.2727	0.6212	0.7258	0.4951	0.8545	0.9938
10	0.8108	0.4458	0.8067	0.7874	0.9359	0.9439	0.7850	0.6329	0.6588	0.1416
11	0.8695	0.0984	0.2850	0.6106	0.6538	0.4530	0.6385	0.7042	0.5296	0.0007
12	0.8845	0.1036	0.3122	0.8744	0.3285	0.2007	0.7726	0.2429	0.6557	0.7386

Lower Cell Limit	Upper Cell Limit	Frequency
0.0	0.1	7
0.1	0.2	12
0.2	0.3	11
0.3	0.4	11
0.4	0.5	9
0.5	0.6	5
0.6	0.7	11
0.7	0.8	11
0.8	0.9	17
0.9	1.0	6

FIGURE 2-16 Frequency Distribution of Random Numbers in Figure 2-15

though the distribution is generally uniform. We may conduct various statistical tests to verify that the random number generation process is indeed uniform and independently distributed. The appendix to this chapter also addresses these issues.

Create a worksheet for generating 500 random numbers and construct a frequency distribution using the *FREQUENCY* function approach. Does the frequency distribution show more uniformity than the distribution of 100 numbers in Figure 2-16? Press the F9 key several times to generate new sets of numbers and frequency distributions. What do you observe?

GENERATING RANDOM VARIATES FROM DISCRETE DISTRIBUTIONS

In both the Dave's Candies and Mantel Manufacturing examples, we need to generate outcomes from a discrete probability distribution. Specifically, we need to generate a random variate, x, from the probability distribution:

x	$f(x)$
1	1/6
2	1/6
3	1/6
4	1/6
5	1/6
6	1/6

Two properties of discrete probability distributions that allow us to use random numbers to generate random variates easily are (1) the probability of any outcome is always between 0 and 1 and (2) the sum of the probabilities of all outcomes adds to 1. We can, therefore, divide the range from 0 to 1 into intervals that correspond to the probabilities of the discrete outcomes. Any random number, then, must fall within one of these intervals. To see this more clearly, first construct the cumulative probability distribution, $F(x)$.

x	$p(x)$	$F(x)$
1	1/6	1/6 (.1667)
2	1/6	1/3 (.3333)
3	1/6	1/2 (.5000)
4	1/6	2/3 (.6667)
5	1/6	5/6 (.8333)
6	1/6	1

The cumulative probability distribution partitions the interval from 0 to 1 into subintervals whose size are equal to the probabilities of the associated outcomes, as shown in Figure 2-17. For instance, the interval from 0 up to but not including 1/6 has a length of 1/6 and corresponds to the probability that the outcome $x = 1$; the interval from 1/6 up to but not including 1/3 also has a length of 1/6 and corresponds to the probability that $x = 2$; and so on. (For consistency, we do not include the upper limit of an interval in the interval to make them mutually exclusive.)

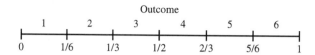

FIGURE 2-17 Assigning Outcomes to the Interval $(0, 1)$

To generate a random variate from this distribution, all we need to do is to select a random number and determine the interval into which it falls. Suppose we use the first row in Figure 2-15.

The first random number is 0.2014. This falls between 1/6 and 1/3; thus, the corresponding random variate is $x = 2$. The second random number is 0.0600. This number falls between 0 and 1/6, generating a random variate $x = 1$. What would the next three random variates be? We have developed a numerical method for rolling dice!

We can generalize this approach to generate outcomes from *any* discrete distribution. For example, suppose we have the distribution:

x	$f(x)$	$F(x)$
5	.1	.1
20	.4	.5
30	.3	.8
50	.2	1.0

The ranges of random numbers corresponding to each outcome are as follows:

Random Number Range	*Outcome*
0–.1	5
.1–.5	20
.5–.8	30
.8–1.0	50

Note that the probability of falling within a particular random number range corresponds to the relative frequency of the associated outcome. To generate an outcome, we select a random number, determine the range into which it falls, and select the corresponding outcome. If this is done repeatedly, the frequency of occurrence of each outcome should be proportional to the random number range because random numbers are uniformly distributed.

USING EXCEL'S VLOOKUP FUNCTION

To implement the procedure of generating discrete random variates in Excel, we use the function

$$=\text{VLOOKUP}(lookup\_value, table\_array, col\_index\_num)$$

This function compares the *lookup_value* to the values in the first column of *table_array* (which must be in ascending order) until it finds the largest value less than or equal to *lookup_value*. Then it returns the value in the column defined by *col_index_num*. We will illustrate the use of this function by applying it to the previous example.

Figure 2-18 shows a spreadsheet and selected cell formulas. The random number range and outcome array are specified in cells A2:C5. We generate random numbers in column A, rows 8 through 12. In cells B8 through B12, we use the *VLOOKUP* function to select an outcome. In this example, the first random number is 0.0328. *VLOOKUP* searches the first column of the array until it finds the largest value less than or equal to 0.0329; this is 0, found in cell A2. Because *col_index_num* is defined as 3 in the function, the function returns the value in the third column of this row of the array, namely, 5. You can see that this procedure correctly generates outcomes based on this probability distribution. Note that column B of the array is not really necessary. However, we included it so that the random number ranges are clearly defined.

SKILLBUILDER EXERCISE

Create a worksheet for generating random variates from two experiments: rolling one die and rolling two dice, using the *VLOOKUP* function. (Make sure you have the correct probabilities for the two-dice case.)

USING EXCEL'S RANDOM NUMBER GENERATION TOOL

Excel allows you to generate random variates from discrete distributions (as well as others, as we shall see in Chapter 3) without using the *VLOOKUP* function. From the main toolbar, select *Tools/Data Analysis/Random Number Generation*. Actually, the name "Random Number Generation" is a misnomer, because the tool generates random variates; Excel improperly uses the term "random number" the way we use "random variate," so don't get confused. The Random Number Generation dialog box, shown in Figure 2-19, asks you to specify the upper-left cell reference of the output table that will store the outcomes, the number of variables (columns of values you want generated), number of "random numbers" (the number of outcomes you want generated for each variable), and the type of distribution. The default distribution is the discrete distribution. A discrete distribution must contain two columns: The left column contains the outcomes, and the right column contains the probabilities associated with the outcomes

	A	B	C
1	Random number range		Outcome
2	0	0.2	5
3	0.2	0.6	20
4	0.6	0.9	30
5	0.9	1	50
6			
7	Random no.	Outcome	
8	0.0328	5	
9	0.4799	20	
10	0.9047	50	
11	0.6323	30	
12	0.3540	20	

(a)

FIGURE 2-18 Spreadsheet Showing Selected Cell Formulas

7	Random no.	Outcome
8	=RAND()	=VLOOKUP(A8,$A$2:$C$5,3)
9	=RAND()	=VLOOKUP(A9,$A$2:$C$5,3)
10	=RAND()	=VLOOKUP(A10,$A$2:$C$5,3)
11	=RAND()	=VLOOKUP(A11,$A$2:$C$5,3)
12	=RAND()	=VLOOKUP(A12,$A$2:$C$5,3)

(b)

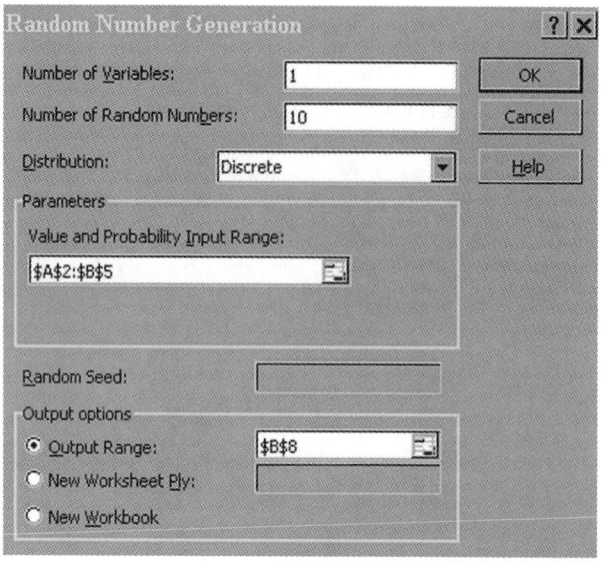

FIGURE 2-19 Random Number Generation Dialog Box

(which must sum to 1.0). Note that this is quite different from the way we set up the data to use the *VLOOKUP* function.

Figure 2-20 shows the spreadsheet used with the dialog box in Figure 2-19 for generating outcomes from the example distribution. The outcomes generated are found in cells B8 through B17. Like the Histogram tool, the Random Number Generation tool produces output that is not linked to the original data; thus, any changes in the random number ranges or outcomes will not be reflected automatically in the output. However, one advantage the Random Number Generation tool does have, as seen in the dialog box in Figure 2-19, is the option of specifying a random number seed. A **random number seed** is a value from which a stream of random numbers is generated. By specifying the same seed, you can produce the same random numbers at a later time. This is desirable when we wish to reproduce an identical sequence of "random" events in a simulation in order to test the effects of different policies or decision variables under the same circumstances. To do this using the *VLOOKUP* function, you would have to save the outcomes using the *Paste . . . Special . . . Values* command. Nevertheless, the Ran-

	A	B
1	Outcome	Probability
2	5	0.1
3	20	0.4
4	30	0.3
5	50	0.2
6		
7	Trial	Outcome
8	1	20
9	2	20
10	3	30
11	4	50
12	5	50
13	6	50
14	7	5
15	8	20
16	9	50
17	10	20

FIGURE 2-20 Results from the Random Number Generation Tool

dom Number Generation tool is generally cumbersome to use; therefore, we do not recommend it for simulation modeling.

Monte-Carlo Simulation on Spreadsheets

We now have all the fundamentals that we need to perform a Monte-Carlo simulation using an Excel spreadsheet model. The process for performing Monte-Carlo simulations with spreadsheets is as follows:

1. Develop the spreadsheet model, paying particular attention to the format for displaying the output results.
2. Generate random variates for each probabilistic variable according to its probability distribution and use the results in the appropriate formulas in the simulation model.
3. Repeat step 2 a sufficient number of times to create a distribution of results. The number of trials to use depends on the statistical precision you wish to obtain; we discuss this issue in Chapter 3.
4. Compute summary statistics and collect output data in a frequency distribution or histogram for analysis.

We will illustrate this approach for both the Dave's Candies and Mantel Manufacturing examples.

SIMULATION ANALYSIS OF DAVE'S CANDIES

Figure 2-21 shows a portion of the spreadsheet for the Dave's Candies problem. We have modified the worksheet in Figure 2-9 to generate demands using the *VLOOKUP* function. The complete spreadsheet replicates the simulation for 100 trials. Figure 2-22 shows the remainder of the worksheet that displays the summary results. We defined the bin range to include all possible values of profit for any order quantity so that the frequency distribution and histogram would be correct for any choice. We used the *FREQUENCY* function to calculate the frequency distribution so that the summary results would be linked to the simulation results if we repeat the simulation or change order quantities.

To compare different order quantities, we could resimulate the entire spreadsheet by changing the order quantity in cell B15. Using this approach, however, would require us to record or copy the summary statistics each time the order quantity is changed. An alternative is to redesign the spreadsheet to evaluate different order quantities together. Excel provides a convenient way to do this using a one-way data table. A *one-way data table* evaluates a formula for different values of inputs. This is shown in the lower left-hand portion of Figure 2-22.

To create a one-way data table, first create a column of inputs (order quantity) to evaluate (the range H27:H32 in Figure 2-22). In the cell immediately to the right and one cell up from the top of the list (cell I26), copy the formula for the output; in this case, the average profit. Thus, cell I26 has the formula =I22. Next, select the data table

	A	B	C	D	E	F
1	Dave's Candies Simulation			Simulation Results		
2				Trial	Demand	Profit
3	Selling price	$ 12.00		1	70	$315.00
4	Cost	$ 7.50		2	40	$135.00
5	Discount price	$ 6.00		3	50	$195.00
6				4	50	$195.00
7	Demand	Probability		5	50	$195.00
8	40	1/6		6	50	$195.00
9	50	1/6		7	80	$315.00
10	60	1/6		8	70	$315.00
11	70	1/6		9	70	$315.00
12	80	1/6		10	60	$255.00
13	90	1/6		11	60	$255.00
14				12	60	$255.00
15	Order Quantity	70		13	70	$315.00
16				14	50	$195.00
17				15	70	$315.00
18	Random Number Range		Demand	16	50	$195.00
19	0	1/6	40	17	40	$135.00
20	1/6	1/3	50	18	80	$315.00
21	1/3	1/2	60	19	80	$315.00
22	1/2	2/3	70	20	80	$315.00
23	2/3	5/6	80	21	70	$315.00
24	5/6	1	90	22	90	$315.00

FIGURE 2-21 Modified Spreadsheet for Dave's Candies Problem

range—the smallest rectangular block that includes both the formula and all the values in the input range (H26:I32). Select *Table* from the *Data* menu on the Excel menu bar, and specify the location of the column input cell in your model as the cell in the model that corresponds to the order quantity (B15). Use the column input cell because the values we wish to evaluate are arranged in a column. (If they were in a row, you would place this cell location in the row input cell box.) In this case the row input cell is left blank. Click OK, and Excel evaluates the average profit for each order quantity based on the random demands in column E as shown in the range I27:I32.

One advantage of this approach versus changing the order quantity in cell B15 in Figure 2-21 and resimulating the entire spreadsheet is that the average profits in the data table are evaluated for the same stream of demands in column E. This provides a more valid comparison. (This is similar to the notion of using the same random number seed to generate the same stream of demands for each order quantity as we discussed earlier.)

SKILLBUILDER EXERCISE

Open the file *Daves Candies Monte Carlo Simulation.xls.* Modify the demand distribution and lookup table for demands of 50, 55, 60, 65, 70, 75, and 80 with probabilities 0.05, 0.10, 0.20, 0.30, 0.20, 0.10, and 0.05, respectively. Construct a new data table for order quantities equal to the range of demands. What is the best order quantity? What happens if the discount price drops to $3.00?

SIMULATION ANALYSIS OF MANTEL MANUFACTURING

Figure 2-23 shows a portion of a modified spreadsheet for the Mantel Manufacturing example that includes a lookup table for generating demands and simulates 260 days (assuming a 5-day workweek, this represents about 1 year). Because this is a dynamic

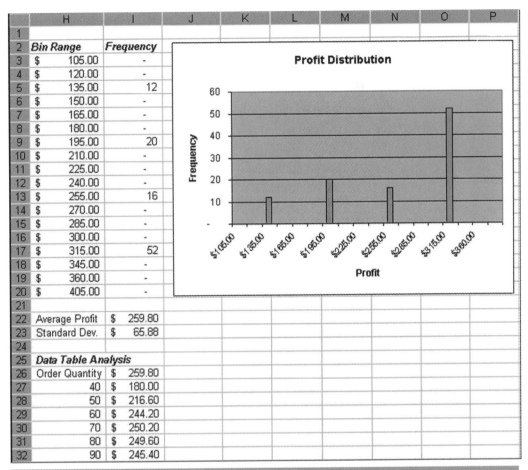

	Bin Range	Frequency
2	Bin Range	Frequency
3	$ 105.00	-
4	$ 120.00	-
5	$ 135.00	12
6	$ 150.00	-
7	$ 165.00	-
8	$ 180.00	-
9	$ 195.00	20
10	$ 210.00	-
11	$ 225.00	-
12	$ 240.00	-
13	$ 255.00	16
14	$ 270.00	-
15	$ 285.00	-
16	$ 300.00	-
17	$ 315.00	52
18	$ 345.00	-
19	$ 360.00	-
20	$ 405.00	-
21		
22	Average Profit	$ 259.80
23	Standard Dev.	$ 65.88
24		
25	**Data Table Analysis**	
26	Order Quantity	$ 259.80
27	40	$ 180.00
28	50	$ 216.60
29	60	$ 244.20
30	70	$ 250.20
31	80	$ 249.60
32	90	$ 245.40

FIGURE 2-22 Summary Results for Dave's Candies Monte-Carlo Simulation

simulation, the 260 days represents only *one* trial, not 260 trials. Thus, the results provide no information about the distribution of the number of additional shifts that might occur over an arbitrary 260-day period; they only show the result for 1 year. If the company management needed to set a budget, they would want to know the expected number of additional shifts as well as the risk that the number of additional shifts might exceed 15, for instance.

We can apply Monte-Carlo principles to this simulation model by replicating the entire 260-day simulation using a one-way data table, as shown in Figure 2-24. First, create a column representing the number of trials, say 100. Immediately to the right and one up from the top value in this column, copy the formula for the number of additional shifts. Select the data table range as before; then select *Table* from the *Data* menu to bring up the data table dialog box. In this case, however, the column input cell does not correspond to any model input. Simply click on any blank cell in the worksheet as the column input cell value. When you click OK, Excel will recalculate the worksheet for each row of the data table. Because the demand lookup function uses the RAND( ) function, each trial will be based on different values of demand. In this fashion, we have easily replicated the entire simulation.

Figure 2-24 also shows a portion of the results. Again, we used the *FREQUENCY* function to construct a frequency distribution and the Chart Wizard for the histogram. Because we may be unsure of the actual range of the data, we add one additional row

	A	B	C	D	E	F	G	H
1	**Mantel Manufacturing**					**Simulation Results**		
2								
3	*Demand*	*Probability*			Average ending inventory		95.62	
4	80	1/6			Number of additional shifts		16	
5	90	1/6						
6	100	1/6			*Beginning*			*Ending*
7	110	1/6		*Day*	*Inventory*	*Demand*	*Production*	*Inventory*
8	120	1/6		1	100	90	100	110
9	130	1/6		2	110	120	100	90
10				3	90	110	100	80
11	Initial Inventory			4	80	130	100	50
12	100			5	50	130	100	20
13				6	20	90	200	130
14				7	130	120	100	110
15	*Random Number Range*		*Demand*	8	110	80	100	130
16	0	1/6	80	9	130	90	100	140
17	1/6	1/3	90	10	140	120	100	120
18	1/3	1/2	100	11	120	130	100	90
19	1/2	2/3	110	12	90	110	100	80
20	2/3	5/6	120	13	80	80	100	100
21	5/6	1	130	14	100	90	100	110
22				15	110	90	100	120

FIGURE 2-23 Modified Spreadsheet for Mantel Manufacturing Problem

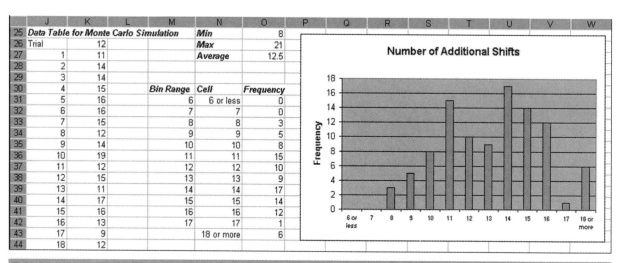

	J	K	L	M	N	O
25	*Data Table for Monte Carlo Simulation*			*Min*	8	
26	Trial	12		*Max*	21	
27	1	11		*Average*	12.5	
28	2	14				
29	3	14				
30	4	15		*Bin Range*	*Cell*	*Frequency*
31	5	16		6	6 or less	0
32	6	16		7	7	0
33	7	15		8	8	3
34	8	12		9	9	5
35	9	14		10	10	8
36	10	19		11	11	15
37	11	12		12	12	10
38	12	15		13	13	9
39	13	11		14	14	17
40	14	17		15	15	14
41	15	16		16	16	12
42	16	13		17	17	1
43	17	9			18 or more	6
44	18	12				

FIGURE 2-24 Summary Results for Mantel Manufacturing Monte-Carlo Simulation

to the bin range in the *FREQUENCY* function. Thus, the bin range in this example is defined as M31:M43. The last bin is the overflow bin that includes any values greater than 17. Also note that the first bin would contain any values 6 or less. To make this clear, we have added an additional column N labeled "Cell" and used this to label the *x* axis in the histogram. We could not use this column to define the bin range because it contains nonnumerical text.

We see considerable variation in the number of shifts that might occur over a 1-year period—from 6 to 19—even though the average is 12.5. The company would face con-

siderable risk in establishing a budget based only on the average value and would probably want to plan for a larger number of additional shifts.

SIMULATING CASINO GAMES

Because the Monte-Carlo simulation was named after the famed casino, we probably
would be remiss not to apply simulation to some popular casino games! Actually, simulating casino games reinforces the fundamental principles of simulation and introduces
some new approaches to spreadsheet simulation. It also shows that simulation can be
applied to recreational activities and be a lot of fun.

Our first example is the game of roulette. Roulette is played at a table similar to the
one in Figure 2-25. A wheel with the numbers 1 through 36 (evenly distributed with the
colors red and black) and two green numbers 0 and 00 rotates in a shallow bowl with a
curved wall. A small ball is spun on the inside of the wall and drops into a pocket corresponding to one of the numbers. Players may make 11 different types of bets by placing chips on different areas of the table. These include bets on a single number, two
adjacent numbers, a row of three numbers, a block of four numbers, two adjacent rows
of six numbers, and the five number combinations of 0, 00, 1, 2, and 3; bets on the numbers 1–18 or 19–36; the first, second, or third group of 12 numbers; a column of 12 numbers; even or odd; and red or black. Payoffs differ by bet. For instance, a single-number
bet pays 35 to 1 if it wins; a three-number bet pays 11 to 1; a column bet pays 2 to 1; and
a color bet pays even money.

To simulate roulette, we observe that the outcome of each of the 38 numbers is
equally likely and that we need to know both the number and its color. Figure 2-26
shows the lookup table used to draw a random variate from this distribution. A portion
of the remainder of the worksheet is shown in Figure 2-27. We will assume that the player
decides to bet on a single number, one column, and a color on every spin. These values
are entered in cells I4, J4, and K4, respectively. For each trial, we select a random number using the RAND( ) function and determine the number and its color. Using *IF* functions, we then determine whether any of the bets win and enter the payoff (or loss of $1
per bet) in each column. In column L, we sum the payoff for each trial and compute the
average payoff per trial in cell L3, based on 100 trials.

The cell formulas are too long to show in Figure 2-27; however, an example of the
formula in cell J5 is:

```
=IF(OR(AND($J$4=1,OR(G5=1,G5=4,G5=7,G5=10,G5=13,G5=16,G5=19,G5=22,G5=25,G5=28,
G5=31,G5=34)),AND($J$4=2,OR(G5=2,G5=5,G5=8,G5=11,G5=14,G5=17,G5=20,G5=23,G5=26,
G5=29,G5=32,G5=35)),AND($J$4=3,OR(G5=3,G5=6,G5=9,G5=12,G5=15,G5=18,G5=21,G5=24,
G5=27,G5=30,G5=33,G5=36))),$J$1,-1)
```

FIGURE 2-25 Layout of a Typical Roulette Table

	A	B	C	D
4	**Random Number Range**		**Number**	**Color**
5	0	1/38	1	Red
6	1/38	1/19	2	Black
7	1/19	3/38	3	Red
8	3/38	2/19	4	Black
9	2/19	5/38	5	Red
10	5/38	3/19	6	Black
11	3/19	7/38	7	Red
12	7/38	4/19	8	Black
13	4/19	9/38	9	Red
14	9/38	5/19	10	Black
15	5/19	11/38	11	Black
16	11/38	6/19	12	Red
17	6/19	13/38	13	Black
18	13/38	7/19	14	Red
19	7/19	15/38	15	Black
20	15/38	8/19	16	Red
21	8/19	17/38	17	Black
22	17/38	9/19	18	Red
23	9/19	1/2	19	Red
24	1/2	10/19	20	Black
25	10/19	21/38	21	Red
26	21/38	11/19	22	Black
27	11/19	23/38	23	Red
28	23/38	12/19	24	Black
29	12/19	25/38	25	Red
30	25/38	13/19	26	Black
31	13/19	27/38	27	Red
32	27/38	14/19	28	Black
33	14/19	29/38	29	Black
34	29/38	15/19	30	Red
35	15/19	31/38	31	Black
36	31/38	16/19	32	Red
37	16/19	33/38	33	Black
38	33/38	17/19	34	Red
39	17/19	35/38	35	Black
40	35/38	18/19	36	Red
41	18/19	37/38	0	Green
42	37/38	1	00	Green

FIGURE 2-26 Lookup Table for Simulating Roulette Outcomes

Although a bit messy, the formulas are not complicated and simply check what number is hit. The results of the simulation in Figure 2-27 show that the player loses $0.39 for each $3 bet (cell L3). For 1,000 bets, this results in an average total loss of $390.

SKILLBUILDER EXERCISE

Open the file *Roulette Simulation.xls*. Examine the formulas in columns I, J, and K to understand how the results are computed. Add a new column to allow a bet on the five numbers 0, 00, 1, 2, or 3. If any of these numbers hits, the payoff is 6 to 1. If a player adds this bet to the other three, how does the average payoff compare? (Hint: Compute both the average payoff for the original three bets and for the four bets using the same simulated outcomes.)

Roulette is an example of a pure Monte-Carlo simulation model because each trial is independent. A more complicated casino game to simulate is the dice game of craps. A new player rolls two dice. If the roll (sum of the dice) is 7 or 11, the player wins. If the roll is 2, 3, or 12, the player loses. Any other number becomes the *point*. The player continues to roll until he or she either rolls the point before 7, in which case the player wins, or rolls a 7 before the point, resulting in a loss. The payoff for winning is even money. Many other bets can be placed during the course of a game, but these are too complicated to describe here.

	E	F	G	H	I	J	K	L
1				*Payoff*	$ 35.00	$ 2.00	$ 1.00	
2					*Single*			*Average Payoff*
3		*Random*			*Number*	*Column*	*Color*	$ (0.39)
4	*Trial*	*Number*	*Winner*		13	2	Black	*Total Payoff*
5	1	0.0278	2	Black	$ (1.00)	$ 2.00	$ 1.00	$ 2.00
6	2	0.4689	18	Red	$ (1.00)	$ (1.00)	$ (1.00)	$ (3.00)
7	3	0.5389	21	Red	$ (1.00)	$ (1.00)	$ (1.00)	$ (3.00)
8	4	0.6708	26	Black	$ (1.00)	$ 2.00	$ 1.00	$ 2.00
9	5	0.6679	26	Black	$ (1.00)	$ 2.00	$ 1.00	$ 2.00
10	6	0.5751	22	Black	$ (1.00)	$ (1.00)	$ 1.00	$ (1.00)
11	7	0.1431	6	Black	$ (1.00)	$ (1.00)	$ 1.00	$ (1.00)
12	8	0.4386	17	Black	$ (1.00)	$ 2.00	$ 1.00	$ 2.00
13	9	0.0207	1	Red	$ (1.00)	$ (1.00)	$ (1.00)	$ (3.00)
14	10	0.1896	8	Black	$ (1.00)	$ 2.00	$ 1.00	$ 2.00
15	11	0.7199	28	Black	$ (1.00)	$ (1.00)	$ 1.00	$ (1.00)
16	12	0.9018	35	Black	$ (1.00)	$ 2.00	$ 1.00	$ 2.00
17	13	0.0453	2	Black	$ (1.00)	$ 2.00	$ 1.00	$ 2.00
18	14	0.2290	9	Red	$ (1.00)	$ (1.00)	$ (1.00)	$ (3.00)
19	15	0.0515	2	Black	$ (1.00)	$ 2.00	$ 1.00	$ 2.00
20	16	0.9151	35	Black	$ (1.00)	$ 2.00	$ 1.00	$ 2.00
21	17	0.2546	10	Black	$ (1.00)	$ (1.00)	$ 1.00	$ (1.00)
22	18	0.1054	5	Red	$ (1.00)	$ 2.00	$ (1.00)	$ -
23	19	0.9410	36	Red	$ (1.00)	$ (1.00)	$ (1.00)	$ (3.00)

FIGURE 2-27 Simulation Model and Results for Roulette

What makes craps more difficult than roulette to simulate is that we never know how many rolls might be needed until the player wins or loses. Thus, a simulation model is dynamic, like the Mantel Manufacturing example, because we need to keep track of the rolls to make a determination of winning or losing. Figure 2-28 shows an Excel model for this game. You have undoubtedly already computed the probabilities of the dice outcomes in an elementary statistics course. These are converted into a lookup table to generate the appropriate random variate. Columns D, E, and F simulate one game. Because we do not know the length of the game ahead of time, we assume some maximum length, in this case, 100 rolls. In cell F5, we determine the outcome from the initial roll—either WIN, LOSE, or CONTINUE. If the player wins or loses on the first roll, the outcome WIN or LOSE is carried down the remaining rows. If a point is rolled, each subsequent row examines whether the player wins, loses, or continues rolling based on the point. However, once a player wins or loses, the final outcome is carried down the remaining rows. Thus, the outcome in the last row represents the outcome of the game and is copied to cell F2. To replicate the game for a number of trials, we use a data table in columns H and I. The table results in a value of 1 for a win and 0 for a loss. Based on 1,000 trials, we see that the winning percentage is 48.4 percent. The mathematical winning percentage is actually 49.3 percent.

SKILLBUILDER EXERCISE

Open the file *Craps Simulation.xls*. In craps, a player may place a bet on one of five numbers (7, 11, 2, 3, or 12) for one roll. A win on 7 is paid 4 to 1; 11 or 3 is paid 14 to 1; and a win on 2 or 12 is paid 29 to 1. Modify the spreadsheet to determine the outcome of a bet on 7, 3, and 12. Assume the player bets on these numbers on the first roll of the dice. Construct data tables to determine the average payoff for each bet. Simulate 500 trials of the game.

	A	B	C	D	E	F	G	H	I
1	Simulation of Craps								
2				Game Outcome		LOSE		Wining %	0.484
3	Outcome	Probability							
4	2	1/36		Roll	Dice Value	Result		Trial	0
5	3	1/18		1	5	CONTINUE		1	1
6	4	1/12		2	9	CONTINUE		2	1
7	5	1/9		3	3	CONTINUE		3	1
8	6	5/36		4	6	CONTINUE		4	1
9	7	1/6		5	4	CONTINUE		5	0
10	8	5/36		6	6	CONTINUE		6	1
11	9	1/9		7	9	CONTINUE		7	1
12	10	1/12		8	10	CONTINUE		8	0
13	11	1/18		9	9	CONTINUE		9	1
14	12	1/36		10	7	LOSE		10	0
15				11	7	LOSE		11	1
16	Random Number Range		Outcome	12	12	LOSE		12	0
17	0	1/36	2	13	6	LOSE		13	1
18	1/36	1/12	3	14	7	LOSE		14	0
19	1/12	1/6	4	15	10	LOSE		15	1
20	1/6	5/18	5	16	9	LOSE		16	1
21	5/18	5/12	6	17	8	LOSE		17	0
22	5/12	7/12	7	18	10	LOSE		18	0
23	7/12	13/18	8	19	9	LOSE		19	1
24	13/18	5/6	9	20	10	LOSE		20	1
25	5/6	11/12	10	21	6	LOSE		21	0
26	11/12	35/36	11	22	10	LOSE		22	0
27	35/36	1	12	23	10	LOSE		23	1
28				24	6	LOSE		24	0
29				25	10	LOSE		25	1

(a)

(b)

	F
4	Result
5	=IF(OR(E5=7,E5=11),"WIN",IF(OR(E5=2,E5=3,E5=12),"LOSE","CONTINUE"))
6	=IF(AND(E6=7,F5="CONTINUE"),"LOSE",IF(AND(E6=$E$5,F5="CONTINUE"),"WIN",IF(F5="WIN","WIN",IF(F5="LOSE","LOSE","CONTINUE"))))
7	=IF(AND(E7=7,F6="CONTINUE"),"LOSE",IF(AND(E7=$E$5,F6="CONTINUE"),"WIN",IF(F6="WIN","WIN",IF(F6="LOSE","LOSE","CONTINUE"))))
8	=IF(AND(E8=7,F7="CONTINUE"),"LOSE",IF(AND(E8=$E$5,F7="CONTINUE"),"WIN",IF(F7="WIN","WIN",IF(F7="LOSE","LOSE","CONTINUE"))))
9	=IF(AND(E9=7,F8="CONTINUE"),"LOSE",IF(AND(E9=$E$5,F8="CONTINUE"),"WIN",IF(F8="WIN","WIN",IF(F8="LOSE","LOSE","CONTINUE"))))
10	=IF(AND(E10=7,F9="CONTINUE"),"LOSE",IF(AND(E10=$E$5,F9="CONTINUE"),"WIN",IF(F9="WIN","WIN",IF(F9="LOSE","LOSE","CONTINUE"))))
11	=IF(AND(E11=7,F10="CONTINUE"),"LOSE",IF(AND(E11=$E$5,F10="CONTINUE"),"WIN",IF(F10="WIN","WIN",IF(F10="LOSE","LOSE","CONTINUE"))))
12	=IF(AND(E12=7,F11="CONTINUE"),"LOSE",IF(AND(E12=$E$5,F11="CONTINUE"),"WIN",IF(F11="WIN","WIN",IF(F11="LOSE","LOSE","CONTINUE"))))

FIGURE 2-28 Excel Model for Simulating the Game of Craps

The third, and most challenging, game to simulate on a spreadsheet is blackjack. In blackjack, a player is initially dealt two cards. The dealer also receives two cards, one of which is face up. A two-card total of 21 (an ace and a ten) is "blackjack," and the player wins (unless the dealer also has blackjack, which usually results in a tie or "push"). For any other amount, the player has a choice of receiving as many additional cards as he or she might want (with an ace counting as 11 or 1). If the total value of the player's hand exceeds 21 (called a "bust"), the player automatically loses. In the basic strategy, a player holds at 17 or better. When the player stops, the dealer makes some decisions on drawing additional cards according to casino rules. If the dealer does not bust, the player wins if his or her hand exceeds the dealer's; otherwise, the player loses, except for ties.

Because of its complexity, we will not simulate the entire game, but only examine some basic statistical results. Specifically, we will be interested in the distribution of the initial two cards, and the percentage of time that a player will bust on the third card if he or she draws to a 16 or less. What makes the simulation more difficult is that not only must we maintain a record of the cards that are played, but the probabilities of drawing a card change, based upon the previous cards that have been drawn. Figure 2-29 shows an Excel model for this situation. In the range A3:D14, we maintain a record of the num-

Card Value	Number at start	Number after first	Number after second		Simulation Model	
11	4	4	4		Card 1	7
10	16	16	16		Card 2	8
9	4	4	4		Total	15
8	4	4	3			
7	4	3	3		Card 3	10
6	4	4	4		Total	25
5	4	4	4		Bust?	1
4	4	4	4			
3	4	4	4			
2	4	4	4			

Card 1 Lookup Table

0	1/13	11
1/13	5/13	10
5/13	6/13	9
6/13	7/13	8
7/13	8/13	7
8/13	9/13	6
9/13	10/13	5
10/13	11/13	4
11/13	12/13	3
12/13	1	2

Card 2 Lookup Table

0	4/51	11
4/51	20/51	10
20/51	8/17	9
8/17	28/51	8
28/51	31/51	7
31/51	35/51	6
35/51	13/17	5
13/17	43/51	4
43/51	47/51	3
47/51	1	2

Card 3 Lookup Table

0	2/25	11
2/25	2/5	10
2/5	12/25	9
12/25	27/50	8
27/50	3/5	7
3/5	17/25	6
17/25	19/25	5
19/25	21/25	4
21/25	23/25	3
23/25	1	2

(a)

(b)

Number after first	Number after second		Simulation Model	
=IF($G$4=A5,B5-1,B5)	=IF($G$5=A5,C5-1,C5)	Card 1	=VLOOKUP(RAND(),$A$17:$C$26,3)	
=IF($G$4=A6,B6-1,B6)	=IF($G$5=A6,C6-1,C6)	Card 2	=VLOOKUP(RAND(),$E$17:$G$26,3)	
=IF($G$4=A7,B7-1,B7)	=IF($G$5=A7,C7-1,C7)	Total	=IF(AND(G4=11,G5<6),1+G5, IF(AND(G4<6,G5=11),1+G4,IF(G4+G5=22,2,G4+G5)))	
=IF($G$4=A8,B8-1,B8)	=IF($G$5=A8,C8-1,C8)	Card 3	=IF(G6<=16,VLOOKUP(RAND(),$A$29:$C$38,3),"")	
=IF($G$4=A9,B9-1,B9)	=IF($G$5=A9,C9-1,C9)	Total	=IF(G8<>"",IF(AND(G6>10,G8=11),G6+1,G6+G8),"")	
=IF($G$4=A10,B10-1,B10)	=IF($G$5=A10,C10-1,C10)	Bust?	=IF(AND(G9<>"",G9>21),1,"")	
=IF($G$4=A11,B11-1,B11)	=IF($G$5=A11,C11-1,C11)			
=IF($G$4=A12,B12-1,B12)	=IF($G$5=A12,C12-1,C12)			
=IF($G$4=A13,B13-1,B13)	=IF($G$5=A13,C13-1,C13)			
=IF($G$4=A14,B14-1,B14)	=IF($G$5=A14,C14-1,C14)			

FIGURE 2-29 Excel Model for Simulating Blackjack

ber of each card in the deck at the start and after the first and second cards have been dealt to the player. The probabilities of drawing each card are converted to lookup tables given below this range. The actual simulation model is at the top of columns F and G. Card 3 is drawn only if the first two cards total 16 or less. If the three-card total exceeds 21, a "1" is entered in cell G10, indicating that the player busts.

To simulate 100 trials of this process, we use a one-way data table with *two* output columns—one for the first two-card total and one for whether or not the player busts, as shown in Figure 2-30. As before, we use a blank cell for the column input cell in the data table because the trial index is not an input value in the model itself. Based on these results, we construct a frequency distribution of the two-card total in columns L and M, using the *FREQUENCY* function, and compute the percentage of blackjacks, hands with value 20 (almost a sure win), hands with value 17–19, hands with value 12–16, and

	I	J	K	L	M
1					
2	*Simulation Results*				
3				*First Two Cards*	
4		2-card total	Bust if 1	*Bin range*	*Frequency*
5	Trial	18		4	8
6	1	6		5	2
7	2	20		6	3
8	3	21		7	1
9	4	14	1	8	3
10	5	19		9	2
11	6	12	1	10	1
12	7	9		11	5
13	8	8		12	8
14	9	18		13	7
15	10	18		14	6
16	11	17		15	10
17	12	18		16	5
18	13	17		17	5
19	14	18		18	16
20	15	14		19	7
21	16	12		20	8
22	17	13	1	21	3
23	18	12		22	0
24	19	18			
25	20	18		*Summary Statistics*	
26	21	15	1	*Blackjack*	3.00%
27	22	6		*Twenty*	8.00%
28	23	16	1	*17-19*	28.00%
29	24	4		*12-16*	36.00%
30	25	15	1	*Bust on 3rd card*	55.56%
31	26	15			

FIGURE 2-30 Results from Blackjack Simulation

percentage of busts for initial hands of 12 to 16. Based on 100 trials, we see in Figure 2-30 that blackjack was dealt 3 percent of the time. Over half of the time, the player busted on the third card after being dealt 12–16 in the first two cards.

SKILLBUILDER EXERCISE

Open the file *Blackjack Simulation.xls*. Modify the spreadsheet to deal a fourth card. Assume that the player will draw the fourth card if he or she has a 16 or less after 3 cards. Simulate the game to determine the likelihood of busting on the fourth card whenever it is taken.

CONCLUDING REMARKS

Through these examples, we introduced the basics of developing Monte-Carlo simulation models to develop a distribution of model outputs using the capabilities of Excel. Several theoretical and practical issues remain, however. For instance, how many trials are "sufficient" to obtain useful results? What do we do when the input data are more appropriately described by some continuous probability distribution such as a normal, exponential, or triangular distribution, for example, instead of a simple discrete distribution? How do we even know what distribution to use to model various inputs? These questions—which revolve around concepts in probability and statistics—are discussed in the next chapter.

We also see that developing Monte-Carlo simulation models using Excel can be tedious because of the need to copy formulas or use data tables to repeat trials, and the need to create frequency distributions and histograms. Although doing this has helped you to gain a fundamental understanding of Monte-Carlo simulation, there are, fortunately, more efficient means of doing this. We will address this in Chapter 4.

Simulation in Practice

RISK/RETURN FINANCIAL MODELING AND ANALYSIS

A common financial analysis for investment decisions involves the trade-off between risk and return. Economic theory of rational decision making assumes that investors select those options they think have the highest net return, given their consideration of risk. An **efficient frontier** consists of those options that have the highest return among all options available at a given level of risk (as measured by the variance of return). Conversely, those options with the lowest risk for a particular rate of return would also lie on the efficient frontier. For example, suppose that three investment options are available: option A with an expected return of 10 percent and a variance of 100; option B with an expected return of 5 percent and a variance of 110; and option C with an expected return of 4 percent and a variance of 50. Options A and C would be on the efficient frontier; B would not be because it is dominated by option A.

Though the concept of the efficient frontier helps explain why safe investments are made, it presents a practical problem in that an individual's preference between risk and return must be identified. Some people are willing to take high risks in order to gain higher returns, while others prefer more safety. An individual's risk/return trade-off function is usually nonlinear, which means that identification of the best portfolio of options for a specific investor will require solution of nonlinear optimization models.

Portfolios of investments provide investors with ways to hedge, in the sense that they can consider the covariance of risk across investments. If two investment options are positively correlated, their returns would tend to change in the same direction. For instance, most stocks have a positive correlation with the Dow Jones average, which measures the overall average trend in stock prices. Investing in stocks that are positively correlated means higher overall risk, because if one stock performs poorly, the other stocks that have positive correlations with it are also likely to perform poorly. On the other hand, some stocks have negative correlations of changes in return. For example, agricultural-producing organizations do well when the price of food rises; grocery outlets tend to do worse under the same conditions, and vice versa. Therefore, a risk-averse investor would be interested in investing in negatively correlated investment options. The following two examples are real applications of simulation to risk/return modeling that demonstrate these concepts.

Pacific Financial Asset Management Company[1]   The Pacific Financial Asset Management Company developed a PC-based system to aid investors in identifying portfolios of long-term investments. The system can support both individuals and institutions that might have policy or legal constraints on particular risk features. The model considers both assets and liabilities, seeking to maximize surplus, which is defined by the market value of assets minus the present value of liability cash flow, and using the market discount rate to evaluate cash flows. The surplus is used to measure the strength of

[1]J. M. Mulvey, "An Asset-Liability Investment System," *Interfaces,* Vol. 24, no. 3, 1994, pp. 22–33.

pension plans. Plans that are in good shape have strong positive surpluses, whereas weak plans have small surpluses relative to assets, or even deficits.

In past approaches, projected liabilities were treated as deterministic point values. However, this assumption does not consider the uncertainty of future liabilities. The system developed considers this uncertainty, as well as the covariances between liabilities and returns.

The system also includes information about the preference function of specific investors. The expected utility for an investor's trade-off between risk and return is optimized over a network model reflecting available assets (to include cash), transaction costs of switching assets to other investment options, and constraints reflecting legal or policy limits. (For example, many pension funds are restricted to keep investment in international investments below 10 or 20 percent.) The system identifies the optimal plan for a given preference function.

Simulation is used as a means to take the mix of investments generated by the system, as well as the current portfolio, and generate possible outcomes over an extended time into the future, for example 20 to 30 years. Investors would examine the distribution of returns using a graphical representation (see Figure 2-31). This presents a clearer picture of relative risk and helps them to either reduce or increase their risks based on the pattern of contributions and the associated probabilities. In fact, investor selection from among these alternative investment mixes described by simulation results proved to be a more accurate means of eliciting preference functions than the lottery trade-offs obtained prior to running the model. An interesting feature is that investors were better able to express risk preference by selecting from descriptions of expected outcome based on simulation output than they were through conventional preference assessment methods.

Institutional Analysis of Investment Risk at Fannie Mae[2]   The Federal National Mortgage Association (Fannie Mae) is a government-sponsored agency providing fi-

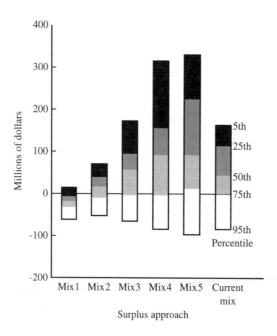

FIGURE 2-31 Simulation Profile of Possible Asset and Wealth Outcomes in 10 Years

Source: Reprinted by permission. J. M. Mulvey, "An Asset-Liability Investment System," *Interfaces,* Vol. 24, no. 3, 1994, pp. 22–33. Copyright 1994, The Institute of Management Sciences and the Operations Research Society of America (currently INFORMS), 2 Charles Street, Suite 300, Providence, RI 02904 USA.

[2]M. R. Holmer, "The Asset-Liability Management Strategy System at Fannie Mae," *Interfaces,* Vol. 24, no. 3, 1994, pp. 3–21.

nancial products and services to increase the availability and affordability of housing. Fannie Mae is shareholder-owned, but regulated by the government. It has two major products: guaranty of mortgages and mortgage investment.

The mortgage guaranty business is a type of insurance. Fannie Mae collects a small fraction of monthly interest payments to cover costs of insuring against mortgagor delinquency or default (and return on equity). This business started in the early 1980s and has grown to over $300 billion of insured mortgages. The mortgage investment business buys mortgage securities as long-term investments, financing them with debt and equity. This business has been operating since the 1940s and reached $117 billion in mortgage principal by the end of 1990. Most securities are individual mortgages, but they in turn are guaranteed so that credit risk is borne by the mortgage guaranty business.

Risk arises from prepayment of fixed-rate mortgages. If Fannie Mae obtains money at 9 percent to cover a 10 percent mortgage and the rate of housing interest drops, mortgage holders will usually refinance at a lower rate, leaving Fannie Mae with a note that is not prepayable for 9 percent without the expected income to cover it. On the other hand, if housing interest should rise, mortgage holders will hang onto the mortgage at the fixed rate, and Fannie Mae will have to get money at higher new rates.

In 1988, Fannie Mae asked that a system to manage their mortgage portfolio be developed. This asset-liability management strategy (ALMS) system optimizes the asset-liability component of Fannie Mae's portfolio to obtain the most desirable risk/return properties. A preference function is required in order to know what risk/return characteristics to optimize. A survey of the risk attitudes of between 10 and 20 top managers at Fannie Mae was used as the basis for this preference function. Simulation is used to estimate returns for fixed-income securities. The simulated security return distributions are then evaluated using the risk-aversion function in order to maximize Fannie Mae's utility.

The system was placed on a distributed processing network, and it has had heavy usage. The primary application has been portfolio-hedging. It has been a mechanism to move toward more callable debt financing of mortgage investments. However, the system has also proven useful in pricing debt securities, for which Fannie Mae had not had prior experience, and in pricing negotiations. It also allows Fannie Mae to estimate the market value of both their mortgage guaranty business and their mortgage investment business.

A DETERMINISTIC SPREADSHEET SIMULATION MODEL FOR PRODUCTION SCHEDULING AT WELCH'S[3]

Welch's, owned by the National Grape Cooperative Association, Inc., produces and distributes bottled juice, frozen juice concentrates, and spreads in three plants in Pennsylvania, Michigan, and Washington. Demand is not typically seasonal but has very pronounced peaks resulting from product promotions. This made it difficult for the company to determine desired inventory levels and to schedule production.

To help with these decisions, a deterministic spreadsheet simulation model was developed that predicts future production and inventory levels based on several user-defined parameters, including customer service level and cases of finished product produced per shift. The model incorporates the runout of current inventory and the scheduling of future production for replenishment of each individual SKU based upon

---

[3]Adapted from Edmund W. Schuster and Byron J. Finch, "A Deterministic Spreadsheet Simulation Model for Production Scheduling in a Lumpy Demand Environment," *Production and Inventory Management Journal,* Vol. 31, no. 1, First Quarter, 1990.

TABLE 2-1	Simulation Output Sample of Manufacturing and Inventory Cycles						
Day	*Inventory*	*Demand*	*Balance*	*Queue*	*Reorder Point*	*Production*	*Lot Size*
1	4.00	0.99	3.02	3.02	5.54	1	7
2	3.02	0.99	2.03	9.03	5.54	0	0
3	2.03	0.99	1.05	8.05	5.54	0	0
4	1.05	0.99	0.05	7.06	3.70	0	0
5	7.06	0.99	6.08	6.08	1.85	0	0

TABLE 2-2	Simulation Output Sample of a Production Schedule								
	Week 1	*Week 2*	*Week 3*	*Week 4*	*Week 5*	*Week 6*	*Week 7*	*Week 8*	*Week 9*
Planned Production	0	7	7	0	7	0	0	7	0
Inventory	9.50	3.02	5.38	8.79	5.21	8.54	5.11	1.77	5.43
Shifts/week	0	1	1	0	1	0	0	1	0
Demand/week	5.50	4.93	3.58	3.58	3.58	3.58	3.34	3.34	3.34

reorder points. Table 2-1 shows an example of the basic simulation output (all numbers are in thousands). The inventory represents the available inventory at the beginning of each day. The next column is the forecast demand for day 1; this is deducted from the inventory to arrive at the balance for the end of the day. This balance plus any previously scheduled production is equal to the queue in the next column. If the queue is less than the reorder point, production is indicated by a 1 in the production column. The last column specifies the number of thousands of cases required. This cycle can be repeated indefinitely to estimate production timing and inventory levels for future months. The model employed at Welch Foods projects a 9-week production schedule, as shown by the sample output in Table 2-2.

The simulation approach allows the user to test different service levels and immediately see the resulting effects on inventory levels. This makes it possible to see a direct cost/benefit relationship between holding costs and service levels. When large swings in demand caused by promotions occur, the forecasts can be used to help ensure that enough production is available to meet the promotion. Production-line load profiles can be constructed from the expected production quantities provided by the simulation. This provides capacity requirements resulting from reorder points that are directly linked to forecasted demand.

The model was implemented on a personal computer and consists of one main spreadsheet along with other small data files. Approximately 300 products are modeled for the entire company, and the model is run weekly, with reports distributed to the three plants electronically. Welch's director of logistics noted that the model predicts production as good or better than manual calculations in much less time, serves as the master schedule for the plants, and drives the central material requirements planning system.

Questions and Problems

1. Develop a spreadsheet model for the magazine inventory (problem 10) in Chapter 1. Enter the results from a manual, dice-driven simulation to verify the correctness of your model.
2. Develop a spreadsheet to replicate the results of the stock investment problem (problem 14) in Chapter 1. Compare results with the results obtained manually using dice.

3. A warehouse manager currently has one truck for local deliveries and is considering the purchase of an additional one to handle occasional high volumes. Currently, she rents additional trucks as needed for $300 per day. The truck that the company currently owns is charged off at a rate of $200 per day whether it is used or not. A new one would have a charge-off rate of $250 per day. Historical records show that the number of trucks needed each day has the following distribution:

Number Needed	Probability
0	0.20
1	0.30
2	0.40
3	0.10

Develop a spreadsheet model to evaluate the alternatives of renting any additional trucks required, as opposed to purchasing another truck and renting a third, as necessary.

4. A simple profit model is:

$$\text{Profit} = (\text{unit price} - \text{unit cost})(\text{quantity sold}) - \text{fixed costs}$$

Suppose that the unit price is $100 and that the other variables have the following distributions:

Unit Cost	Probability	Quantity Sold	Probability	Fixed Costs	Probability
$45	0.15	100	0.20	$5,000	0.30
$55	0.45	200	0.60	$6,500	0.50
$65	0.25	300	0.20	$8,000	0.20
$75	0.15				

Develop a spreadsheet model that would evaluate profitability. Identify the average profit, as well as the number of times there was a loss, and the maximum loss. Enter some values for the uncertain variables to verify the correctness of your model.

5. A firm produces radios for the consumer market. Their profit function is:

$$\text{Profit} = (\text{unit price} - \text{unit cost})(\text{quantity sold}) - \text{fixed costs}$$

Suppose that the unit price is $200 per radio, and that the other variables have the following probability distributions:

Unit Cost	Probability	Quantity Sold	Probability	Fixed Costs	Probability
$80	0.20	1,000	0.10	$50,000	0.40
$90	0.40	2,000	0.60	$65,000	0.30
$100	0.30	3,000	0.30	$80,000	0.30
$110	0.10				

Develop a spreadsheet model to evaluate expected profitability. Identify the average profit, the number of times there was a loss, and the maximum loss. Enter some values for the uncertain variables to verify your model, and run the simulation for 100 trials.

6. A garage band is planning a rock concert. They will sell tickets for $10. The cost for the facility will be $50,000 plus 10 percent of revenue from both tickets and concessions.

Crowd	Probability
10,000	0.10
15,000	0.30
20,000	0.30
25,000	0.20
30,000	0.10

Concession Spent/Individual	Probability
0	0.20
$5	0.30
$10	0.30
$15	0.20

Develop a spreadsheet model, and construct a frequency distribution of results based on 100 trials.

7. A firm produces guava juice in distinctive 1-gallon jugs. The profit function is:

$$\text{Profit} = (\$10 - \text{variable cost}) \times \text{gallons sold} - \text{fixed cost}$$

The current plant capacity is 5 million gallons per year. The firm can expand the plant by an additional 1 million or 2 million gallons per year, with the following cost impacts:

	Plant Capacity (gallons/year)	Fixed Cost	Variable Cost
Current	5 million	$5 million	$5/gallon
Add 1 million	6 million	$6 million	$4.90/gallon
Add 2 million	7 million	$7 million	$4.80/gallon

Annual demand is probabilistic, with the following distribution:

Demand	Probability
3 million	0.1
4 million	0.2
5 million	0.3
6 million	0.3
7 million	0.2

Develop a spreadsheet model for each of the capacity options, using the same stream of random numbers to generate results for each. Simulate 100 trials and identify average profit for each alternative.

8. Develop a spreadsheet model for a 3-year financial analysis of total profit based on the following data and information. Sales volume in the first year is estimated to be 100,000 units and is projected to grow at a rate of 7 percent per year. The selling price is $10 and will increase by $0.50 each year. Per unit variable costs are $3, and annual fixed costs are $200,000. Per unit costs are expected to increase by 5 percent per year. Fixed costs are expected to increase by 10 percent per year. What components of your model would probably be uncertain? Develop a spreadsheet model and discuss how simulation might be applied to this model.

9. Most firms maintain a minimum amount of cash on hand to meet their daily obligations or as a requirement from the firm's bank. A maximum amount may also be specified to reflect the trade-off between the transaction cost of investing in liquid assets and the cost of lost interest if the cash is not invested. The Miller-Orr model for cash management computes the spread between the minimum and maximum cash balance limits as:

Spread =
$3 \times (.75 \times \text{transaction cost} \times \text{variance of daily cash flows/daily interest rate})^{.333}$

The maximum cash balance is the spread plus the minimum cash balance, which is assumed to be known. The *return point* is defined as the minimum cash balance plus spread/3. Whenever the cash balance hits the maximum, the firm should invest the difference between the maximum and the return point; if the minimum is reached, sufficient securities should be sold to bring it up to the return point. Develop a spreadsheet model for computing these decision points if the minimum cash balance is $100,000; the variance of cash flows is $6,250,000; the interest rate is 9 percent annually; and the transaction cost is $20. What variables in this model would be uncertain? Discuss how simulation might be applied to this model.

10. A hotel wants to analyze its room-pricing structure prior to a major remodeling effort. Currently, rates and average number of room sold are as follows:

Room Type	Rate	Average Sold/Day
Standard	$ 85	250
Gold	$ 98	100
Platinum	$139	50

Each market segment has its own elasticity of demand to price. The elasticity values are standard, −1.5; gold, −2; platinum, −1. These mean, for example, if the price of a standard room is reduced by 1 percent, the number of rooms sold each day is expected to increase by 1.5 percent. The projected number of rooms sold can be determined using the following formula:

Projected number sold = elasticity $\times$ (price change in $) $\times$ (average daily sales/room rate)

where the average daily sales and room rates are those in the preceding table. Develop a spreadsheet model to compute the projected revenue for each market segment and the total projected revenue for any set of price changes. Apply your model to room rates of $78, $90, and $145 for standard, gold, and platinum rooms, respectively. What aspects of this problem would be uncertain? How might simulation help in the analysis?

11. Define the term *random number*. How does the literal interpretation differ from a *pseudorandom number?*

12. Generate 10, 30, 50, and 100 random numbers on a spreadsheet and construct a frequency distribution for each sample using 10 cells. What do you observe?

13. The number of replacement halogen headlight bulbs sold each week at an auto parts store has the following probability distribution:

Number of Bulbs	Probability
0	.15
1	.20
2	.35
3	.15
4	.10
5	.05

Show how to use random numbers to simulate outcomes from this distribution.

14. Use the roulette simulation spreadsheet to play 500 games. Select a number you want to play in cell I4 and a color in cell K4. In two blank columns, accumulate running totals for both bets. Identify the maximum loss and the maximum gain over the course of 500 games, and when they occur. What conclusions might you draw about the results?

15. Use the craps simulation spreadsheet to estimate the average winning percentage based on samples of 10, 100, and 500 games. Use the F9 key to repeat each experiment 10 times. What do you observe about the results?

16. The blackjack simulation spreadsheet is set up to draw a third card when the total of the first two cards is 16 or less (see cell G8). Modify the blackjack simulation spreadsheet to evaluate the following strategies:

    a. Draw a third card only if the total of the first two cards is 15 or less.

    b. Draw a third card only if the total of the first two cards is 14 or less.

    How do these strategies compare to the original policy of drawing to a 16 or less, based on the percentage of times a player busts?

17. What is a *random number seed?* What is it used for?

References

Carroll, J. M. *Simulation Using Personal Computers.* Reston, 1987.

Law, A. M. and W. D. Kelton. *Simulation Modeling and Analysis,* 2d ed. McGraw-Hill, 1991.

L'Ecuyer, P. "Efficient and Portable Combined Random Number Generators," *Communications of the ACM,* Vol. 31, no. 6, 1988, pp. 742–774.

Meyerson, Mark D. "Random Numbers," UMAP Modules and Monographs in Undergraduate Mathematics and Its Applications Project, COMAP/UMAP/Suite No. 4, 271 Lincoln St., Lexington, MA 02173.

Park, S. K. and K. W. Miller. "Random Number Generators: Good Ones Are Hard to Find," *Communications of the ACM,* Vol. 31, no. 10, 1988, pp. 1192–1201.

Pritsker, A. A. B. *Introduction to Simulation and SLAM II,* 3d ed. New York: Halsted Press, 1986.

Sheel, Atul. "Monte-Carlo Simulations and Scenario Analysis—Decision-Making Tools for Hoteliers," *Cornell Hotel and Restaurant Administration Quarterly,* Vol. 36, no. 5, 1995, pp. 18–26.

Smith, G. N. *Lotus 1-2-3 Quick,* 2d ed. Boston: Boyd & Fraser Publishing Co., 1990.

Thommes, M. C., *Proper Spreadsheet Design.* Boston: Boyd & Fraser Publishing Co., 1992.

Random Number Generation

Random number generation has been an interesting practical as well as research issue since computers were first built. The ability to generate a reproducible stream of independent random numbers is crucial to successful simulation studies. Some of the earliest attempts to generate random numbers included using a Geiger counter to count the number of rays in a short time span or to rely on the computer's internal clock to generate a random sequence. Although these methods can produce random numbers, a major shortcoming is their inability to repeat the same sequence, which we saw in Chapter 2 is an important feature for simulation applications. Thus, numbers that were generated would have to be saved in large tables if we needed to use the same sequence in other simulation experiments. The methods in use today focus on simple mathematical algorithms that generate a new random number from the previous one. Although a mathematical algorithm is deterministic (meaning that we could predict the next number in the stream if we know the algorithm), the sequence of random numbers appear to be random. This means they possess three important properties:

1. All numbers are uniformly distributed between 0 and 1.
2. Numbers in the stream have no serial correlation.
3. The random number stream has a long cycle (that is, the length of the random number stream before some number repeats itself).

A uniform distribution ensures that all numbers between 0 and 1 are equally likely. We should not expect find random numbers following a bell-shaped or any other distribution. **Serial correlation** refers to the relationship between any random number in the stream and the n subsequent random numbers. If serial correlation exists in a random number stream, then some pattern exists in the sequence, and the numbers are not independent. For example, consider the sequence .1, .2, .3, .4, .5, .6, .7, .8, .9, 0. These 10 numbers are uniformly distributed between 0 and 1. However, each successive number is precisely 0.1 greater than the one before it. Knowing this, we could predict the next number in the sequence, so the numbers do not appear to be random. The value of n can be 1 or any other positive integer, so we might look for patterns between every other number, every third, and so on, although usually smaller degrees of serial correlation are of more interest.

The **cycle** of a random number stream is determined by how many numbers can be generated before the sequence repeats itself. For example, the cycle of the sequence .250, .750, .875, .500, .625, .250 is 5. Notice that because the algorithm generates the next number from the previous one, once .250 appears the second time, the sequence will continually repeat itself. Only *five* possible numbers between 0 and 1 are generated! All random number generation algorithms have a cycle of some length, but we would like the cycle to be long enough that no one would ever notice.

If these properties do not hold, then simulation may produce biased and erroneous results. In fact, after the first personal computers were introduced, researchers found that the methods they used to generate random numbers were seriously flawed and even declared them to be unusable for simulation purposes.[4]

Random numbers should have some practical features as well. First, algorithms to generate them should be fast. Despite the ever-improving speed of today's computers, some simulations require that millions of random numbers be generated. Second, random numbers should be reproducible; that is, we should be able to generate the same sequence on demand. In addition to being able to generate identical conditions to evaluate alternative policies, different computer systems (such as an IBM mainframe, an Apple PC, or a Compaq) will generally use different internal random number generators. If you want to compare systems using the same program, but on different computer systems, you will need to be able to reproduce the same sequence. Finally, random number generators should require minimal computer storage.

Methods for Random Number Generation

Many algorithms for generating random numbers have been devised. One of the first approaches that was developed was called the **midsquare technique.** This method takes a number, called the **random number seed**—because it starts the stream of random numbers—squares it, and then takes the set of middle digits as the next random number. For example, suppose we use 7143 as the seed. Because the square of any 4-digit number will have at most 8 digits, we will always add leading zeros to create an 8-digit number. Thus, digits 3, 4, 5, and 6 represent the middle set. The sequence of random numbers that would be generated are shown in bold below:

$$7143^2 = \quad 51\mathbf{0224}49$$
$$0224^2 = \quad 00\mathbf{0501}76$$
$$0501^2 = \quad 00\mathbf{2510}01$$
$$2510^2 = \quad 06\mathbf{3001}00$$
$$3001^2 = \quad 09\mathbf{0060}01$$

and so forth.

The midsquare technique was one of the first numerical methods to generate reproducible random numbers on early computers. The random number seed determines the sequence of all subsequent random numbers in the stream, eliminating the need to save them. But this approach had a limitation as well. If we continue the sequence started above, we obtain:

$$0060^2 = \quad 00\mathbf{0036}00$$
$$0036^2 = \quad 00\mathbf{0012}96$$
$$0012^2 = \quad 00\mathbf{0001}44$$
$$0001^2 = \quad 00\mathbf{0000}01$$
$$0000^2 = \quad 00\mathbf{0000}00$$

After this, the rest of the random number stream is woefully predictable. When the stream converges to a constant value (often 0) or to a short cycle of values, the stream is basically useless as a random number generator.

Congruential random number generators are transportable and efficient methods of generating random numbers. To discuss these, we need to review the mathematical function **mod.** If m is a positive integer (called the *modulus*), then $x \bmod m$ is the

---

[4]Doan T. Modianos, Robert C. Scott, and Larry W. Cornwell, "Random Number Generation on Microcomputers," *Interfaces,* Vol. 14, no. 4, July–August 1984, pp. 81–87.

remainder after dividing m into x as many times as possible. For example, 7 mod 3 = 1, since we can divide 3 into 7 two times, leaving a remainder of 1. Similarly, 10 mod 2 = 0, and 231 mod 17 = 10. Note that the value of $x \bmod m$ will always be an integer between 0 and $m - 1$. The Excel function =MOD(x, m) performs this operation.

Congruential random number generators have the form $z_{i+1} = f(z_i) \bmod m$, where $f(z_i)$ is some function. Typically, $f(z_i)$ is either multiplicative: $f(z_i) = az_i$, or linear: $f(z_i) = az_i + c$. The initial seed is z_0. Congruential generators generate a sequence of values z_{i+1} that will be integers between 0 and $m - 1$. Then, by dividing z_{i+1} by m, we have a number $R_{i+1} = z_{i+1}/m$ between 0 and 1; that is, a random number!

To illustrate this, suppose that we use a multiplicative congruential generator with $m = 13$ and $a = 6$ and start with a seed $z_0 = 1$. An Excel spreadsheet, *Congruential generator.xls,* for computing a stream of random numbers is shown in Figure A-1. This generator has a cycle of 12. Observe that every positive integer less than m is generated before the sequence repeats itself. We call such a generator a **full-cycle generator.** However, note that having only 12 different random numbers hardly constitutes a continuous uniform distribution. To increase the totality of possible random numbers, it is desirable to make the modulus as large as possible. This is the largest integer that the computer can store. For a 32-bit computer, the maximum integer the computer can store is $2^{bit-1} - 1$, or $2^{31} - 1 = 2147483647$. For a 36-bit computer, the maximum integer is 34359738337. Both are large enough to serve the practical needs of simulation with respect to large random number generator cycles. You may experiment with the Excel congruential generator worksheet to build your own random number generators.

However, not all congruential generators are "good." An example is the linear congruential generator with $a = 4$, $c = 1$, and $m = 7$, and using a seed for $z_0 = 3$:

i	z_{i+1}	R_{i+1}
0	6	0.8571
1	4	0.5714
2	3	0.4286
3	6	0.8571
4	4	0.5714

This is not a full-cycle generator, producing only three values before repeating.

Mathematicians have proven that it is always possible to have a full-cycle generator for any m and any seed if and only if c ends in the digit 1, 3, 7, or 9 and a ends in 01, 21, 41, 64, or 81. It is best to use prime numbers for a modulus. Of course, this will not guarantee that the stream lacks serial correlation. You should take heart that most random number generators embedded in software packages today have been thoroughly tested. Some of the common statistical tests are described in the next section.

Testing Random Number Generators

To test for uniformity, you could apply the Chi-Square and/or Kolmogorov-Smirnov tests that you may have studied in a statistics course. (We review these procedures in Chapter 3.) To test for serial correlation, the Chi-Square test could be applied to pairs of numbers and numbers n units later. Because serial correlation is the same as autocorrelation, the Durbin-Watson test used to identify autocorrelation in time series forecasting could also be used to test for serial correlation, treating the stream of random numbers the same as a string of forecasts.

	A	B	C	D	E
1	Multiplicative Congruential Generator				
2					
3					
4	*a*	6			
5	*m*	13			
6					
7	*Index*	*z*	*az*	*az mod(m)*	*R*
8	0	1	6	6	0.461538
9	1	6	36	10	0.769231
10	2	10	60	8	0.615385
11	3	8	48	9	0.692308
12	4	9	54	2	0.153846
13	5	2	12	12	0.923077
14	6	12	72	7	0.538462
15	7	7	42	3	0.230769
16	8	3	18	5	0.384615
17	9	5	30	4	0.307692
18	10	4	24	11	0.846154
19	11	11	66	1	0.076923
20	12	1	6	6	0.461538
21	13	6	36	10	0.769231
22	14	10	60	8	0.615385
23	15	8	48	9	0.692308

(a)

	A	B	C	D	E
7	*Index*	*z*	*az*	*az mod(m)*	*R*
8	0	1	=B8*$B$4	=MOD(C8,$B$5)	=D8/$B$5
9	1	=D8	=B9*$B$4	=MOD(C9,$B$5)	=D9/$B$5
10	2	=D9	=B10*$B$4	=MOD(C10,$B$5)	=D10/$B$5
11	3	=D10	=B11*$B$4	=MOD(C11,$B$5)	=D11/$B$5
12	4	=D11	=B12*$B$4	=MOD(C12,$B$5)	=D12/$B$5
13	5	=D12	=B13*$B$4	=MOD(C13,$B$5)	=D13/$B$5
14	6	=D13	=B14*$B$4	=MOD(C14,$B$5)	=D14/$B$5
15	7	=D14	=B15*$B$4	=MOD(C15,$B$5)	=D15/$B$5
16	8	=D15	=B16*$B$4	=MOD(C16,$B$5)	=D16/$B$5
17	9	=D16	=B17*$B$4	=MOD(C17,$B$5)	=D17/$B$5
18	10	=D17	=B18*$B$4	=MOD(C18,$B$5)	=D18/$B$5
19	11	=D18	=B19*$B$4	=MOD(C19,$B$5)	=D19/$B$5
20	12	=D19	=B20*$B$4	=MOD(C20,$B$5)	=D20/$B$5
21	13	=D20	=B21*$B$4	=MOD(C21,$B$5)	=D21/$B$5
22	14	=D21	=B22*$B$4	=MOD(C22,$B$5)	=D22/$B$5
23	15	=D22	=B23*$B$4	=MOD(C23,$B$5)	=D23/$B$5

(b)

FIGURE A-1 Spreadsheet Implementation of Multiplicative Congruential Generator

Random number streams also can be tested for runs. A **run** is defined as the number of consecutive increases (or decreases) in a stream of random numbers. The number of increasing runs should be approximately the same as the number of decreasing runs. This does not mean that every other random number must go up or down. That would violate the random property as much as a constant increase. However, the occurrence of runs of various lengths should match the theoretical probability of occurrence.

Several other statistical tests can be applied to random number streams. These include the *poker test* (concerned with the number of pairs, three of a kind, etc., observed relative to their theoretical probability), *gap tests* (concerned with the number of digits drawn before a number repeats), the *coupon collector test* (seeing how long it would take to get a full set of digits), and many other tests. Details of these tests can be found in textbooks on nonparametric statistics. The overall point is that random numbers should be as close to truly random as possible, including the possibility that a number such as 0.123456789 is followed by 0.123456789 (which should happen once in a billion[2])!

Random Number Generation in Computer Programming Languages

You might find yourself writing a simulation program in some high-level computer programming language such as BASIC, FORTRAN, or C. These languages typically use linear congruential generators, but each language uses different syntax to call random numbers.

FORTRAN AND PASCAL

The command to generate a random number in FORTRAN or PASCAL is RAND. This function returns a uniform random number between 0 and 1 that is not replicable. Use of RAND(x) where *x* is a seed value will return a random number that can be replicated because the seed is specified.

A good linear congruential generator subroutine in FORTRAN is shown here. The call RND will return the random number with seed IXX.

```
FUNCTION RND(IXX)
INTEGER A,P,IXX,B15,B16,XHI,XALO,LEFTLO,FHI,K
DATA A/16807/,B15/32768/,B16/65536/,P/2147483647/
XHI=IXX/B16
XALO=(IXX-XHI*B16)*A
LEFTLO=XALO/B16
FHI=XHI*A+LEFTLO
K=FHI/B15
IXX=(((XALO-LEFTLO*B16)-P)+(FHI-K*B15)*B16)+K
IF(IXX.LT.0) IXX=IXX+P
RND=FLOAT(IXX)*4.656612875E-10
RETURN
END
```

BASIC AND C

In BASIC or C, a random number can be obtained by RND(#), which provides a random number with options. If the seed # is less than 0, the same number is returned every time. If # is greater than 0, or if no # is used (typing simply RND), a new random number is returned. If # = 0, the most recently generated number is returned. Given an initial seed, a random number sequence can be replicated because each successive call uses the previous number generated as a seed for the next number in the sequence.

The command RANDOMIZE provides a random seed based on the system timer and asks the user for a seed number between −32768 and 32767. The following code generates random numbers using the system clock as a generator (from Carroll, 1987):

```
10 CLS
20 TIME=VAL(RIGHT$(TIME$,2))+VAL(MID$(TIME$,4,2))+VAL(LEFT$(TIME$,2))
30 RANDOMIZE TIME
40 FOR I=1 TO 100
50 PRINT RND
60 NEXT I
```

We do warn you, however, that if you do experiment with writing random number generators, you test them thoroughly using valid statistical procedures.

Questions and Problems

1. What properties should a sequence of random numbers possess?
2. What is *serial correlation?*
3. Why is understanding the cycle of a random number generator important?
4. Generate 5 random numbers using the midsquare technique, beginning with the seed 1,659.
5. Generate 10 random numbers using a linear congruential generator with $m = 13$, $a = 7, c = 5$ and seed $(z) = 8$.
6. Generate 10 random numbers using a linear congruential generator with $m = 15$, $a = 4, c = 3$, and seed $(z) = 5$.
7. Describe methods for testing random number generators.

Probability and Statistics in Simulation

Chapter Outline

A working knowledge of probability and statistics is fundamental to effective simulation modeling and analysis. One application of probability and statistics is in building simulation models. In Monte-Carlo simulation models, for instance, we must make assumptions about the uncertainty of key inputs, such as future sales, growth rates, and inflation factors. We characterize this uncertainty by specifying probability distributions for these model inputs. In systems simulation models, we must also be able to describe the random variation in various elements of the system. For example, in a communication system, we may need to characterize the time between message arrivals to the system and the length of time of each message by a probability distribution. Likewise, in manufacturing models, times of job arrivals, job types, processing times, times between machine breakdowns, and repair times are also random variables. Choosing appropriate probability distributions for such input data can often be done by analyzing empirical or historical data and fitting these data to a distribution. At other times, such data are not available, and the modeler must select an appropriate distribution and its parameters judgmentally.

As we have seen, a simulation is only a sample from an infinite population of possible results. After a simulation model is built, we need to select the number of trials to run in a Monte-Carlo simulation, or the length of time to simulate a systems model. These decisions are statistical issues because they will influence the quality of the results. Questions such as How precise should estimates of the output measures be? and What sample size should be used to guarantee a specified precision? need to be addressed. Answers to such questions require statistics.

A third important use of statistics in simulation is to analyze the results of simulation experiments. In order to interpret the results properly, we must be able to characterize the variability in the results. This might include computing basic descriptive statistical measures or developing confidence intervals for population parameters of interest. Finally, statistical methods are also needed to validate simulation models and design simulation experiments.

In this chapter, we focus on the role of probability and statistics in simulation. We describe the more common probability distributions used in simulation models, describe approaches to identifying and modeling input distributions, discuss how to generate random variates from many of these common probability distributions, and examine various statistical issues in simulation related to interpreting output data and choosing sample sizes. We assume that you have had a basic course in probability and statistics; thus, some of this chapter should be a review, primarily to establish notation and consistent terminology, although many concepts will more than likely be new.

Basic Statistical Concepts

In this section, we summarize some essential concepts in statistics that are important in simulation.

DESCRIPTIVE STATISTICS

A **statistic** is a summary measure of sample data used to draw inferences about population parameters. It is common practice in statistics to use Greek letters to represent population parameters and Roman letters to represent sample statistics. Several different statistical measures are used to describe the results from simulations. Common measures of central tendency with which you should be quite familiar are the **mean, median,** and **mode.** If a population consists of N observations $x_1, \ldots x_N$, the **population mean,** μ, is

$$\mu = \frac{\sum_{i=1}^{N} x_i}{N}$$

The **sample mean** of a sample with n observations is

$$\bar{x} = \frac{\sum\limits_{i=1}^{n} x_i}{n}$$

The **population variance** measures dispersion about the mean, and is computed as:

$$\sigma^2 = \frac{\sum\limits_{i=1}^{N} (x_i - \mu)^2}{N}$$

where x_i is the value of the ith item, N is the number of items in the population, and μ is the population mean. The **standard deviation,** σ, is the square root of the variance. For sample data, we calculate the **sample variance** by the formula:

$$s^2 = \frac{\sum\limits_{i=1}^{n} (x_i - \bar{x})^2}{n - 1}$$

where n is the number of items in the sample, and $\bar{x}$ is the sample mean. The square root of the sample variance is the sample standard deviation. It may seem peculiar to use a different "average" for populations and samples, but statisticians have shown that the sample variance provides a more accurate representation of the true population variance when computed in this way.

Skewness describes the degree to which a distribution is or is not symmetric. Negatively skewed distributions tail off to the left (most of their density is to the right); positively skewed distributions tail off to the right (most of their density is to the left). The **coefficient of skewness** is a measure of skewness and is computed as:

$$CS = \frac{\sum\limits_{i=1}^{n} (x_i - \mu)^3}{\sigma^3}$$

(For sample data, we replace the population mean and standard deviation with the corresponding sample statistics.) If the sign of CS is positive, the distribution is positively skewed; if negative, the distribution is negatively skewed. The magnitude of the skewness can be assessed by the closeness of CS to zero. The closer CS is to zero, the less the degree of skewness (that is, the more symmetry) in the distribution. A distribution whose coefficient of skewness greater than 1 or less than -1 is highly skewed. A value between .5 and 1 or $-.5$ and -1 is moderately skewed. Coefficients between .5 and $-.5$ are fairly symmetric. A normal distribution has a coefficient of skewness equal to 0.

Kurtosis refers to the peakedness (i.e., high, narrow) or flatness (i.e., short, flat-topped) of the distribution. The **coefficient of kurtosis** measures the degree of kurtosis of a distribution, and is computed as:

$$CK = \frac{\sum\limits_{i=1}^{n} (x_i - \mu)^4}{\sigma^4}$$

(Again, for sample data, use the sample statistics instead of the population parameters.) A normal distribution has a coefficient of kurtosis of 3. If the value of CK is less than 3, the distribution is relatively flat with a very wide degree of dispersion. As the values of CK get farther from 3, the distribution becomes more peaked with less dispersion.

The **coefficient of variation** (sometimes called the **coefficient of variability**) is the ratio of the standard deviation to the mean:

$$CV = \sigma/\mu$$

The coefficient of variation for sample data is $s/\bar{x}$. Because it is independent of the units of the variable, it can be used to compare the variability of different variables, even when their scales differ. For example, a variable with a small mean might exhibit relatively more variation than one with a large mean, even though the value of standard deviation may be smaller.

Correlation is a measure of the strength of linear relationship between two variables, X and Y, and is measured by the (population) **correlation coefficient:**

$$\rho_{x,y} = \frac{\mathrm{cov}(X, Y)}{\sigma_x \sigma_y}$$

The numerator is called the **covariance,** and is the average of the products of deviations of each observation from its respective mean:

$$\mathrm{cov}(X, Y) = \frac{\sum_{i=1}^{N} (x_i - \mu_x)(y_i - \mu_y)}{N}$$

Similarly, the sample correlation coefficient is computed as:

$$r = \frac{\sum_{i=1}^{n} (x_i - \bar{x})(y_i - \bar{y})}{(n - 1)s_x s_y}$$

Correlation coefficients will range from -1 to $+1$. A correlation of 0 indicates that the two variables have no linear relationship to each other. Thus, if one changes, we cannot reasonably predict what the other variable might do. A correlation coefficient of $+1$ indicates a perfect positive relationship; as one variable increases, the other will also increase. A correlation coefficient of -1 also shows a perfect relationship, except that as one variable increases, the other decreases. In economics, for instance, a perfectly price-elastic product has a correlation between price and sales of -1; as price increases, sales decrease, and vice versa.

Excel provides a useful tool for basic data analysis, which is available in the *Data Analysis* option from the *Tools* menu. The Descriptive Statistics tool provides a variety of statistical measures for one or more data sets: the mean, standard error, median, mode, standard deviation, sample variance, kurtosis, skewness, range, minimum, maximum, sum, count, kth largest and smallest values (for any value of k you specify), and the confidence level for the mean. If you select *Descriptive Statistics,* the dialog box shown in Figure 3-1 will appear. You need only enter the range of the data, which must be in a single row or column. If the data are in a matrix form, the tool treats each row or column as a separate data set, depending on which you specify.

SKILLBUILDER EXERCISE

Open the file *Random Numbers.xls,* which was shown in Figure 2-15. Use the Descriptive Statistics tool to find the summary statistics for this data set.

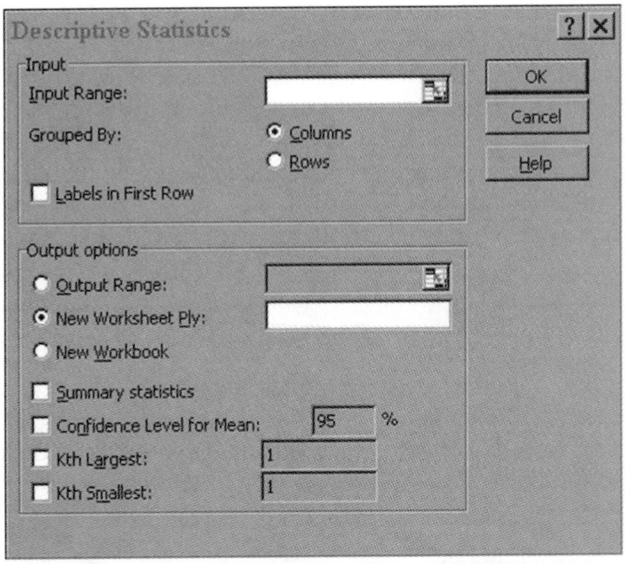

FIGURE 3-1 Excel Descriptive Statistics Dialog Box

SAMPLING DISTRIBUTIONS

Simulation is essentially a sampling experiment whose results represent a sample from some generally unknown probability distribution. Usually the goal is to estimate a population parameter, such as the mean. An important statistical question is How good is the estimate obtained from the sample? Suppose that we simulate a Monte-Carlo model for n trials and repeat the simulation many times, each time computing the mean of the n trials. For example, in Figure 2-22, we replicated the Dave's Candies simulation 100 times and computed the mean profit in cell I22. If we repeated this simulation again, using different random numbers to generate a different stream of demands, we would expect a somewhat different value for the mean profit. The means of all possible simulation results will form a distribution, which we call the **sampling distribution of the mean.** The sampling distribution of the mean characterizes the population of means of *all possible samples* of a given size. As the sample size increases, the variance of the sampling distribution decreases. This suggests that the estimates we obtain from larger numbers of trials provide greater accuracy in estimating the true population mean.

We would like to have some measure of the variability of the sampling distribution of the mean to assess how accurate our estimate of the sample mean is to the true population mean. We would expect the variability of this distribution to be higher if we replicated the simulation, say only 25 times instead of 100. On the other hand, we would expect the variability to be lower if we had replicated it 500 times. A measure of variability of the sample mean is called the **standard error of the mean** (or simply, **standard error**) and is computed as:

$$\text{Standard error of the mean} = \sigma/\sqrt{n}$$

where σ is the standard deviation of the distribution of individual observations and n is the sample size. The standard error is essentially the standard deviation of the sampling distribution of the mean. From this formula, we see that as n increases, the standard error decreases.

What about the shape of the sampling distribution of the mean? Statisticians have shown that if the population is normal, then the sampling distribution of the mean will also be normal for any sample size, and that the mean of the sampling distribution will be the same as that of the population. Furthermore, the **Central Limit Theorem,** one of the most important practical results in statistics, states that if the sample size is large enough, the sampling distribution of the mean can be approximated by a normal distribution, regardless of the shape of the population distribution. As the sample size and the number of samples increase, the distribution will become closer in shape to a normal distribution. Another important statistical concept often useful in simulation is the **strong law of large numbers.** Essentially, this states that if you take a large enough number of samples, the sample mean will be very close to the population mean.

SKILLBUILDER EXERCISE

Copy the *RAND( )* function into a matrix of 30 rows by 50 columns. This represents 50 samples of size $n = 30$. Compute the mean of each column, and construct a histogram of the 50 means. Although the distribution of the individual values is uniform, does the distribution of the sample means look like a normal distribution?

CONFIDENCE INTERVALS

An interval estimate provides a range within which we believe the true population parameter falls. A **confidence interval (C.I.)** is an interval estimate that also specifies the likelihood that the interval contains the population parameter. This probability is called the *level of confidence,* and it is usually expressed as a percentage. For example, we might state that "a 90 percent C.I. for the mean is 10 ± 2." The value 10 is the point estimate calculated from the sample data, and 2 can be thought of as a margin for error. Thus, the interval estimate is [8, 12]. The level of confidence, denoted by $1 - \alpha$, is 0.90. This means that, based on the sample data, the probability that the confidence interval contains the true population parameter is 0.90. Equivalently, the value of α represents the probability that a confidence interval will *not* include the true population parameter. Commonly used confidence levels are 90, 95, and 99 percent; the higher the confidence level, the more assurance we have that the interval contains the population parameter. As the confidence level increases, the confidence interval becomes larger.

The sample mean, $\bar{x}$, is a point estimate for the population mean μ. We can use the central limit theorem to quantify the sampling error in $\bar{x}$. Recall that the central limit theorem states that no matter what the underlying population, the distribution of sample means is approximately normal with mean μ and standard deviation $\sigma_{\bar{x}} = \sigma/\sqrt{n}$. The value of α represents the proportion of sample means expected to be outside the confidence interval. Therefore, $100(1 - \alpha)\%$ of sample means (for all possible samples of size n) would fall within $\mu \pm z_{\alpha/2}(\sigma/\sqrt{n})$.

Since we do not know μ but estimate it by $\bar{x}$, a $100(1 - \alpha)\%$ confidence interval for the population mean μ is

$$\bar{x} \pm z_{\alpha/2}(\sigma/\sqrt{n})$$

The value $z_{\alpha/2}$ may be found from a standard normal table (Appendix A), or may be computed in Excel using the function *NORMSINV(1 − α/2).* For the most common confidence levels used, we have

Confidence Level	$z_{\alpha/2}$
90%	1.645
95%	1.96
99%	2.575

The confidence interval we developed, however, assumes that we know the standard deviation. When the standard deviation is unknown, as would almost always be the case, we need to use the **Student-t** (or **t**) **distribution** to compute a confidence interval. The t-distribution is actually a family of probability distributions with a shape similar to the standard normal distribution. Different t-distributions are distinguished by an additional parameter, **degrees of freedom (d.f.).** The t-distribution has a larger variance than the standard normal, but as the number of degrees of freedom increases, it converges to the standard normal distribution. When sample sizes get to be as large as 120, the distributions are virtually identical; even for sample sizes as low as 30 to 35, it becomes difficult to distinguish between the two. Thus, for large sample sizes, many people use z-values to establish confidence intervals even when the standard deviation is unknown. Because the t-distribution explicitly accounts for the effect of the sample size in estimating the population variance, it is the proper one to use for any sample size. However, for large samples the difference between t- and z-values is very small.

Using the t-distribution, a $100(1 - \alpha)\%$ confidence interval for the population mean μ is

$$\bar{x} \pm t_{\alpha/2, n-1}(s/\sqrt{n})$$

where $t_{\alpha/2, n-1}$ is the value from the t-distribution with $n - 1$ degrees of freedom giving an upper-tail probability of $\alpha/2$. We may find t-values in Appendix B or by using the Excel function *TINV(probability, degrees of freedom)*. However, take careful note that "probability" in the function *TINV* refers to the probability in *both* tails of the distribution. Thus, $t_{\alpha/2, n-1}$ in Appendix B is equivalent to $TINV(\alpha, d.f.)$.

In simulation, we often wish to compare the difference between two means (see Chapter 7). A $100(1 - \alpha)\%$ confidence interval for the difference in means with sample sizes n_1 and n_2, respectively, is given by the following formula:

$$\bar{x}_1 - \bar{x}_2 \pm z_{\alpha/2}\sqrt{\frac{s_1^2}{n_1} + \frac{s_2^2}{n_2}}$$

Random Variables and Probability Distributions

A good knowledge of probability distributions is critical to applying simulation successfully. A **random variable** is a numerical description of the outcome of some experiment. Performing a simulation is an example of an experiment. Outcomes might be profit values, failure times, numbers of customer arrivals, or service times. Random variables may be discrete or continuous and generally are denoted by capital letters such as X or Y. For example, the outcome of rolling a die and the number of customers who respond to a telemarketing campaign are discrete random variables; the time to repair a failed machine is continuous. A discrete random variable can have only a countable number of outcomes. For the experiment of rolling one die, the possible outcomes are the numbers 1 through 6. For the telemarketing campaign, the outcomes are nonnegative integers up to the number of customers contacted. Two random variables x and y are **independent** if the probability of both x and y occurring equals the product of the two probabilities; that is $p(x, y) = p(x)p(y)$. For example, the random variables associated with the experiment of throwing two dice are independent.

A **probability distribution** is a characterization of the possible values that a random variable may assume along with the probability of assuming these values. Discrete probability distributions are defined by probability mass functions, like the examples we used in Chapter 2. Continuous probability distributions are defined by their probability density functions. We will use the notation $f(x)$ to denote the appropriate probability mass function or density function for discrete or continuous random variables respectively, and $F(x) = P(X \le x)$ to denote the cumulative distribution function, which specifies the probability that the random variable X will assume a value less than or equal to a specified value, x. A probability mass function has the properties that the probability of each outcome must be between 0 and 1, and the sum of all probabilities must add to 1; that is

$$0 \le f(x_i) \le 1$$

$$\sum_i f(x_i) = 1$$

A continuous random variable assumes outcomes over a continuous range of real numbers. The probabilities of various outcomes are characterized by a **probability density function, $f(x)$.** A density function has the property that the total area under the function is 1. The cumulative distribution function is denoted $F(x)$. The probability that the random variable will assume a value between a and b is given by the area under the function between a and b, and is equal to the difference of the cumulative distribution function evaluated at these two points. Mathematically,[1]

$$\text{Area between } a \text{ and } b = \int_a^b f(x)dx = F(b) - F(a)$$

The **expected value** of a random variable corresponds to the notion of the mean, or average. For a discrete random variable, the expected value is the weighted average of the outcomes, where the weights are the probabilities:

$$E[X] = \sum_{j=1}^{\infty} x_j p(x_j)$$

For example, in rolling one die, the expected value $= \sum_{j=1}^{6} j(1/6) = 21/6 = 3.5$. In the continuous case, the expected value is defined using calculus as $E[X] = \int_{-\infty}^{\infty} xf(x)dx$.

Probability density and mass functions depend on one or more *parameters*. Many continuous distributions can assume different shapes and sizes, depending on the value of the parameters. There are three basic types of parameters. A **shape parameter** controls the basic shape of the distribution. For certain distributions, changing the shape parameter will cause major changes in the form of the distribution. For others, the changes will be less severe. A **scale parameter** controls the unit of measurement within the range of the distribution. Changing the scale parameter either contracts or expands the distribution along the horizontal axis. Finally, a **location parameter, L,** specifies the location of the distribution relative to zero on the horizontal axis. L may represent the midpoint or the lower endpoint of the range of the distribution. All distributions will not have all three parameters; some may have more than one shape parameter. Understanding the effects of these parameters is important in selecting distributions as inputs to simulation models.

In this section, we review some of the more common types of probability distributions that are used in simulation modeling; discuss how shape, scale, and location pa-

---

[1]The mathematics of continuous random variables depends on a knowledge of calculus. However, calculus is not required to follow the discussions or in any application of simulation in this book.

rameters affect the distributions; and describe typical situations for which each distribution often applies.

USEFUL CONTINUOUS DISTRIBUTIONS

Uniform Distribution   The uniform distribution characterizes a random variable for which all outcomes between some minimum and maximum value are equally likely. For a uniform distribution with a minimum value a and a maximum value b, the density function is

$$f(x) = \frac{1}{b - a} \quad \text{if } a \le x \le b$$

and is shown in Figure 3-2. The distribution function is

$$F(x) = \begin{cases} 0 & \text{if } x < a \\ \dfrac{x - a}{b - a} & \text{if } a \le x \le b \\ 1 & \text{if } b < x \end{cases}$$

The mean of the uniform distribution is $(a + b)/2$, and the variance is $(b - a)^2/12$. Note that a is a location parameter because it controls the location of the distribution along the horizontal axis. The difference, $b - a$, is a scale parameter. Increasing $b - a$ elongates the distribution; decreasing $b - a$ compresses it. There is no shape parameter because any uniform distribution is flat.

Random numbers generated by computer algorithms provide an example of a uniform distribution with $a = 0$ and $b = 1$. The uniform distribution is often used when little knowledge about a random variable is available; the parameters a and b are chosen judgmentally to reflect a modeler's best guess about the range of the random variable.

Normal Distribution   The normal distribution is described by the familiar bell-shaped curve. The normal distribution is symmetric and has the property that the median equals the mean. Although the range of x is unbounded, most of the density is close to the mean. It is characterized by two parameters: the mean, μ (the location parameter), and the variance, σ^2 (the scale parameter). The probability density function for the normal distribution is

$$f(x) = \frac{e^{-(x-\mu)^2/2\sigma^2}}{\sqrt{2\pi\sigma^2}}$$

No closed form of the distribution function exists. Figure 3-3 provides a sketch of the **standard normal distribution,** the random variable Z having $\mu = 0$ and $\sigma = 1$. Appendix A provides a tabular calculation of the cumulative distribution function $F(Z) = P(Z \le z)$.

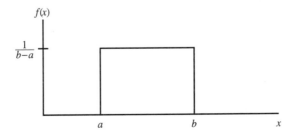

FIGURE 3-2 Uniform Probability Density Function

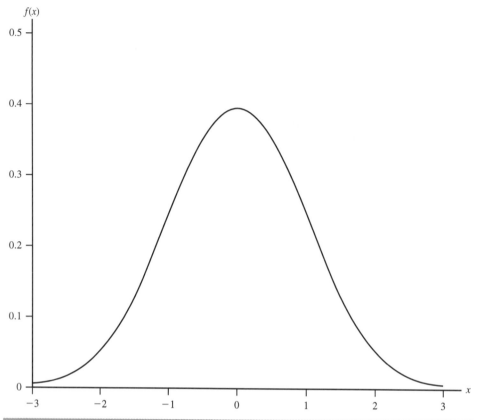

FIGURE 3-3 Normal Density Function with Mean 0 and Standard Deviation 1

Any normal distribution with an arbitrary mean μ and standard deviation σ may be transformed to a standard normal distribution by applying the following formula:

$$z = \frac{x - \mu}{\sigma}$$

This makes it easy to use Appendix A to compute probabilities associated with the normal distribution as you have undoubtedly done in your statistics classes.

Two Excel functions are used to compute normal probabilities: *Normdist(x, mean, standard_deviation, cumulative)* and *Normsdist(z)*. For the cumulative distribution, the last argument of the function, *cumulative,* must be set to *TRUE. Normsdist(z)* generates the same values as can be found in Appendix A. The Excel function *Standardize (x, mean, standard_deviation)* can be used to compute z-values within a spreadsheet; this function essentially computes z-values using the formula $z = (x - \text{mean})/\text{standard deviation}$. Thus, *Standardize(700, 750, 100)* $= -0.5$.

The normal distribution is observed in many natural phenomena. Errors of various types, such as deviation from specifications of machined items, often are normally distributed. Processing times in some service systems can be well-approximated by a normal distribution. Also, as a consequence of the Central Limit Theorem, the mean of batches of random variables having any distribution will also be approximated by a normal distribution.

Triangular Distribution   The triangular distribution is defined by three parameters: the minimum, *a*; maximum, *b*; and most likely, *c*. Outcomes near the most likely value have a higher chance of occurring than those at the extremes. By varying the position of the most likely value relative to the extremes, the triangular distribution can be symmetric or skewed in either direction as shown in Figure 3-4. The probability density function is given by:

$$f(x) = \begin{cases} \dfrac{2(x-a)}{(b-a)(c-a)} & \text{if } a \le x \le c \\[2mm] \dfrac{2(b-x)}{(b-a)(b-c)} & \text{if } c < x \le b \\[2mm] 0 & \text{otherwise} \end{cases}$$

From Figure 3-4, you can see that *a* is the location parameter, $(b-a)$ is the scale parameter, and *c* is the shape parameter. The distribution function is

$$F(x) = \begin{cases} 0 & \text{if } x < a \\[2mm] \dfrac{(x-a)^2}{(b-a)(c-a)} & \text{if } a \le x \le c \\[2mm] 1 - \dfrac{(b-x)^2}{(b-a)(b-c)} & \text{if } c < x \le b \\[2mm] 1 & \text{if } b < x \end{cases}$$

FIGURE 3-4 Examples of the Triangular Distribution

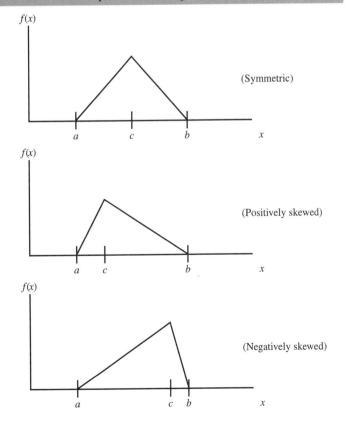

The mean is computed as $(a+b+c)/3$ and the variance is $(a^2+b^2+c^2-ab-ac-bc)/18$.

The triangular distribution is often used as a rough approximation of other distributions, such as the normal, or in the absence of more complete data. For instance, a manager might be able to state that a job usually takes c hours, and if everything goes well, it can be done in a hours. However, when things go wrong, it can take as long as b hours. Without any other data, a triangular distribution can be assumed. Because it depends on three simple parameters and can assume a variety of shapes, it is very flexible in modeling a wide variety of assumptions. One drawback, however, is that it is bounded by the parameters a and b, thereby eliminating the possibility of extreme outlying values that might possibly occur.

Exponential Distribution   The exponential distribution models events that recur randomly over time. Thus, it is often used to model the time between customer arrivals to a service system and the time to failure of machines, lightbulbs, and other mechanical or electrical components. A key property of the exponential distribution is that it is *memoryless;* that is, the current time has no effect on future outcomes. For example, the length of time until a machine failure has the same distribution no matter how long the machine has been running.

The exponential distribution has the density function

$$f(x) = \lambda e^{-\lambda x} \qquad x \geq 0$$

and distribution function

$$F(x) = 1 - e^{-\lambda x} \qquad x \geq 0$$

The mean of the exponential distribution $= 1/\lambda$ and the variance $= (1/\lambda)^2 = (\text{mean})^2$. The exponential distribution has no shape or location parameters; λ is the scale parameter. Figure 3-5 provides a sketch of the exponential distribution. The exponential distribution has the properties that it is bounded below by 0, has its greatest density at 0, and a declining density as x increases.

The Excel function *Expondist(x, lambda, cumulative)* can be used to compute exponential probabilities. Note that the mean is *not* the same as the parameter lambda; thus, in the Excel function, lambda is set equal to the reciprocal of the mean.

Lognormal Distribution   If the natural logarithm of a random variable X is normal, then X has a lognormal distribution. The probability density function is

$$f(x) = \begin{cases} \dfrac{1}{x\sqrt{2\pi\sigma^2}} e^{-[\ln(x)-\mu]^2/2\sigma^2} & \text{if } x > 0 \\ 0 & \text{otherwise} \end{cases}$$

The mean is $e^{\mu+\sigma^2/2}$ and the variance is $e^{2\mu+\sigma^2}(e^{\sigma^2}-1)$. The parameter σ is the shape parameter and must be greater than 0. Figure 3-6 presents lognormal density functions with $\mu = 1$ and $\sigma = 1/2, 1$, and $3/2$. When σ takes on larger values, the distribution takes on a very spiked value starting from 0, and then tails off towards 0 as x approaches ∞. For values of $\sigma < 1$, the skewness of the distribution decreases, and the distribution takes on an almost normal shape, with slight skewness with the tail to the right. The parameter μ is the scale parameter and can assume any real value.

The lognormal distribution is positively skewed and bounded below by zero. Thus, it finds applications in modeling phenomena that have low probabilities of large values and cannot have negative values, such as the time to complete a task. Other common examples include stock prices, and real estate prices. The lognormal distribution is also

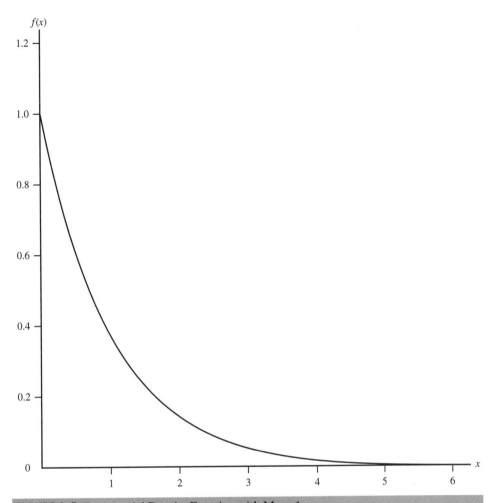

FIGURE 3-5 Exponential Density Function with Mean 1

often used for "spiked" service times; that is, when the probability of zero is very low but the most likely value is just greater than zero. Quantities that are the product of a large number of other quantities are also approximately lognormally distributed.

Note that μ and σ are the mean and standard deviation, respectively, of the *normal* distribution, not the lognormal distribution. If we know the mean μ_L and standard deviation σ_L of the lognormal distribution, we can use the following formulas to convert the parameters:

$$\mu = \ln(\mu_L^2/\sqrt{\sigma_L^2 + \mu_L^2})$$
$$\sigma^2 = \ln[(\sigma_L^2 + \mu_L^2)/\mu_L^2]$$

The Excel function *Lognormdist(x, mean, standard_dev)* returns the value for lognormally distributed data with given mean and standard deviation (i.e., μ_L and σ_L). The Excel function *Loginv(probability, mean, standard_dev)* returns the numeric value for a given cumulative probability.

Truncating Distributions   Many continuous distributions extend to infinity in either one or both directions. For many models, this might not be a reasonable assumption. For

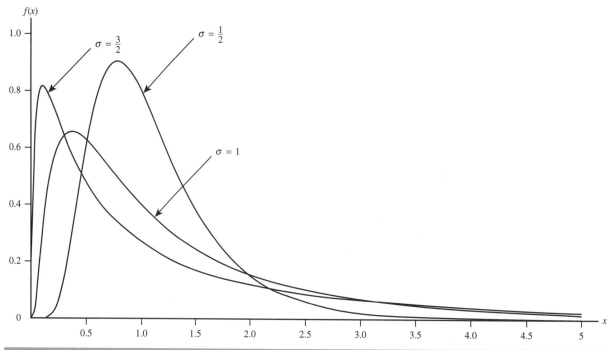

FIGURE 3-6 Lognormal Density Functions with Mean 1 and Standard Deviations of $\frac{1}{2}$, 1, and $\frac{3}{2}$

Source: A. Law and D. Kelton, *Simulation Modeling Analysis,* McGraw-Hill, 1991. Material is produced with permission of The McGraw Hill Companies.

instance, suppose you wanted to model an Internet auction. You might assume that the selling price for a particular item is normally distributed with a mean of $1,000 and standard deviation of $200, but that the minimum bid is $500. Thus, you would want to cut off any values below $500, resulting in a distribution that looks like the one in Figure 3-7. When you do this, however, the actual mean and standard deviation of the distribution are no longer $1,000 and $500, respectively, so you must be careful in how you use the parameters in calculations.

FIGURE 3-7 Truncated Normal Distribution

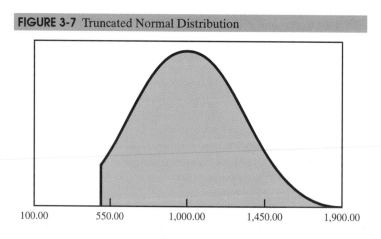

USEFUL DISCRETE DISTRIBUTIONS

Discrete Uniform Distribution   This is a discrete analogy of the continuous uniform distribution. The discrete uniform distribution characterizes a random variable for which all integer outcomes between some minimum value a and maximum value b are equally likely. An example is the distribution of the roll of one die, which we used in Chapter 2. The density function is

$$f(x) = \frac{1}{b - a + 1} \qquad \text{if } a \leq x \leq b$$

where a, b, and x are integers, and the distribution function is

$$F(x) = \begin{cases} 0 & \text{if } x < a \\ \dfrac{x - a + 1}{b - a + 1} & \text{if } a \leq x \leq b \\ 1 & \text{if } b < x \end{cases}$$

The mean of the discrete uniform distribution is $(a + b)/2$, and the variance is $[(b - a + 1)^2 - 1]/12$. As with the continuous uniform distribution, a is a location parameter and $b - a$ is a scale parameter.

Bernoulli and Binomial Distributions   The Bernoulli distribution characterizes a random variable with two possible outcomes with constant probabilities of occurrence. Typically, these outcomes represent "success" ($x = 1$) or "failure" ($x = 0$). The probability mass function is

$$f(x) = \begin{cases} 1 - p & \text{if } x = 0 \\ p & \text{if } x = 1 \end{cases}$$

where p represents the probability of success. A Bernoulli distribution might be used to model whether an individual responds positively to a telemarketing promotion.

The binomial distribution models n independent replications of a Bernoulli trial with probability p of success on each trial. The random variable x represents the number of successes in these n trials. The probability mass function is

$$f(x) = \begin{cases} \dbinom{n}{x} p^x (1 - p)^{n-x} & \text{if } x = 1, 2, \ldots, n \\ 0 & \text{otherwise} \end{cases}$$

The mean is np, and the variance is $np(1 - p)$. A binomial distribution might be used to model the results of sampling inspection in a production operation or the effects of drug research on a sample of patients. Binomial probabilities are cumbersome to compute by hand but can be computed in Excel using the function *Binomdist(number_s, trials, probability_s, cumulative),* where *cumulative* should be set equal to *TRUE* for the cumulative distribution function and *FALSE* for the probability mass function.

Poisson Distribution   The Poisson distribution is a discrete distribution used to model the number of occurrences in some unit of measure; for example, the number of events occurring in an interval of time, number of items demanded per customer from an inventory, or the number of errors per line of software code. The Poisson distribution assumes no limit on the number of occurrences, that occurrences are independent, and that the average number is constant. The probability mass function is

$$p(x) = \begin{cases} \dfrac{e^{-\lambda}\lambda^{x}}{x!} & \text{if } x = 0,1,2,\ldots \\ 0 & \text{otherwise} \end{cases}$$

where the mean is λ and the variance also is equal to λ. Poisson probabilities are cumbersome to compute by hand. Many books have tables, but probabilities can easily be computed in Excel by using the function *Poisson(x, mean, cumulative)*.

The Poisson distribution bears a close relationship to the exponential distribution. If X has a Poisson distribution, then $1/X$ has an exponential distribution. This property is often used in simulation models. For example, if the *number of arrivals* to a system is Poisson, then the *time between arrivals* would be exponentially distributed.

OTHER USEFUL DISTRIBUTIONS

Many other probability distributions find application in simulation studies, especially those distributions that assume a wide variety of shapes. Such distributions provide a great amount of flexibility in representing empirical data that may be available. Although these are somewhat complicated mathematically, they are available in software packages such as Crystal Ball and in many other commercial simulation languages.

Gamma Distribution   The gamma distribution is a family of distributions defined by a shape parameter α, a scale parameter β, and a location parameter L. L is the lower limit of the random variable X; that is, the gamma distribution is defined for $x > L$. The density function depends on some advanced mathematical functions and thus is not included here. Figure 3-8 shows gamma distributions for various values of α and $\beta = 1$.

Gamma distributions are often used to model the time to complete a task, such as customer service or machine repair. It is used to measure the time between the occurrence of events when the event process is not completely random. It also finds application in inventory control and insurance risk theory.

A special case of the Gamma distribution when $\alpha = 1$ and $L = 0$ is called the *Erlang distribution*. The Erlang distribution can also be viewed as the sum of k independent and identically-distributed exponential random variables. Its probability density function is

$$f(x) = \dfrac{\dfrac{1}{\lambda}\left(\dfrac{x}{\lambda}\right)^{k-1} e^{-x/\lambda}}{(k-1)!} \qquad \text{if } x \geq 0$$

where k is a positive integer. The mean is k/λ, and the variance is k/λ^2. When $k = 1$, the Erlang is identical to the exponential distribution. For $k = 2$, the distribution is highly skewed to the right. For larger values of k, this skewness decreases, until for $k = 20$, the Erlang distribution looks similar to a normal distribution. One common application of the Erlang distribution is for modeling the time to complete a task when it can be broken down into independent tasks, each of which has an identical exponential distribution.

Weibull Distribution   The Weibull distribution is another probability distribution capable of taking on a number of different shapes, as shown in Figure 3-9. The density function for the Weibull distribution is

$$f(x) = \begin{cases} \left(\dfrac{\beta}{\alpha}\right)\left(\dfrac{x-L}{\alpha}\right)^{\beta-1} e^{-\left(\frac{x-L}{\beta}\right)^{\beta}} & \text{if } x \geq L \\ 0 & \text{otherwise} \end{cases}$$

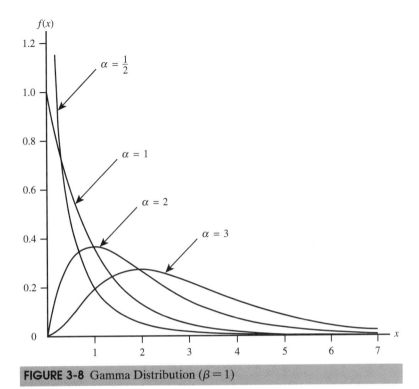

FIGURE 3-8 Gamma Distribution ($\beta = 1$)

Source: A. Law and D. Kelton, *Simulation Modeling Analysis,* McGraw-Hill, 1991.
Material is produced with permission of The McGraw Hill Companies.

The parameter α is the scale parameter, and β is the shape parameter. Both α and β must be greater than zero. When $L = 0$ and $\beta = 1$, the Weibull distribution is the same as the exponential distribution with $\lambda = 1/\alpha$. By choosing the scale parameter L different from 0, you can model an exponential distribution that has a lower bound different from zero. When $\beta = 3.25$, the Weibull approximates the normal distribution. The mathematical formulas needed to compute the mean and variance of this distribution are too advanced for this book. Weibull distributions are often used to model results from life and fatigue tests, equipment failure times, and times to complete a task.

Beta Distribution   One of the most flexible distributions for modeling variation over a fixed interval from 0 to a positive value s is the beta. The density function is

$$
f(x) = \begin{cases} \dfrac{(\alpha + \beta - 1)!}{(\alpha - 1)!(\beta - 1)!} \left(\dfrac{x}{s}\right)^{\alpha-1}\left(1 - \dfrac{x}{s}\right)^{\beta-1} & \text{if } 0 < x < s \\ 0 & \text{otherwise} \end{cases}
$$

The beta distribution is a function of two shape parameters, α and β, both of which must be positive. The mean of the beta distribution is

$$
\frac{\alpha}{\alpha + \beta}
$$

and the variance is

$$
\frac{\alpha\beta}{(\alpha + \beta)^2(\alpha + \beta + 1)}
$$

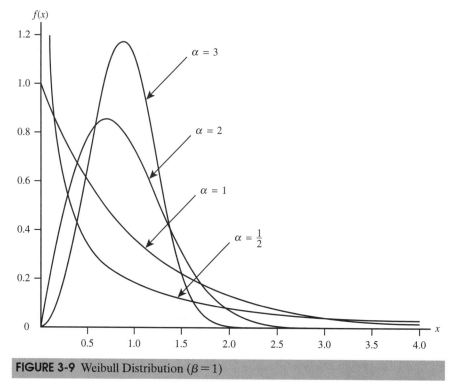

FIGURE 3-9 Weibull Distribution ($\beta = 1$)

Source: A. Law and D. Kelton, *Simulation Modeling Analysis,* McGraw-Hill, 1991. Material is produced with permission of The McGraw Hill Companies.

The parameter *s* is the scale parameter. Note that *s* defines the upper limit of the distribution range. Figure 3-10 shows various beta density functions. If α and β are equal, the distribution is symmetric. If either parameter is 1.0 and the other is greater than 2, the distribution is in the shape of a "J." If α is less than β, the distribution is positively skewed; otherwise, it is negatively skewed. These properties can help you to select appropriate values for the shape parameters.

Lesser-Used Distributions   Several other distributions that may be used in special applications include:

- *Geometric distribution*—a sequence of Bernoulli trials that describes the number of trials until the first success. An example would be the number of parts manufactured until a defect occurs, assuming that the probability of a defect is constant for each trial.
- *Negative binomial distribution*—similar to the geometric distribution, it models the distribution of the number of trials until the *r*th success (for example, the number of sales calls needed to sell 10 orders).
- *Hypergeometric distribution*—similar to the binomial, except that it applies to sampling without replacement. The hypergeometric distribution is often used in quality control inspection applications.
- *Logistic distribution*—commonly used to describe growth of a population over time.
- *Pareto distribution*—describes phenomena in which a small proportion of items accounts for a large proportion of some characteristic. For example, a small number of cities constitutes a large proportion of the population.

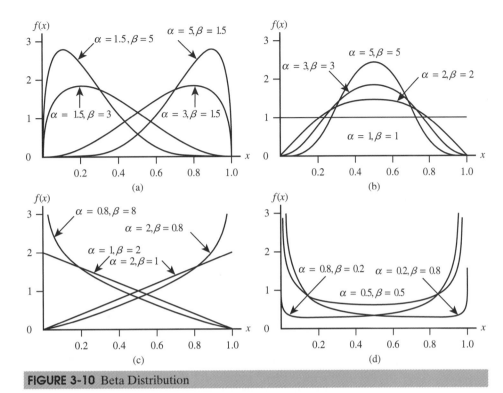

FIGURE 3-10 Beta Distribution

Source: A. Law and D. Kelton, *Simulation Modeling Analysis,* McGraw-Hill, 1991. Material is produced with permission of The McGraw Hill Companies.

Other examples include the size of companies, personal incomes, and stock price fluctuations.

- *Extreme value distribution*—describes the largest value of a response over a period of time, such as rainfall, earthquakes, and breaking strengths of materials.

Generating Random Variates

In Chapter 2, we introduced the notion of a random variate and described how to generate random variates from discrete distributions using random numbers and the *VLOOKUP* function in Excel. Here we describe some techniques for generating random variates from many of the other distributions discussed in this chapter. All of these approaches involve some transformation of random numbers. For example, because a random number is uniformly distributed between 0 and 1, it is quite easy to transform it into a random variate from an arbitrary uniform distribution with parameters a and b. Consider the formula:

$$U = a + (b - a)*R$$

where R is some random number. Note that when $R = 0$, $U = a$, and when $R = 1$, $U = b$. For any other value of R between 0 and 1, $(b - a)*R$ represents the same proportion of the interval (a,b) as R does of the interval $(0,1)$. Thus, all real numbers between a and b can occur. Because R is uniformly distributed, so also is U. However, it is certainly not obvious how to generate random variates from other distributions such as a normal or exponential. One approach that has wide applicability is called the *inverse transformation method*.

INVERSE TRANSFORMATION METHOD

The inverse transformation method exploits the properties of random numbers and the cumulative distribution function, $F(x)$, of a random variable X. Because $F(x)$ is a *function*, every value of x has a unique value $F(x)$ associated with it. Because $F(x)$ is non-decreasing, the inverse function exists, at least in principle. That is, for every value of $F(x)$, we have associated a unique value of x. If we choose a value of $F(x)$ randomly, we can often find the associated value of x, either explicitly or by using some computational algorithm. The x-values are random variates from the probability distribution $f(x)$.

We may choose a value of $F(x)$ randomly by realizing that the cumulative distribution function of any random variable X has the property that $0 \leq F(x) \leq 1$. Random numbers have the same property. Therefore, we need only generate a random number R, set it equal to the cumulative distribution function $F(x)$, and solve for x as shown in Figure 3-11(a). Another way of looking at this is to realize that R represents the area from negative infinity to x under the density function. The inverse transformation method finds the value of x associated with a "random" area under the density, as illustrated in Figure 3-11(b).

The application of the *VLOOKUP* function on a spreadsheet is actually an application of the inverse transformation method for discrete distributions. Refer to the example shown in Figure 2-18. The cumulative probability function is shown in Figure 3-12. Note that this forms a natural partition of the interval $[0,1]$ into the random number range that we defined in the spreadsheet. Figure 3-13 shows that the *VLOOKUP* approach is essentially an application of the inverse transformation method.

Let us see how the inverse transformation method applies to a triangular distribution. Figure 3-14 shows a triangular distribution with minimum 0, mode 3, and maximum 4. We will apply some simple geometry to implementing the inverse transformation method. The distribution has two triangles, the left one extending from $x = 0$ to $x = 3$, and the right one extending from $x = 3$ to $x = 4$ with a common side of length H. The combined areas of these triangles must be 1.0, because it is a density function. Therefore,

$$3H/2 + H/2 = 1$$

Thus, $H = 0.5$, with the left triangle having an area of 0.75 and the right triangle having an area of 0.25.

In order to convert a random number R into a random variate x from this triangular distribution, we first determine whether x occurs under the left or right triangle. We will demonstrate both cases. Assume that the first random number is 0.4, falling to the left of the mode. We seek the distance x as shown in Figure 3-15. The area of the small triangle bounded by x and h is $xh/2 = 0.4$. We know by proportion that $H/3 = h/x$. Therefore, the area of the small triangle must be $(x/6)(x)/2 = .4$. Solving this equation for x yields $x = 2.1909$. This is a random variate from the triangular distribution we have defined.

Now suppose that the random number is 0.8, as shown in Figure 3-16 on page 86. By proportion, $H/1 = h/(4 - x)$; thus, $h = .5(4 - x)$. The area of the triangle to the right of x is $0.2 = (4 - x)h/2$. Substituting for h, we have

$$0.2 = (4 - x)(.5)(4 - x)/2 = (4 - x)^2/4$$

Solving for x, we get $x = 3.1056$ as the random variate.

If we know the mathematical form of the cumulative distribution function, we can often use this explicitly to solve the equation $F(x) = R$ explicitly for x. Earlier in this chapter, we saw that the cumulative distribution function for the triangular distribution is

$$F(x) = \begin{cases} 0 & \text{if } x < a \\ \dfrac{(x-a)^2}{(b-a)(c-a)} & \text{if } a \le x \le c \\ 1 - \dfrac{(b-x)^2}{(b-a)(b-c)} & \text{if } c < x \le b \\ 1 & \text{if } b < x \end{cases}$$

FIGURE 3-11 Inverse Transformation Method

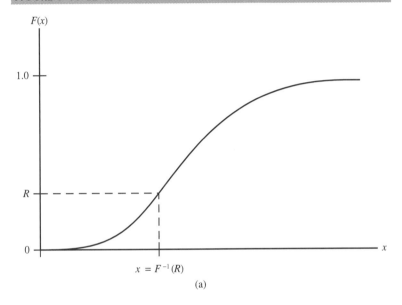

(a)

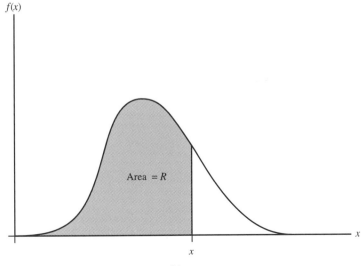

(b)

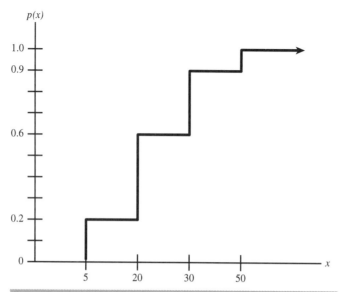

FIGURE 3-12 Cumulative Distribution $P(x)$

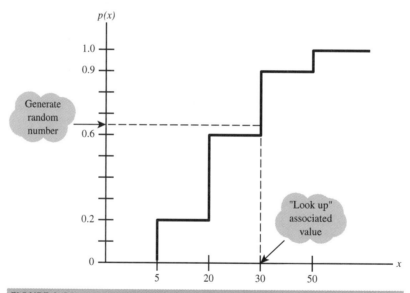

FIGURE 3-13 Inverse Transformation Method for Discrete Distributions

Thus, when $R = .4$, we have

$$F(x) = \frac{(x - 0)^2}{(4 - 0)(3 - 0)} = 0.4$$

or, $x^2/12 = .4$. Solving for x, we obtain $x = 2.1909$, as we did earlier.

As another example, the cumulative distribution function for the exponential distribution is $F(x) = 1 - e^{-\lambda x}$. Setting this equal to a random number R gives

$$R = 1 - e^{-\lambda x}$$

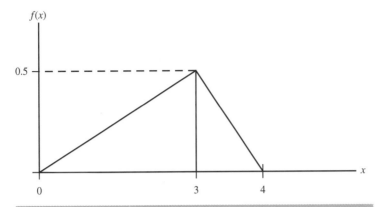

FIGURE 3-14 Triangular Distribution

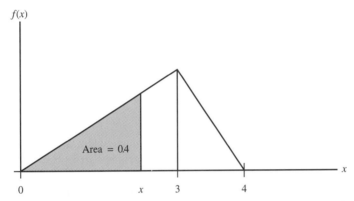

FIGURE 3-15 Inverse Transformation Method Applied to the
Triangular Distribution, $R = 0.4$

Solving for x, we have

$$e^{-\lambda x} = 1 - R$$
$$-\lambda x = \ln(1 - R)$$
$$x = -(1/\lambda)\ln(1 - R)$$

Note that $1 - R$ is a random number, so we may simplify this expression as:

$$x = -(1/\lambda)\ln(R)$$

The inverse transformation principle will work for any distribution, *provided* that we can somehow solve for x after setting $F(x) = R$. For the normal distribution, for example, it is impossible to develop a closed-form mathematical expression for $F(x)$. However, we may use a standardized normal table to apply the inverse transformation method. Suppose we have a normal distribution with a mean of 5 and a variance of 2. From Appendix A at the end of this book, the z-value corresponding to an area of, for example, $R = 0.8$ is approximately 0.84. Thus, solving the equation $z = (x - \mu)/\sigma$ for x, we have $0.84 = (x - 5)/2$, and $x = 6.68$.

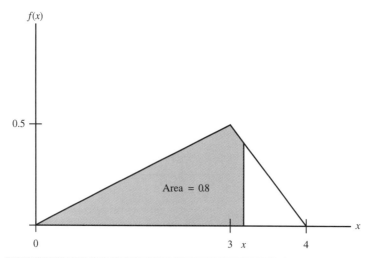

FIGURE 3-16 Inverse Transformation Method Applied to Triangular Distribution, $R = 0.8$

SPECIAL METHODS

The inverse transformation method for generating random variates cannot be applied to every probability distribution, particularly when a closed-form mathematical solution does not exist, as in the case of the normal distribution. In such cases, special methods have been developed.

The **convolution approach** defines a random variate as a sum of other independent, identically distributed random variables. An Erlang random variate can be generated by using this approach. Recall that an Erlang random variable is the sum of k independent and identically distributed exponential random variables. To generate an Erlang random variate with mean k/λ, we could first generate k independent exponential random variates, each with mean $1/\lambda$, and add them together.

A special method for generating normally distributed random variates is called the *Box-Muller method.* This approach uses two random numbers, R_1 and R_2, and generates two standard normal random variates as $X = \sqrt{-2\ln(R_1)}\cos(2\pi R_2)$ and $Y = \sqrt{-2\ln(R_1)}\sin(2\pi R_2)$. An alternative approach for normal random variates, called the *polar method,* relies on a general principle for generating random variates called the **acceptance-rejection method.** This method is useful when most of a distribution can be transformed by formula, but parts cannot. The method proceeds as follows.

> Step 1: Draw two random numbers, R_1 and R_2. Compute $V_1 = 2R_1 - 1$, $V_2 = 2R_2 - 1$, and $W = V_1^2 + V_2^2$.
>
> Step 2: If $W > 1$, discard these results and return to step 1. Otherwise, compute $Y = \sqrt{-2\ln(W)/W}$. Then $X_1 = V_1 Y$ and $X_2 = V_2 Y$ are standard normal random variates.

These approaches are useful when normal variates must be generated in higher level programming languages such as FORTRAN, BASIC, or C++. As we have seen, the Excel function *NORMINV* is easy to use within a spreadsheet environment.

GENERATING RANDOM VARIATES IN EXCEL AND CRYSTAL BALL

In Chapter 2, we discussed how to generate outcomes from discrete probability distributions in Excel using the (misnamed) Random Number Generation tool. This tool also has the capability of providing random variates from several other common probability distributions. On the main toolbar, select *Tools . . . Data Analysis . . . Random Number Generation*. From the Random Number Generation dialog box, you may select from seven distributions: uniform, normal, Bernoulli, binomial, Poisson, and patterned, as well as discrete, which we illustrated in Chapter 2. The patterned distribution is characterized by a lower and upper bound, a step, a repetition rate for values, and a repetition rate for the sequence. The dialog box prompts you for the distribution's parameters; other features in the dialog box are discussed in Chapter 2. However, as noted, the output from this tool is not dynamically linked to the spreadsheet and must be repeated whenever new random variates are needed.

Excel also has several "inverse functions" for probability distributions that can be used to generate random variates. These are useful because they can be embedded within formulas and are updated dynamically as the spreadsheet is changed. For example, the Excel function *NORMINV(probability, mean, standard_deviation)* can be used to generate a normal random variate. All you need to do is to replace *probability* with a random number as in the function *NORMINV(RAND( ), mean, standard_deviation)*. Other inverse functions are

- Standard normal distribution: *NORMSINV(probability)*
- Beta distribution: *BETAINV(probability, alpha, beta, A, B)*. In this function, A and B represent lower and upper bounds, which default to 0 and 1 respectively if omitted from the function.
- Gamma distribution: *GAMMAINV(probability, alpha, beta)*
- Lognormal distribution: *LOGINV(probability, mean, standard_dev)*. The mean and standard deviation refer to a random variable whose natural logarithm is normally distributed.

Excel also has a function for generating random variates from a discrete uniform distribution between integer values *a* and *b*: *RANDBETWEEN(a, b)*.

SKILLBUILDER EXERCISE

Open the file *Daves Candies Monte Carlo Simulation.xls*. Modify the spreadsheet to generate demands from (1) a discrete uniform distribution between 40 and 90 and (2) a normal distribution with mean 65 and standard deviation 10. How do these affect the best order quantity from the data table analysis?

Crystal Ball is an Excel add-in that is provided with this book. Although the principal use of Crystal Ball is to perform Monte-Carlo simulations, as we will discuss in the next chapter, it has some useful capabilities purely from a probability and statistics viewpoint. "Extras" provided by Crystal Ball are functions similar to the Excel inverse functions for all of the distributions described in the first section of this chapter: uniform, normal, triangular, binomial, Poisson, geometric, hypergeometric, lognormal, exponential, Weibull, beta, gamma, logistic, Pareto, extreme value, negative binomial, as well as user-defined custom distributions. These are listed in Table 3-1. For example, to generate a uniform random variate between 10 and 20, you would enter the formula:

```
=CB.UNIFORM(10,20)
```

into the appropriate cell. The CB.CUSTOM function is used for discrete distributions much in the same fashion as we used lookup tables (although the syntax is quite different). The *Cell Range* argument of the function allows for a variety of distributions, including individual values, continuous ranges, and discrete uniform ranges. For a typical discrete distribution, use a two-column range, with the first column containing the outcomes and the second column containing the probabilities (*not* the cumulative probabilities). For a continuous range, use a three-column range where the first two columns give the (uniform) ranges and the third column the probabilities. For example:

$$
\begin{array}{ccc}
10 & 20 & .4 \\
20 & 30 & .5 \\
50 & 60 & .1
\end{array}
$$

Here, the function would select one of the ranges according to the probabilities in the third column, and then generate a value uniformly within the selected range. If you define a four-column range and place a 1 in the fourth column of the range, the function would generate a discrete uniform distribution between the values in the first two columns if that range were chosen. We will see how many of these functions can be used in simulation models in Chapter 6.

SKILLBUILDER EXERCISE

Install Crystal Ball on your computer. Open the file *Mantel Manufacturing Monte Carlo Simulation.xls*. Using the CB functions, modify the demand to be (1) normally distributed with a mean of 105 and standard deviation 15 and (2) triangular with a lower limit of 80, most likely value of 95, and upper limit of 130. In both cases, truncate the decimal to give integer values.

TABLE 3-1   Random Variate Functions Available in Crystal Ball

CB.Beta(Alpha, Beta, Scale)

CB.Binomial(Probability, Trials)

CB.Custom(Cell Range)

CB.Exponential(Rate)

CB.ExtremeValue(Mode, Scale) (for maximum)

CB.ExtremeValue2(Mode, Scale) (for minimum)

CB.Gamma(Location, Scale, Shape)

CB.Geometric(Probability)

CB.Hypergeometric(Probability, Trials, Population)

CB.Logistic(Mean, Scale)

CB.Lognormal(Mean, Standard Deviation)

CB.NegBinomial(Probability, Trials)

CB.Normal(Mean, Standard Deviation)

CB.Pareto(Location, Shape)

CB.Poisson(Rate)

CB.Triangular(Minimum, Likeliest, Maximum)

CB.Uniform(Minimum, Maximum)

CB.Weibull(Location, Scale, Shape)

Applications in Financial Analysis and Management

Simulation is used extensively in financial analysis and financial management. We present several examples, drawing from the topics introduced in this chapter.

SIMULATING STOCK PRICES

Financial analysts often use the following lognormal model to characterize changes in stock prices:

$$P_t = P_o e^{(\mu - 0.5\sigma^2)t + \sigma Z\sqrt{t}}$$

where P_o = current stock price

P_t = price at time t

μ = mean (logarithmic) change of the stock price per unit time

σ = (logarithmic) standard deviation of price change

Z = standard normal random variable

This model assumes that the logarithm of a stock's price is a normally distributed random variable (see the discussion of the lognormal distribution and note that the first term of the exponent is the mean of the lognormal distribution). Using historical data, one can estimate values for μ and σ. For example, Figure 3-17 shows the daily closing prices during 1999 for Cisco Systems (Nasdaq: CSCO). In column C we take the logarithm of the daily prices, and compute the change from one day to the next. The logarithmic mean daily change and standard deviation are computed in cells D3 and D4. Now suppose that we wish to simulate future stock prices based on these historical estimates. We may use the previous formula, replacing Z by a standard normal random variate, such as the Excel function NORMSINV(RAND ()). This is shown in columns F and G of Figure 3-17.

OPTIONS PRICING

An approach to protecting stock investments in volatile markets is to purchase options.[2] A *call option* gives the holder of the option the right to buy a stock at a given price, called the *strike price,* on or before a certain (expiration) date. Thus, if the stock rises above the strike price, you would in effect purchase the stock at the strike price and sell it at the current price, pocketing the difference (less the cost of the option). A *put option* is the opposite of a call; it allows you to sell a stock at the strike price, effectively gambling that the stock price will fall. There are two types of options: an American option allows you to exercise the option anytime before the expiration date; a European option may only be exercised on the expiration date. For example, suppose the current price of a stock is $20 and you purchase a European option to buy at a strike price of $24 on a certain date. If the stock price is $30 on this date, you would make $6 per share less the cost of the option. However, if the price is only $22, you would not buy it at $24, but would only lose the price you paid for the option.

Call options trade at their net present values. We may use simulation to estimate the net present value of the option, thus determining a fair price for the option. We have extended the Cisco stock price simulation example to do this as shown in Figure 3-18. We start with the closing price on December 31 and select a strike price of $60 in 50 days. We will assume that the risk-free rate (what you can earn in the money market, for example) is 8 percent and use this in the net present value calculation. Using the price

---

[2]David G. Luenberger, *Investment Science,* New York: Oxford University Press, 1998.

	A	B	C	D	E	F	G
1	CSCO Stock Price Simulation						
2							
3			Average Daily Change	0.003227			
4			Standard Deviation	0.026154			
5						Price Simulation	
6	Date	Closing Price	Ln(Closing Price)	Change		Time (days)	Price
7	01/04/1999	23.828	3.17086136			0	53.563
8	01/05/1999	24.234	3.187756606	0.016895		1	51.64392
9	01/06/1999	24.938	3.216392745	0.028636		2	53.45594
10	01/07/1999	25.906	3.254474602	0.038082		3	54.46299
11	01/08/1999	26.672	3.283614326	0.02914		4	55.71758
12	01/11/1999	26.172	3.264690137	-0.01892		5	52.56872
13	01/12/1999	24.531	3.199937624	-0.06475		6	54.8451
14	01/13/1999	23.969	3.176761329	-0.02318		7	48.95794
15	01/14/1999	24.094	3.181962847	0.005202		8	52.53837
16	01/15/1999	25.422	3.235614941	0.053652		9	53.7544
17	01/19/1999	26.594	3.280685626	0.045071		10	52.14764
18	01/20/1999	26.531	3.278313861	-0.00237		11	52.83773
19	01/21/1999	25.328	3.231910503	-0.0464		12	51.78305
20	01/22/1999	25.703	3.246607717	0.014697		13	59.14699
21	01/25/1999	25.859	3.252658703	0.006051		14	51.54203
22	01/26/1999	26.781	3.287692681	0.035034		15	55.86757

(a)

	F	G
5	Price Simulation	
6	Time (days)	Price
7	0	=B258
8	1	=$G$7*EXP(($D$3-0.5*$D$4^2)*F8+$D$4*NORMSINV(RAND())*SQRT(F8))
9	2	=$G$7*EXP(($D$3-0.5*$D$4^2)*F9+$D$4*NORMSINV(RAND())*SQRT(F9))
10	3	=$G$7*EXP(($D$3-0.5*$D$4^2)*F10+$D$4*NORMSINV(RAND())*SQRT(F10))
11	4	=$G$7*EXP(($D$3-0.5*$D$4^2)*F11+$D$4*NORMSINV(RAND())*SQRT(F11))
12	5	=$G$7*EXP(($D$3-0.5*$D$4^2)*F12+$D$4*NORMSINV(RAND())*SQRT(F12))
13	6	=$G$7*EXP(($D$3-0.5*$D$4^2)*F13+$D$4*NORMSINV(RAND())*SQRT(F13))
14	7	=$G$7*EXP(($D$3-0.5*$D$4^2)*F14+$D$4*NORMSINV(RAND())*SQRT(F14))
15	8	=$G$7*EXP(($D$3-0.5*$D$4^2)*F15+$D$4*NORMSINV(RAND())*SQRT(F15))

(b)

FIGURE 3-17 Stock Price Simulation Model

simulation in columns F and G, we find the simulated price on the expiration date using a lookup table. We then determine if we have made a profit, and then compute the net present value of the profit. However, this is only a single simulated trial. As before, we may use a data table to replicate the simulation for several trials and compute the average NPV of the profit to estimate the fair price of the option. This is shown in columns L and M.

SKILLBUILDER EXERCISE

Open the file *1999 CSCO Stock Prices.xls*. Expand the data table to simulate 500 trials for the options pricing model. What proportion of time will the option yield a positive return?

	I	J	K	L	M
1					
2	**European Call Option Pricing Model**				NPV
3	Initial price	$53.56		Trial	$ -
4	Option strike price	$ 60.00		1	$ -
5	Expiration date (days)	50		2	$ 0.10
6	Risk-free rate	8%		3	$ 4.31
7				4	$ 22.61
8	**Model Results**			5	$ -
9				6	$ -
10	Stock price at expiration date	$ 51.77		7	$ 51.83
11	Profit	$ -		8	$ 8.49
12	Net present value of profit	$0.00		9	$ 7.39
13				10	$ -
14	Average NPV	$ 6.45		11	$ 3.39
15				12	$ 15.72
16				13	$ 2.51
17				14	$ 2.40
18				15	$ -

(a)

	I	J
1		
2	**European Call Option Pricing Model**	
3	Initial price	=G7
4	Option strike price	60
5	Expiration date (days)	50
6	Risk-free rate	0.08
7		
8	**Model Results**	
9		
10	Stock price at expiration date	=VLOOKUP(J5,$F$7:$G$107,2)
11	Profit	=MAX(J10-J4,0)
12	Net present value of profit	=NPV(J6*J5/365,J11)
13		
14	Average NPV	=AVERAGE(M4:M103)

(b)

FIGURE 3-18 Call Option Simulation Model

CASH MANAGEMENT SIMULATION

The *Miller-Orr model* in finance addresses the problem of a firm's managing its cash position by purchasing or selling securities at a transaction cost in order to lower or raise its cash position. That is, the firm needs to have enough cash on hand to meet its obligations, but does not want to maintain too high a cash balance because it loses the opportunity for earning higher interest by investing in other securities. The Miller-Orr model assumes that the firm will maintain a minimum cash balance, m, a maximum cash balance, M, and an ideal level, R, called the *return point*. Cash is managed using a decision rule that states that whenever the cash balance falls to m, $R - m$ securities are sold to bring the balance up to the return point. When the cash balance rises to M, $M - R$ securities are purchased to reduce the cash balance back to the return point. Using some advanced mathematics, the return point and maximum cash balance levels are shown to be:

$$R = m + Z$$
$$M = R + 2Z$$

where

$$Z = \left(\frac{3C_0\sigma^2}{4r}\right)^{1/3}$$

σ^2 = variance of the daily cash flows, and

r = average daily rate of return corresponding to the premium associated with securities.

For example, if the premium is 4 percent, $r = .04/365$. To apply the model, note that we do not need to know the actual demand for cash, only the daily variance. Essentially, the Miller-Orr model determines the decision rule that minimizes the expected costs of making the cash-security transactions and the expected opportunity costs of maintaining the cash balance based on the variance of the cash requirements.

Figure 3-19 shows a spreadsheet for implementing the model and for simulating the results. In the simulation model, we begin with a cash level equal to the return point. The next day's requirement is randomly generated in column F as a normal random variate with mean 0 and variance given in cell B7. The decision rule is applied in column G. If the cash balance for the current day (cell E4) is less than or equal to the minimum

FIGURE 3-19 Miller-Orr Cash Management Simulation Model

	A	B	C	D	E	F	G
1	Miller-Orr Model				Simulation	Next Day's	Transaction
2				Day	Cash Balance	Requirement	Amount
3	*Model Inputs*			0	$ 9,931.30	$ (5.36)	$ -
4				1	$ 9,925.93	$ 327.10	$ -
5	Transaction cost	$ 35.00		2	$ 10,253.03	$ 119.89	$ -
6	Interest rate premium of securities	4%		3	$ 10,372.92	$ (0.75)	$ -
7	Variance of daily cash flows	$ 60,000.00		4	$ 10,372.17	$ 173.96	$ -
8	Required minimum balance	$ 7,500.00		5	$ 10,546.13	$ 675.86	$ -
9				6	$ 11,221.99	$ 171.55	$ -
10	*Model Results*			7	$ 11,393.54	$ 352.78	$ -
11				8	$ 11,746.31	$ 37.82	$ -
12	Return point	$ 9,931.30		9	$ 11,784.13	$ (166.51)	$ -
13	Upper limit	$ 14,793.89		10	$ 11,617.62	$ 500.03	$ -
14				11	$ 12,117.65	$ 230.98	$ -
15				12	$ 12,348.63	$ 108.20	$ -
16	Cash Balance			13	$ 12,456.83	$ 143.40	$ -
17				14	$ 12,600.23	$ 253.76	$ -
18			M	15	$ 12,853.99	$ (710.93)	$ -
19				16	$ 12,143.05	$ (64.06)	$ -
20				17	$ 12,078.99	$ 62.25	$ -
21				18	$ 12,141.24	$ 245.26	$ -
22			R	19	$ 12,386.50	$ (467.78)	$ -
23				20	$ 11,918.71	$ (459.86)	$ -
24			m	21	$ 11,458.85	$ (270.31)	$ -
25				22	$ 11,188.55	$ (419.09)	$ -
26	Day			23	$ 10,769.45	$ 288.03	$ -
27				24	$ 11,057.48	$ (40.80)	$ -
28				25	$ 11,016.68	$ (427.63)	$ -

(a)

	B	C	D	E	F	G
1				Simulation	Next Day's	Transaction
2			Day	Cash Balance	Requirement	Amount
3			0	=B12	=NORMINV(RAND(),0,SQRT($B$7))	0
4			1	=E3+F3+G3	=NORMINV(RAND(),0,SQRT($B$7))	=IF(E4<=$B$8,$B$12-E4,IF(E4>=$B$13,-E4+$B$12,0))
5	35		2	=E4+F4+G4	=NORMINV(RAND(),0,SQRT($B$7))	=IF(E5<=$B$8,$B$12-E5,IF(E5>=$B$13,-E5+$B$12,0))
6	0.04		3	=E5+F5+G5	=NORMINV(RAND(),0,SQRT($B$7))	=IF(E6<=$B$8,$B$12-E6,IF(E6>=$B$13,-E6+$B$12,0))
7	60000		4	=E6+F6+G6	=NORMINV(RAND(),0,SQRT($B$7))	=IF(E7<=$B$8,$B$12-E7,IF(E7>=$B$13,-E7+$B$12,0))
8	7500		5	=E7+F7+G7	=NORMINV(RAND(),0,SQRT($B$7))	=IF(E8<=$B$8,$B$12-E8,IF(E8>=$B$13,-E8+$B$12,0))
9			6	=E8+F8+G8	=NORMINV(RAND(),0,SQRT($B$7))	=IF(E9<=$B$8,$B$12-E9,IF(E9>=$B$13,-E9+$B$12,0))
10			7	=E9+F9+G9	=NORMINV(RAND(),0,SQRT($B$7))	=IF(E10<=$B$8,$B$12-E10,IF(E10>=$B$13,-E10+$B$12,0))
11			8	=E10+F10+G10	=NORMINV(RAND(),0,SQRT($B$7))	=IF(E11<=$B$8,$B$12-E11,IF(E11>=$B$13,-E11+$B$12,0))
12	=B8+(3*B5*B7/(4*B6/365))^(1/3)		9	=E11+F11+G11	=NORMINV(RAND(),0,SQRT($B$7))	=IF(E12<=$B$8,$B$12-E12,IF(E12>=$B$13,-E12+$B$12,0))
13	=B12+2*(3*B5*B7/(4*B6/365))^(1/3)		10	=E12+F12+G12	=NORMINV(RAND(),0,SQRT($B$7))	=IF(E13<=$B$8,$B$12-E13,IF(E13>=$B$13,-E13+$B$12,0))

(b)

level (cell B8), we sell B12-E4 dollars of securities to bring the balance up to the return point. Otherwise, if the cash balance exceeds the upper limit (B13), we buy enough securities (i.e., subtract an amount of cash) to bring the balance back down to the return point. If neither of these conditions hold, there is no transaction and the balance for the next day is simply the current value plus the net requirement.

SKILLBUILDER EXERCISE

Open the file *Miller-Orr.xls*. The model is designed to simulate 365 days. Modify the spreadsheet to compute a count (hint: use the COUNTIF function) of the number of securities transactions per year and the total annual cost of securities transactions. Construct a data table to simulate the model for 100 trials.

Modeling Probabilistic Inputs

Given the wide variety of probability distributions that we described in this chapter, it is not always easy to select the most appropriate one. However, the choice of an appropriate assumption for a simulation model is a critical one. To illustrate the importance of identifying the correct distribution in simulation as well as in other quantitative models, we discuss the following example in advertising.[3]

The amount that companies spend on the creative component of advertising (that is, making better ads) is traditionally quite small relative to the overall media budget. One expert noted that the expenditure on creative development was about 5 percent of that spent on the media delivery campaign. A packaged-goods company executive concluded that of the advertising his firm conducted, roughly 50 percent was wasted entirely, 40 percent was remembered by consumers, and only 10 percent actually motivates consumers to buy the product. Of the total, only about 5 percent is extremely effective.

Whatever money is spent on creative development is usually directed through a single advertising agency. However, one theory that has been proposed is that more should be spent on creative ad development, and the expenditures should be spread across a number of competitive advertising agencies. In research studies of this theory, the distribution of advertising effectiveness was assumed to be normal. In reality, data collected on the response to consumer product ads shows that this distribution is actually quite skewed, and therefore not normally distributed. Using the wrong assumption in any model or application can produce erroneous results. In this situation, the skewness actually provides an advantage for advertisers, making it more effective to obtain ideas from a variety of advertising agencies.

A mathematical model (called Gross's model) relates the relative contributions of creative and media dollars to total advertising effectiveness. This model includes factors of ad development cost, total media spending budget, the distribution of effectiveness across ads (assumed to be normal), and the unreliability of identifying the most effective ad from a set of independently-generated alternatives. The economic decision model used to identify the optimal number of draft ads to purchased uses net discounted cash flow. Gross concluded that large gains were possible if multiple ads were obtained from independent sources, and the best ad is selected.

---

[3]Adapted from G. C. O'Connor, T. R. Willemain, and J. MacLachlan, "The Value of Competition Among Agencies in Developing Ad Campaigns: Revisiting Gross's Model," *Journal of Advertising,* Vol. 25, no. 1, 1996, pp. 51–62.

Because the data observed on ad effectiveness is clearly skewed, other researchers examined ad effectiveness by studying standard industry data on ad recall. This allowed examination of Gross's theory without requiring the assumption of normally-distributed effects. This analysis found that the best of a number of ads was more effective than any single ad. Further analysis revealed that the optimal number of ads to commission can vary significantly depending on the shape of the distribution of effectiveness for a single ad.

The research examined published data on the effectiveness of advertisements in two media. These data revealed strong evidence of skewness. As a result, the researchers developed an alternative to Gross's model. Comparisons were made by simulating 80,000 replications for each value of the number of total ads commissioned. As the number of draft ads was increased, the effectiveness of the best ad also increased. For example, if one ad is used, the average number of retained impressions per dollar was 57, whereas the average for the better of two draft ads was 82, and the average for the best of five ads was 124. The Monte-Carlo simulation results identified the average number of retained impressions per dollar spent on media. Both the optimal number of draft ads and the payoff from creating multiple independent drafts were higher *when the correct distribution was used* than the results reported in Gross's original study. Thus, simulation modelers should choose carefully the assumptions made in simulation models.

There are several approaches to choosing a probability distribution to represent a random variable in a simulation model. We will address two situations: (1) when empirical data exist, and (2) when no data are available.

DISTRIBUTION FITTING OF EMPIRICAL DATA

For many inputs to simulation models, empirical data may be available in historical records or collected through special efforts. For example, maintenance records might provide data on machine failure rates and repair times, or observers might collect data on service times in a bank or post office. Figure 3-20 shows a portion of a spreadsheet (*Airport Service Times.xls*) containing 812 observations of the service times (in seconds) at an express service counter at an airport along with a frequency table in 30-second intervals and a histogram of the data. We could sample from this distribution and use the results as inputs to a simulation model. This can be done easily by using a lookup table to select a cell in the frequency distribution and then assuming that the values within the cell are uniformly distributed. An Excel spreadsheet for this example is shown in Figure 3-21.

Sampling from an empirical distribution has some drawbacks, however. First, the empirical data may not adequately represent the true underlying population because of sampling error. Second, using an empirical distribution precludes sampling values *outside* the range of the actual data. Again, because of sampling error, such values may indeed occur at other times, and any simulation results would not reflect this. A way to overcome these drawbacks is to attempt to fit a theoretical distribution to the data and to verify goodness-of-fit statistically. Once an appropriate distribution and its parameters are selected, we can generate random variates from the distribution in our simulation model.

To select an appropriate theoretical distribution, we might begin by examining a histogram of the data to look for the distinctive shapes of particular distributions. For example, normal data is symmetric, with a peak in the middle. Exponential data is very positively skewed, with no negative values. Lognormal data is also very positively skewed, but the density drops to zero at 0. Various forms of the gamma, Weibull, or beta distributions could be used for distributions that do not seem to fit one of the other

	A	B	C	D	E
1	**Service Times at an Airport Ticketing Counter**				
2					
3	Data (sorted)		Range in Seconds		Number of Passengers
4	9		0	29	58
5	9		30	59	105
6	9		60	89	257
7	10		90	119	84
8	10		120	149	96
9	11		150	179	41
10	12		180	209	52
11	14		210	239	23
12	15		240	269	24
13	15		270	299	18
14	15		300	329	19
15	15		330	359	4
16	15		360	389	8
17	16		390	419	4
18	16		420	449	4
19	16		450	479	4
20	16		480	509	2
21	17		510	539	1
22	18		540	569	1
23	18		570	599	0
24	18		600	629	2
25	19		630	659	0
26	19		660	689	1
27	19		690	719	1
28	19		720	749	2
29	19		750	779	0
30	19		780	809	0
31	19		810	839	0
32	20		840	869	0
33	20		870	899	1

(a)

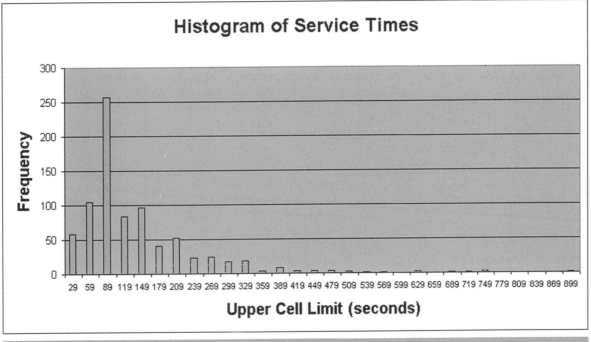

(b)

FIGURE 3-20 Portion of Airport Service Times and Frequency Table

95

	A	B	C	D	E	F	G	H	I
1	**Airport Service Time Simulation**								
2									
3	**Random Number Range**		**Range in Seconds**		**Trial**	**Random Number**	**Selected Range**		**Service Time**
4	0.0000	0.0714	0	29	1	0.5764	90	119	111
5	0.0714	0.2007	30	59	2	0.5066	60	89	62
6	0.2007	0.5172	60	89	3	0.1686	30	59	31
7	0.5172	0.6207	90	119	4	0.7300	120	149	133
8	0.6207	0.7389	120	149	5	0.2131	60	89	85
9	0.7389	0.7894	150	179	6	0.1526	30	59	33
10	0.7894	0.8534	180	209	7	0.2726	60	89	83
11	0.8534	0.8818	210	239	8	0.4275	60	89	61
12	0.8818	0.9113	240	269	9	0.0745	30	59	38
13	0.9113	0.9335	270	299	10	0.6522	120	149	122
14	0.9335	0.9569	300	329	11	0.5272	90	119	114
15	0.9569	0.9618	330	359	12	0.2989	60	89	84
16	0.9618	0.9717	360	389	13	0.4276	60	89	76
17	0.9717	0.9766	390	419	14	0.9852	450	479	469
18	0.9766	0.9815	420	449	15	0.2442	60	89	63
19	0.9815	0.9865	450	479	16	0.0420	0	29	13
20	0.9865	0.9889	480	509	17	0.7906	180	209	199
21	0.9889	0.9901	510	539	18	0.7756	150	179	152
22	0.9901	0.9914	540	569	19	0.3479	60	89	84
23	0.9914	0.9938	600	629	20	0.4327	60	89	78
24	0.9938	0.9951	660	689					
25	0.9951	0.9963	690	719					
26	0.9963	0.9988	720	749					
27	0.9988	1.0000	870	899					

(a)

	E	F	G	H	I
3	**Trial**	**Random Number**	**Selected Range**		**Service Time**
4	1	=RAND()	=VLOOKUP(F4,$A$4:$D$27,3)	=VLOOKUP(F4,$A$4:$D$27,4)	=RANDBETWEEN(G4,H4)
5	2	=RAND()	=VLOOKUP(F5,$A$4:$D$27,3)	=VLOOKUP(F5,$A$4:$D$27,4)	=RANDBETWEEN(G5,H5)
6	3	=RAND()	=VLOOKUP(F6,$A$4:$D$27,3)	=VLOOKUP(F6,$A$4:$D$27,4)	=RANDBETWEEN(G6,H6)
7	4	=RAND()	=VLOOKUP(F7,$A$4:$D$27,3)	=VLOOKUP(F7,$A$4:$D$27,4)	=RANDBETWEEN(G7,H7)
8	5	=RAND()	=VLOOKUP(F8,$A$4:$D$27,3)	=VLOOKUP(F8,$A$4:$D$27,4)	=RANDBETWEEN(G8,H8)
9	6	=RAND()	=VLOOKUP(F9,$A$4:$D$27,3)	=VLOOKUP(F9,$A$4:$D$27,4)	=RANDBETWEEN(G9,H9)
10	7	=RAND()	=VLOOKUP(F10,$A$4:$D$27,3)	=VLOOKUP(F10,$A$4:$D$27,4)	=RANDBETWEEN(G10,H10)
11	8	=RAND()	=VLOOKUP(F11,$A$4:$D$27,3)	=VLOOKUP(F11,$A$4:$D$27,4)	=RANDBETWEEN(G11,H11)
12	9	=RAND()	=VLOOKUP(F12,$A$4:$D$27,3)	=VLOOKUP(F12,$A$4:$D$27,4)	=RANDBETWEEN(G12,H12)
13	10	=RAND()	=VLOOKUP(F13,$A$4:$D$27,3)	=VLOOKUP(F13,$A$4:$D$27,4)	=RANDBETWEEN(G13,H13)
14	11	=RAND()	=VLOOKUP(F14,$A$4:$D$27,3)	=VLOOKUP(F14,$A$4:$D$27,4)	=RANDBETWEEN(G14,H14)
15	12	=RAND()	=VLOOKUP(F15,$A$4:$D$27,3)	=VLOOKUP(F15,$A$4:$D$27,4)	=RANDBETWEEN(G15,H15)
16	13	=RAND()	=VLOOKUP(F16,$A$4:$D$27,3)	=VLOOKUP(F16,$A$4:$D$27,4)	=RANDBETWEEN(G16,H16)
17	14	=RAND()	=VLOOKUP(F17,$A$4:$D$27,3)	=VLOOKUP(F17,$A$4:$D$27,4)	=RANDBETWEEN(G17,H17)
18	15	=RAND()	=VLOOKUP(F18,$A$4:$D$27,3)	=VLOOKUP(F18,$A$4:$D$27,4)	=RANDBETWEEN(G18,H18)
19	16	=RAND()	=VLOOKUP(F19,$A$4:$D$27,3)	=VLOOKUP(F19,$A$4:$D$27,4)	=RANDBETWEEN(G19,H19)
20	17	=RAND()	=VLOOKUP(F20,$A$4:$D$27,3)	=VLOOKUP(F20,$A$4:$D$27,4)	=RANDBETWEEN(G20,H20)
21	18	=RAND()	=VLOOKUP(F21,$A$4:$D$27,3)	=VLOOKUP(F21,$A$4:$D$27,4)	=RANDBETWEEN(G21,H21)
22	19	=RAND()	=VLOOKUP(F22,$A$4:$D$27,3)	=VLOOKUP(F22,$A$4:$D$27,4)	=RANDBETWEEN(G22,H22)
23	20	=RAND()	=VLOOKUP(F23,$A$4:$D$27,3)	=VLOOKUP(F23,$A$4:$D$27,4)	=RANDBETWEEN(G23,H23)

(b)

FIGURE 3-21 Simulating Airport Service Times

common forms. This approach is not, of course, always accurate or valid, and sometimes it can be difficult to apply, especially if sample sizes are small. However, it may narrow the search to a few potential distributions. It is not clear from the histogram in Figure 3-20(b) what the distribution might be. It clearly does not appear to be exponential, but it might be lognormal, or perhaps gamma or beta distributed.

Summary statistics can provide additional clues about the nature of a distribution. The mean, median, standard deviation, and coefficient of variation often provide clues about the nature of the distribution. For instance, normally-distributed data tend to have a fairly low coefficient of variation; however, this may not be true if the mean is small. For normally-distributed data, we would also expect the median and mean to be approximately the same. For exponentially distributed data, however, the median will be less than the mean. Also, we would expect the mean to be about equal to the standard deviation. We could also look at the skewness index. Normal data are not skewed; lognormal and exponential data are positively skewed. Table 3-2 shows these summary statistics for the sample data using Excel's Descriptive Statistics tool. We can see that the mean is not close to being equal to the standard deviation, suggesting that the data are probably not exponential. The data are positively skewed, suggesting that the lognormal distribution might be appropriate. However, this information does not help much in this case.

GOODNESS-OF-FIT TESTS

The examination of histograms and summary statistics might provide some hypotheses; however, these should be verified in a more formal manner. *Goodness-of-fit tests* provide statistical evidence to test hypotheses about the nature of the distribution. Goodness-of-fit is important in selecting the correct input distribution. Two popular statistical tests for goodness-of-fit are the Chi-Square (χ^2) test and the Kolmogorov-Smirnov test. A third, but lesser-used, test is the Anderson-Darling test. These tests are tedious to apply manually for all but the simplest distributions, so we recommend that computerized algorithms be used.

TABLE 3-2 Descriptive Statistics for Airport Service Times	
Mean	126.2783251
Standard error	3.691220698
Median	88
Mode	83
Standard deviation	105.1835991
Sample variance	11063.58952
Kurtosis	8.70752635
Skewness	2.413577454
Range	867
Minimum	9
Maximum	876
Sum	102538
Count	812

Chi-Square Test   The most commonly used goodness-of-fit test is the Chi-Square (χ^2) test. This approach tests the hypothesis

H_0: the sample data come from a specified distribution (e.g., normal, uniform, etc.)

against the alternative hypothesis

H_1: the sample data do not come from the specified distribution

This test is nonparametric and one-sided, and as with any test of hypothesis, you can disprove the null hypothesis (the fit of a particular distribution), but cannot statistically *prove* that data come from the distribution. To use the χ^2 test, the data need to be divided into a finite number of cells, each of which should contain at least five observations and have approximately an equal expected count. The expected frequency in each cell is computed as np_j, where n is the sample size and p_j is the theoretical probability of the jth cell for the hypothesized distribution. For example, suppose the hypothesized distribution is Poisson with a mean of 2.0 and 100 observations are available. Because the Poisson is discrete, each "cell" would correspond to a discrete value. Table 3-3 shows the probabilities and expected frequencies for 100 observations for values of x from 0 to 10. Because the expected frequencies of cells 5 through 10 are less than 5, we would group them together into one cell with an expected frequency of 5.265.

As a general rule, you should have at least 50 observations to perform the test. With n observations, use somewhere between $\sqrt{n}$ and $n/5$ cells. Appendix C provides a table of χ^2 critical values for different degrees of freedom and probability levels of rejecting the hypothesis that the data come from a particular distribution. The number of degrees of freedom equals the number of cells minus the number of distribution parameters estimated with the data minus 1. For an hypothesized normal distribution, for instance, if the sample mean and standard deviation are estimated from the data, the number of degrees of freedom would equal the number of cells minus 3 (because two parameters are estimated). If the hypothesized distribution is the exponential or Poisson, there is only one parameter to estimate, the mean, and therefore the degrees of freedom would be the number of cells minus 2. There are no parameters to estimate for the uniform distribution, and therefore the number of degrees of freedom would be the number of cells minus 1.

TABLE 3-3 Probabilities and Expected Frequencies for 100 Observations of x		
x	$p(x)$	$100p(x)$
0	0.13534	13.5335
1	0.27067	27.0671
2	0.27067	27.0671
3	0.18045	18.0447
4	0.09022	9.0224
5	0.03609	3.6089
6	0.01203	1.2030
7	0.00344	0.3437
8	0.00086	0.0859
9	0.00019	0.0191
10	0.00004	0.0038

The formula for the χ^2 test statistic is

$$\chi^2 = \sum_{i=1}^{k} \frac{(f_{oi} - f_{ei})^2}{f_{ei}}$$

where

$$f_{oi} = \text{observed frequency of cell } i$$
$$f_{ei} = \text{expected frequency of cell } i$$
$$k = \text{number of cells}$$

If χ^2 is less than the critical value in Appendix C, we fail to reject the hypothesis that the data come from the hypothesized distribution.

We will illustrate the Chi-Square test to test uniformity of the random numbers we generated in Figure 2-15. Figure 3-22 shows an enhancement to the spreadsheet in Figure 2-15. The expected frequency in each cell is 10, and column P shows the calculation of χ^2. At a level of significance of 0.05, the Chi-Square value with 9 degrees of freedom is 16.919 (see Appendix C). Thus, we cannot reject the hypothesis that the data come from a uniform distribution.

Kolmogorov-Smirnov Test   The χ^2 test requires a fair amount of data in order to have a reasonable number of cells, each with at least five observations. For small samples, the Kolmogorov-Smirnov test provides a more suitable test for goodness-of-fit. Furthermore, the χ^2 test often requires using the data itself to estimate distribution parameters. The Kolmogorov-Smirnov test is not limited by these factors.

The underlying idea of the Kolmogorov-Smirnov test is to sort the data in ascending order and find the greatest difference between the theoretical value for each ranked observation and that observation's theoretical counterpart. Generally a K-S value less than 0.03 indicates a good fit.

Anderson-Darling Test   This method is similar to the Kolmogorov-Smirnov approach, except that it weights the differences between the two distributions at their tails greater than at their midranges. This approach is useful when you need a better fit at the extreme tails of the distribution. A computed value less than 1.5 generally indicates a good fit. We will not discuss the details of this approach here, but note that it is included in the distribution fitting capability of the Crystal Ball software, which we discuss next.

FIGURE 3-22 Calculation of Chi-Square Statistic

	L	M	N	O	P
1					
2	Lower	Upper			Chi Square statistic
3	Cell Limit	Cell Limit	Frequency	Expected	Calculation
4	0.0	0.1	7	10	0.9
5	0.1	0.2	12	10	0.4
6	0.2	0.3	11	10	0.1
7	0.3	0.4	11	10	0.1
8	0.4	0.5	9	10	0.1
9	0.5	0.6	5	10	2.5
10	0.6	0.7	11	10	0.1
11	0.7	0.8	11	10	0.1
12	0.8	0.9	17	10	4.9
13	0.9	1.0	6	10	1.6
14					
15				Chi Square	10.80

DISTRIBUTION FITTING WITH CRYSTAL BALL

Crystal Ball has a very useful and powerful data-fitting capability. We will illustrate this capability using the airport service time data given in Figure 3-20. Figure 3-23 shows the Excel screen after Crystal Ball has been loaded. Three new menu items appear on the main menu bar: *Cell, Run,* and *CBTools.* A new button bar also provides short-cut ways to invoke Crystal Ball menu commands.

To invoke the distribution fitting capability, click on any cell in the spreadsheet that contains a constant value. Then choose *Cell* from the main menu, and select *Define Assumption . . .* The Crystal Ball distribution gallery (see Figure 3-24) is displayed. Click on *Fit.* The first of two dialog boxes, shown in Figure 3-25, is displayed. If the data are in the front-most spreadsheet, select *Active Worksheet* (Crystal Ball also allows fitting data from a separate text file), and enter the range of the data in the box to the right. Clicking on *Next* displays the second dialog box, shown in Figure 3-26. You may select all continuous distributions in the distribution gallery, any subset, or just the normal distribution. You must also select the type of test; in this example, we select the Chi-Square test. Check the box *Show Comparison Chart and Goodness-of-Fit Statistics* to display comparative results.

Crystal Ball then fits each of the continuous distributions available in the Distribution Gallery to the data set. It then rank orders them by the Chi-Square statistic and p-value. A p-value greater than 0.5 generally indicates a good fit. The best-fitting distribution is then displayed in a Comparison Chart window with distribution parameters and goodness-of-fit scores, shown in Figure 3-27. You may select the *Prefs* button to cus-

FIGURE 3-23 Portion of Excel Screen after Crystal Ball Installation

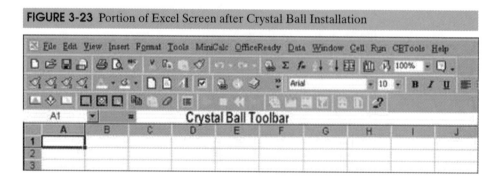

FIGURE 3-24 Crystal Ball Distribution Gallery

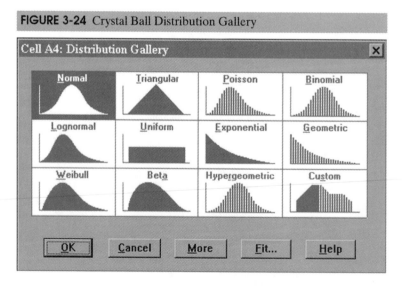

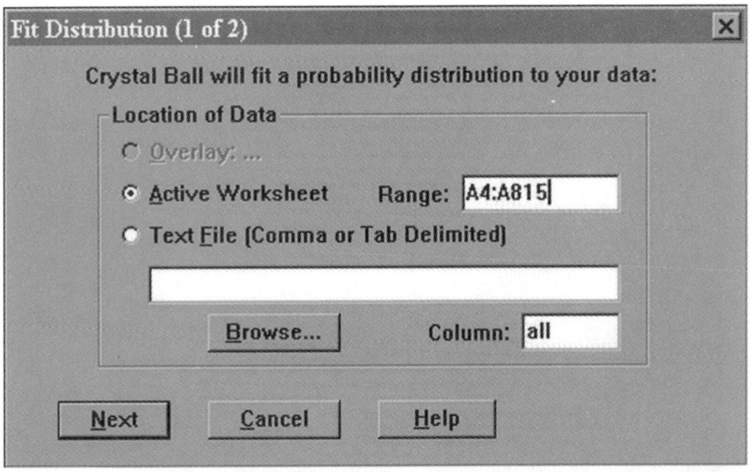

FIGURE 3-25 Crystal Ball Fit Distribution Dialog Box[1]

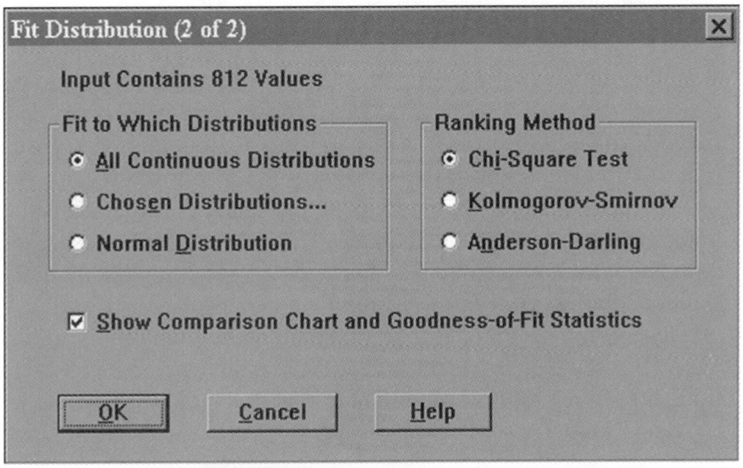

FIGURE 3-26 Crystal Ball Fit Distribution Dialog Box[2]

tomize the chart display type. We see that although the lognormal distribution is the best fit, its Chi-Square value is high and the p-value is essentially zero. Therefore, the fit is not particularly good from a statistical test of hypothesis. By pressing the *Next Distribution* button, the comparison charts for other distributions can be displayed.

SKILLBUILDER EXERCISE

Open the file *Airport Service Times.xls*. Eliminate every other data point between 60 and 89 (the large spike in the histogram in Figure 3-20). Now apply the Crystal Ball Distribution Fit technique to the remaining data. What is the best-fitting distribution now? Does it pass the Chi-Square test?

JUDGMENTAL MODELING

When empirical data are unavailable, modelers must resort to judgment to select an input distribution. This is where knowledge of different distributions is valuable. In describing each distribution, we gave several examples of typical applications. For

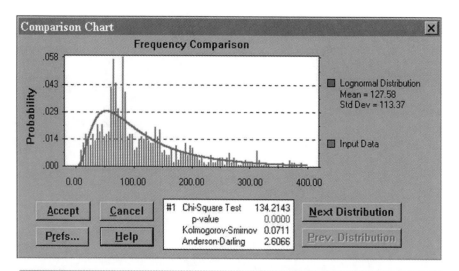

FIGURE 3-27 Crystal Ball Comparison Chart Showing Best-Fitting Distribution

example, in developing a financial simulation model for a new public offering of stock, we might use a lognormal distribution because its properties are similar to those we might expect to observe in a distribution of stock prices. Or, we might use an exponential distribution to model service times in a proposed service system.

In many cases, we might simply have no idea of what distribution to use. Using expert judgment, you can specify an interval $[a, b]$ over which you believe the data should fall. If you have no reason to believe that any value within this interval is more likely than the others, an appropriate choice would be the uniform distribution. If you believe that some value c between a and b is more likely to occur than others, then a triangular distribution could be used. Finally, if you have some reason to estimate an average value, m, in addition to a most likely value c, you might use the beta distribution, with α and β defined by

$$\alpha = \frac{(m - a)(2c - a - b)}{(c - m)(b - a)}$$

$$\beta = \frac{(b - m)\alpha}{m - a}$$

If $m > c$, then the density will be skewed to the right; if $m < c$, it will be skewed to the left.

Statistical Issues in Monte-Carlo Simulation

Two important issues in Monte-Carlo simulation are (1) choosing the number of trials of a simulation and (2) analyzing the results. Both of these issues depend on statistical principles.

SAMPLE SIZE

The number of trials used in a simulation affect the quality of the results. In general, the higher the number of trials, the more accurate will be the characterization of the output distribution and estimates of its parameters, such as the mean. In cell I23 in Figure 2-22, we computed the standard deviation of the 100 profit values in the simulation. If we assume that this value, 65.88, is a good estimate of the true standard deviation of individual profit values, we can compute the standard error of the mean as

$$\text{Standard error} = 65.88/\sqrt{100} = 6.588$$

This means that a 95 percent confidence interval for the mean profit would be $259.80 ± 1.96(6.588) = $259.80 ± 11.13 or $248.67 to $270.93. If we used 500 trials, we would expect the standard error to be approximately (because the actual standard deviation would change) $65.88/\sqrt{500} = 2.95$, yielding a tighter confidence interval.

If we view the width of a confidence interval as the accuracy that we wish to obtain in estimating the mean, we can identify the appropriate number of trials. If we wish our estimate to be accurate within an amount $±A$ with confidence level $100(1 - \alpha)\%$, then the half-width of the confidence interval is set equal to A:

$$\frac{z_{\alpha/2}\sigma}{\sqrt{n}} = A$$

Solving for n, we obtain

$$n = \frac{z_{\alpha/2}^2\sigma^2}{A^2}$$

However, to apply this formula, we need to know the standard deviation, σ, of the individual values of output variable. Generally, this is not known in advance. A common-sense approach would be to take an initial sample of N trials to estimate the standard deviation and compute the accuracy. If the accuracy is not sufficient, use the sample size formula to determine what n should be and repeat the simulation using the larger number of trials. We should use the new value of σ to check whether the new number of replications provided the required accuracy; if not, we might need to repeat the procedure.

For instance, suppose we wish to ensure that our estimate of the mean profit in the Dave's Candies simulation is within $5.00 with a confidence level of at least 0.99 ($z_{.995} = 2.575$). Using the initial sample of 100 trials in Figure 2-22, we compute n as

$$n = \frac{(2.575)^2(65.88)^2}{(5)^2} = 1151.12 \text{ or } 1152$$

Thus, we need to run the simulation for 1,152 trials. (It would make more sense to liberally round up, say, to 1,200—computing power generally is not an issue in today's world. Of course, for very complex simulations that do consume a lot of computing time, we would be more conservative.) We would need to check the new confidence interval, because the standard deviation estimate is likely to change.

SKILLBUILDER EXERCISE

Open the file *Roulette Simulation.xls*. Compute the average payoff and standard error for 50, 100, and 250 trials. What would 95 percent confidence intervals for the mean payoff be in each case?

OUTPUT ANALYSIS

We may express the results of a Monte-Carlo simulation in various ways. We may compute a variety of basic descriptive statistical measures, such as the mean, median, standard deviation, and standard error, as well as some more advanced statistical measures such as skewness, kurtosis, and the coefficient of variability. From this information, we can easily develop a confidence interval for the mean as we had previously illustrated. We would also want to construct a frequency distribution of results and use this information to answer various questions associated with risk. It is also common to express a distribution in terms of **percentiles,** that is, the percentage of outcomes above (or below) any given value. Often, increments of 10 percent (deciles) are used. We might also test various hypotheses about a mean or proportion, or the difference between two alternative system designs. In any case, statistics are critical to the effective use of simulation!

Simulation in Practice

SIMULATION ANALYSIS FOR RED CROSS BLOOD DRIVES[4]

The Red Cross collects over 6 million units of blood per year in the United States, all from volunteer donors. Blood is collected at over 400 fixed and mobile sites, with about 80 percent of blood collected from the mobile sites. The system relies heavily on repeat donors, who give blood on a regular basis. Because these donors are volunteers, their recruitment and retention is difficult. In the early 1990s, the Red Cross was very concerned about how to keep these donors satisfied.

In order to minimize donor discomfort, the Red Cross wanted to minimize donor time—especially waiting time—spent at the donation site. This was also important to sponsors who gave their employees time off for blood donation. The Red Cross told donors and sponsors that donation time was about 1 hour, but for many donors, it took as long as 1 1/2 to 2 hours. This was because arrivals at the donor site were often random, and because of the growing need to conduct blood screening tests.

The blood collection process was examined via simulation to try to develop policies that would reduce the average waiting time and overall time in the system. The first step was to describe the blood-collection process, shown in Figure 3-28. The process consists of

1. registration, after which the donor is directed to a waiting area,
2. completion of a self-administered health history form,
3. measurement of donor's vitals by a staff member: temperature, blood pressure, and pulse as well as simple blood test,
4. a health history interview by a staff member,
5. issuance and explanation of a confidential unit exclusion (CUE) form and a blood bag set used to collect the blood,
6. actual drawing of blood, and finally
7. short recovery period in the "canteen," a waiting room that provides liquid refreshment and a snack.

In Figure 3-28, shaded boxes represent steps that have capacity limitations; the others have unlimited capacity. The time required for each of the major tasks in the current process was identified by gathering data by observation. The actual donation process was broken down into three parts: the time required to prepare the donor and insert the needle, blood-gathering time, and the time required to disconnect the needle and bandage the donor. The times for each step were fit to probability distributions as inputs to the simulation model.

Arrival patterns were collected at six collection sites. The arrival data showed three dominant patterns. For business organizations in which employees were given time off from work to donate, there were two peak arrival times—around mid-morning and mid-afternoon. This accounted for employer-sponsored drives where employees received time off from work to donate. For schools and other organizations where shift work was going on, arrivals peaked in mid- to late-morning. For example, if the workday ended at

[4]John E. Brennan, Bruce L. Golden, and Harold K. Rappaport, "Go with the Flow: Improving Red Cross Bloodmobiles Using Simulation Analysis," *Interfaces* 22(5), September–October, 1992, pp. 1–13.

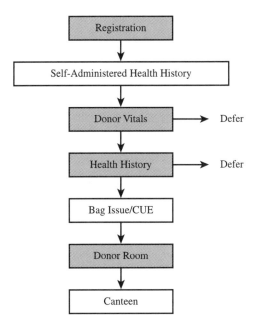

FIGURE 3-28 Blood-Collection Process

Source: Reprinted by permission, J. E. Brennan, B. L. Golden, and H. K. Rappaport, "Go with the Flow: Improving Red Cross Bloodmobiles, Using Simulation Analysis," *Interfaces,* Vol. 22, no. 5, 1992, Copyright 1992, The Institute of Management Sciences and the Operations Research Society of America (currently INFORMS), 2 Charles Street, Suite 300, Providence, RI 02904 USA.

3 P.M., employees usually received time off in the morning. In open community drives where donors came on their own time, the peak donation time was around mid-day. Analysis of the data indicated that a nonstationary Poisson distribution could be used to model arrivals. By nonstationary, we mean that the mean of the Poisson distribution varied over the course of the day.

Service times were collected for each step in the donation process using both manual (e.g., stopwatch) and automated techniques (bar-code identification). A computer routine was used to fit probability distributions to the observed data for all steps of the process. The best-fitting distributions were either lognormal or Weibull, and were verified using a Chi-Square goodness-of-fit test.

Several different policy alternatives were considered, for example, combining some of the screening steps into one workstation, increasing the number of beds served by a blood collector (phlebotomist), and having some staff members float around to wherever a bottleneck occurred. These three alternatives were simulated in various combinations and compared to the existing system. Output statistics included the mean time in the system, mean waiting times, and percentage of donors with no waiting. Confidence intervals were computed for both the transit time and waiting time for each alternative, and compared to a control scenario to estimate the amount of time saved. The alternative combining initial processing activities, using 8 beds per phlebotomist, and having a floating staff member to help with bottleneck activities was adopted. This solution was tested in three real blood drives and resulted in improved performance and greater satisfaction on the part of both donors and sponsors. After the study, the Red Cross initiated a program to disseminate the results to all Red Cross regions.

SIMULATING A MAJOR RUNNING RACE[5]

The Bolder Boulder is a 10-kilometer running race held each Memorial Day in Boulder, Colorado. Since its debut in 1979, it has grown from 2,200 participants to just under 20,000 in 7 years and has become one of America's premier races. As the race grew in

[5]Adapted from Ron Farina, Gary A. Kochenberger, and Tom Obremski, "The Computer Runs the Bolder Boulder: A Simulation of a Major Running Race," *Interfaces,* Vol. 19, no. 2, March–April 1989, pp. 48–55.

size, major concerns centered around crowding in the streets and especially at the finish line. All runners had to be individually tagged as they finished. Because the space, the number of chutes, and other equipment were limited, large queues of runners built up in front of the finish line at certain times, causing many complaints. To address this problem, management implemented an interval start system. Entrants were assigned to one of 24 groups based on ability and released in 1-minute intervals. Although this system helped crowding problems in the streets, it did little to alleviate queues at the finish line.

Race personnel proposed several options:

1. Increase the number of chutes at the finish line.
2. Increase the length of the chutes.
3. Modify the tag procedure to reduce the time it took to tag runners as they crossed the finish line.
4. Increase or modify the block release intervals to spread out the race further.
5. Use some combination of these alternatives.

Because of various restrictions involving facilities, equipment, and personnel, option 4 seemed to be the most promising; however, television coverage dictated that it was necessary to ensure that all blocks were released within in 43 minutes.

To help identify a workable solution, a simulation model was developed. The model was used to identify a sensible chute configuration and a set of block-starting intervals to satisfy all constraints and realize acceptable finish-line behavior. An important input to the model was the distribution of block run times. Figure 3-29 shows a histogram of finish times for the first block of runners in the 1984 race. A preliminary examination of the data suggested that a lognormal distribution was the most appropriate representation for this distribution, and it also applied to other block run times. The historical

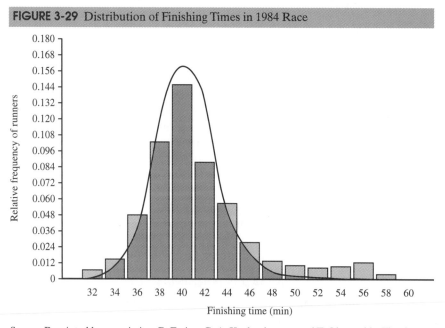

FIGURE 3-29 Distribution of Finishing Times in 1984 Race

Source: Reprinted by permission, R. Farina, G. A. Kochenberger, and T. Obremski, "The Computer Runs the Bolder Boulder; A Simulation of a Major Running Race," *Interfaces,* Vol. 19, no. 2, March–April 1989. Copyright 1989, The Institute of Management Sciences and The Operations Research Society of America (currently INFORMS), 2 Charles Street, Suite 300, Providence, RI 02904 USA.

data was used to identify parameters for the lognormal distributions, which were used to drive the simulation.

The race managers investigated two major alternatives: increasing the number of chutes from 8 to either 12 or 15. Because of severe time constraints in preparing for the upcoming race, it was not possible to conduct a thorough validation of the model. However, simulation results indicated that both alternatives yielded acceptable results, and the 12-chute recommendation was used for the 1985 race. The actual race behavior was very close to the results suggested by the simulation model. No crowded queues appeared at the finish line, and the director of the New York City Marathon stated that he hadn't seen a "better organized race anywhere."

After the 1985 race, the model developers addressed validation and statistical analysis more thoroughly. First, they put actual 1984 race conditions into the model to see if it would yield the same unacceptable conditions (queues at the finish line) that were experienced. The model responded as anticipated and was consistent with estimated queues that race managers had observed. Furthermore, the actual 1985 race results looked very much like that predicted by the model, leading to the conclusion that the model was indeed valid. On the other hand, formal goodness-of-fit tests could not substantiate the lognormal assumption for the run times. Because the tests are highly discriminating, the large number of data points would make it difficult not to reject virtually any distribution. Also, runner behavior, such as running in pairs or small groups or shooting for a goal, invalidates the assumption that finishes are independent. Despite these shortcomings, the histograms did suggest a lognormal distribution, and the results closely resembled actual race conditions. Thus, the modelers decided to keep this assumption for future analyses.

Questions and Problems

1. Describe how statistics are used in simulation modeling and analysis.
2. Using the data in Table 3-4, answer the following. Construct a spreadsheet for doing your calculations and check your results using the Excel *Descriptive Statistics* tool.
   a. Compute the mean and median.
   b. Treat this data set as a sample. Calculate the sample variance and standard deviation.
   c. Calculate the coefficient of skewness and the coefficient of kurtosis for this data set. What do you conclude?

TABLE 3-4	Data For Question 2								
14.23	7.58	18.60	16.89	22.67	16.40	19.58	23.34	14.22	18.81
21.38	20.20	23.47	14.16	18.13	18.94	15.85	22.23	10.87	18.16
20.42	29.07	13.58	28.28	19.90	19.99	17.85	20.26	21.51	22.06
13.18	1.66	21.93	14.00	15.93	23.74	14.54	21.29	24.57	29.61
22.20	25.70	21.22	17.07	21.50	13.50	22.55	23.13	24.24	19.74
8.51	21.54	30.64	18.73	26.19	27.15	21.33	24.94	31.80	18.94
22.06	18.37	18.64	26.17	14.18	20.34	22.43	20.98	24.60	23.64
14.36	16.91	15.50	26.15	20.83	15.64	23.30	14.57	16.65	14.99
24.75	23.63	16.56	9.56	17.51	18.50	20.26	28.26	25.39	24.77
19.38	21.60	21.21	26.07	24.79	23.62	20.19	20.90	13.35	14.57

3. Using the data in Table 3-5, answer the following. Construct a spreadsheet for doing your calculations and check your results using the Excel *Descriptive Statistics* tool.

TABLE 3-5   Data for Question 3

12.41	0.65	38.94	26.73	70.36	23.55	46.61	74.79	12.38	40.60
60.87	51.63	75.60	12.16	35.46	41.57	20.35	67.20	3.39	35.63
53.31	96.51	9.96	95.12	49.22	49.96	33.34	52.11	61.87	66.00
8.63	0.01	65.01	11.50	20.76	77.25	13.74	60.15	81.99	97.27
66.98	87.30	59.63	27.92	61.83	9.68	69.53	73.46	80.16	47.92
1.08	62.07	98.33	40.00	89.20	92.36	60.46	83.86	99.09	41.61
66.02	37.22	39.29	89.12	12.24	52.73	68.67	57.79	82.10	76.66
12.97	26.86	18.39	89.08	56.60	19.15	74.55	13.88	25.17	15.84
82.90	76.61	24.59	1.84	30.89	38.22	52.09	95.08	85.95	83.02
45.04	62.55	59.60	88.76	83.08	76.53	51.48	57.17	9.17	13.88

   a. Compute the mean and median.
   b. Treat this data set as a sample. Calculate the sample variance and standard deviation.
   c. Compute the coefficient of skewness and the coefficient of kurtosis for this data set. What do you conclude?
  4. Using the data in Table 3-6, answer the following. Construct a spreadsheet for doing your calculations and check your results using the Excel *Descriptive Statistics* tool.
   a. Compute the mean and median.
   b. Treat this data set as a sample. Calculate the sample variance and standard deviation.
   c. Identify the coefficient of skewness and the coefficient of kurtosis for this data set. What do you conclude?

TABLE 3-6   Data for Question 4

1.33	0.07	4.93	3.11	12.16	2.69	6.28	13.78	1.32	5.21
9.38	7.26	14.10	1.30	4.38	5.37	2.28	11.15	0.35	4.41
7.62	33.56	1.05	30.19	6.78	6.92	4.06	7.36	9.64	10.79
0.90	0.00	10.50	1.22	2.33	14.81	1.48	9.20	17.14	36.00
11.08	20.63	9.07	3.27	9.63	1.02	11.88	13.26	16.17	6.52
0.11	9.69	40.93	5.11	22.25	25.72	9.28	18.24	46.97	5.38
10.79	4.65	4.99	22.19	1.31	7.49	11.61	8.63	17.21	14.55
1.39	3.13	2.03	22.14	8.35	2.13	13.68	1.49	2.90	1.72
17.66	14.53	2.82	0.19	3.70	4.82	7.36	30.12	19.63	17.73
5.98	9.82	9.06	21.86	17.77	14.50	7.23	8.48	0.96	1.49

  5. Using the data in Table 3-7, calculate the covariance and the correlation coefficient between years with the firm and each of the other three variables.
  6. For the data in Table 3-4, calculate the mean of each of the ten columns (each of which contains ten numbers), and calculate the standard deviation of these ten estimates of the mean. Then calculate the mean of the top five and bottom five of each column (groups of five), and then calculate the standard deviation of these twenty estimates of the mean. Finally, take the average of groups of size 2 (first and second, third and fourth, etc., from each column) to get fifty estimates of the mean, and then calculate the standard deviation of these fifty estimates. Finally, compute the standard error of the mean for each of your sample sizes. What do you observe?
  7. Calculate a 90 percent confidence interval for the mean of data in Table 3-4.

TABLE 3-7	Data for Question 5		
Years with Firm	*Years Education*	*College GPA*	*Age When Hired*
10.0	18	3.01	33
10.0	16	2.78	25
10.0	18	3.15	26
10.0	18	3.86	24
9.6	16	2.58	25
8.5	16	2.96	23
8.4	17	3.56	35
8.4	16	2.64	23
8.2	18	3.43	32
7.9	15	2.75	34
7.6	13	2.95	28
7.5	13	2.50	23
7.5	16	2.86	24
7.2	15	2.38	23
6.8	16	3.47	27
6.5	16	3.10	26
6.3	13	2.98	21
6.2	16	2.71	23
5.9	13	2.95	20
5.8	18	3.36	25

8. Using the data in Table 3-5, assume that each column represents one sample from the same population. Find the mean of each sample by averaging the ten entries in each column. What is the sample variance of these ten means? What is the sample variance of the entire data set? Compare the sample variance of the ten means with the standard error of the mean computed from the entire data set using $n = 10$. What do you find?

9. Explain the role of a shape, scale, and location parameter in characterizing a probability distribution.

10. List some common applications or situations for which the following distributions might be used:
    a. uniform
    b. normal
    c. triangular
    d. exponential
    e. lognormal
    f. gamma
    g. Weibull

11. Draw the graph of the cumulative distribution function for a uniform random variable with $a = 5$ and $b = 13$. What is the mean and variance?

12. Calculate the probability of values lower than 27 for data from a normal distribution with mean of 20 and standard deviation of 5.

13. Calculate the mean and variance of a triangular distribution with minimum 3, mode 7, and maximum 15.

14. Calculate the value from a triangular distribution with minimum 3, mode 7, and maximum 15 that is 30 percent of the distance from the minimum. Repeat for the value that is 70 percent of the distance from the minimum.

15. Find the probability that an exponential random variable with mean 0.7 is less than 1.0.

16. Find the value of an exponential distribution with mean 5 whose cumulative probability is 0.8.
17. Find the probability that lognormally distributed data with a mean of 5 and a standard deviation of 3 falls below 10.
18. For lognormally distributed data with mean $= 7$ and standard deviation $= 4$, find the value that is larger than 40 percent of the data.
19. Calculate the mean and variance of the roll of one die (possible results 1 through 6, equally probable).
20. Given an individual event probability of success of 0.9, find the binomial probability of obtaining 7 successes out of 8 trials.
21. Given an average arrival rate of 12 per hour following the Poisson distribution, find the probability of 8 or less arrivals.
22. Gather real data on arrivals and services at some location, and plot the data using a histogram. Good locations are fast food restaurants, bank teller windows, or post office service windows. By visual inspection, does the data appear more uniform, normal, or exponentially distributed?
23. Gather daily stock closing prices for your favorite stock for a six-month period (which may be found at many Internet sites; do a search on "stock prices"). Re-name the *1999 CSCO Stock Prices.xls* file and save it. Insert your daily stock prices in column B (replacing the Cisco stock prices). Adjust the formulas for mean daily change and standard deviation to reflect the correct range of data. Based on 100 trials, calculate the average NPV of profit.
24. Using the *Miller-Orr.xls* file, identify the number of buys and sells (and amounts) for five simulations, using a discount broker transaction cost of $8 per transaction, an interest rate of 8 percent, a variance of cash flows of $90,000 and a minimum balance of $10,000.
25. Use the Chi-Square test to test the hypothesis at the 0.95 confidence level that data given in Table 3-4 is uniformly distributed over the range 0 through 40.
26. Use the Chi-Square test to test the hypothesis at the 0.95 confidence level that the data in Table 3-5 is uniformly distributed over the range 0 to 100.
27. Use the Chi-Square test to test the hypothesis at the 0.95 confidence level that data in Table 3-6 is uniformly distributed over the range 0 to 50.
28. Construct a histogram for the data given in Table 3-4. Are the data most likely to be normal, exponential, or uniform?
29. Construct a histogram for the data given in Table 3-5. Are the data most likely to be normal, exponential, or uniform?
30. Construct a histogram for the data given in Table 3-6. Are the data most likely to be normal, exponential, or uniform?
31. Compute the parameters of a beta distribution with mean 5, most likely value 6, minimum 2, and maximum 9.
32. Show how to generate random variates using the inverse transformation method for the following distribution:

$$f(x) = 2x, \quad 0 \le x \le 1$$

33. Show how to generate random variates using the inverse transformation method for the following distribution:

$$f(x) = \begin{cases} x & \text{if } 0 \le x \le 1 \\ 2 - x & \text{if } 1 \le x \le 2 \end{cases}$$

34. An Erlang random variate can be generated using the inverse transformation using the following formula:

$$Er = -\frac{1}{\lambda} \ln\left(\prod_{i-1}^{k} (1 - R_i) \right)$$

where the mean of each exponential distribution is $1/k\lambda$.
  a. Show how this formula derives from the inverse transformation function for exponential random variates.
  b. Using a spreadsheet, generate three columns of random numbers. Transform each of these into exponential random variates with mean 5 using the inverse transformation function and sum them to obtain an Erlang random variate. In another column, *using the same random numbers,* generate an Erlang random variate using the formula just given. What do you observe? Is either the convolution approach or inverse transformation approach for generating Erlang random variates better than the other? Why?

35. Generate 100 normal random variates with mean 50 and variance 25 using Excel. Fit these data to each distribution in Crystal Ball. Do the tests support the hypotheses that the data were drawn from a normal distribution?

36. Determine the number of samples required to obtain a 90 percent confidence level and an accuracy within 75, if an estimate of the variance is 60,000.

37. Determine the number of samples required to obtain a 95 percent confidence level and an accuracy within 10, if an estimate of the variance is 750.

38. Suppose you have collected the following service times:

106	29	98	56	128
46	68	105	48	52
173	52	54	69	93
69	84	68	41	128
42	18	33	71	152

Analyze this data and suggest an appropriate distribution to use in a simulation model.

39. The Excel file *SP-CPI.xls* shows 30 years of data for the annual percent change in the Standard & Poor's 500 Index and the Consumer Price Index. Construct histograms of these data, suggest possible probability distributions that might fit the data, and use Crystal Ball to find the best-fitting model. If you were to use your best-fitting model as an input to a simulation model to simulate changes in future years, what would be some of the practical issues that you would have to deal with?

40. Collect real data from one of the following situations, and determine an appropriate distribution to use in a simulation model.
  a. time between arrivals of cars at a gas station
  b. number of cars passing a point on a road or highway during a 1-minute interval
  c. size of parties arriving at a restaurant
  d. time taken to wash *and* remove clothes from washers at a laundromat once they are loaded

References

Berk, K. N., and P. Carey. *Data Analysis with Microsoft Excel 5.0 for Windows,* Cambridge, MA: Course Technology, Inc., 1995.

Decisioneering, Inc., *Crystal Ball Version 4.0 User Manual.*

Evans, M., N. Hastings, and B. Peacock. *Statistical Distributions,* 2d ed. New York: John Wiley & Sons, Inc., 1993.

Law, A. M., and W. D. Kelton. *Simulation Modeling & Analysis,* 2d ed. New York: McGraw-Hill, 1991.

Lawless, J. F. *Statistical Models and Methods for Lifetime Data.* New York: John Wiley & Sons, Inc., 1982.

Middleton, M. R. *Data Analysis Using Microsoft Excel 5.0.* Belmont, CA: Duxbury Press, 1995.

Pritsker, A. A. B., *Introduction to Simulation and SLAM II,* 3d ed. New York: Halsted Press, 1986.

Risk Analysis Using Crystal Ball

Chapter Outline

In Chapter 1, we defined *risk* as the probability of occurrence of an undesirable outcome. Thus, risk is related to the uncertainty associated with things that one cannot control, and the results of this uncertainty. In fact, it is difficult to find any decision—business or personal—that has no risk associated with it. The pharmaceutical industry, for example, operates in a high-risk environment.[1] It costs hundreds of millions of dollars and as much as 10 years to bring a drug to market. Once there, seven of 10 products fail to return the cost of the company's capital. With interest rate and currency rate fluctuations, financial risk becomes even more of a concern. In consumer goods industries, manufacturers maintain "safety stock" inventories of goods to protect against the risk of running out of stock due to uncertain demand. On a personal level, individuals insure their homes, automobiles, vacation packages, and lives because of risk. Investing in the stock market is fraught with risk as the ups and downs of technology stocks have

---

[1]Nancy A. Nichols, "Scientific Management at Merck: An Interview with CFO Judy Lewent," *Harvard Business Review,* January–February 1994, pp. 89–99.

demonstrated over the past several years. To measure risk in evaluating mutual funds, *Fortune* introduced the standard deviation in 1997, as it explains how much of a fund's short-term results vary from its long-term average, noting that the "standard deviation can be an important tool for investors—one that can offer some insight not only into how risky a fund is but even into how it might perform in a given market environment in the future."[2]

The importance of risk in business has long been recognized. The renowned management writer, Peter Drucker, observed in 1974:

> To try to eliminate risk in business enterprise is futile. Risk is inherent in the commitment of present resources to future expectations. Indeed, economic progress can be defined as the ability to take greater risks. The attempt to eliminate risks, even the attempt to minimize them, can only make them irrational and unbearable. It can only result in the greatest risk of all: rigidity.[3]

In recent years, managers have taken a renewed interest in the subject, spawned to a large extent by the ability to analyze risk using spreadsheet models and powerful personal computer software. In this chapter, we provide the details on how to run simulations and conduct risk analysis using the Excel add-in, Crystal Ball.[4]

Risk Analysis

Risk analysis is an approach for developing "a comprehensive understanding and awareness of the risk associated with a particular variable of interest (be it a payoff measure, a cash flow profile, or a macroeconomic forecast)."[5] Hertz and Thomas present a simple scenario to illustrate the concept of risk analysis:

> The executives of a food company must decide whether to launch a new packaged cereal. They have come to the conclusion that five factors are the determining variables: advertising and promotion expense, total cereal market, share of market for this product, operating costs, and new capital investment. On the basis of the "most likely" estimate for each of these variables the picture looks very bright—a healthy 30 percent return indicating a significantly positive expected net present value. This future, however, depends on each of the "most likely" estimates coming true in the actual case. If each of these "educated guesses" has, for example, a 60 percent chance of being correct, there is only an 8 percent chance that all five will be correct (.60 × .60 × .60 × .60 × .60) if the factors are assumed to be independent. So the "expected" return or present value measure is actually dependent on a rather unlikely coincidence. The decision maker needs to know a great deal more about the other values used to make each of the five estimates and about what he stands to gain or lose from various combinations of these values.[6]

[2]David Whitford, "Why Risk Matters," *Fortune,* Dec. 29, 1997, pp. 147–152.

[3]Drucker, P. F., *The Manager and the Management Sciences in Management: Tasks, Responsibilities, Practices,* London: Harper & Row, 1974.

[4]Much of this chapter has been adopted from *Crystal Ball 2000 User Manual,* 1988–2000, Decisioneering, Inc. The *User Manual* provides many other details about Crystal Ball that cannot be described here. Copies of the *User Manual* can be purchased from the Technical Support or Sales Departments of Decisioneering at (303) 534-1515; Toll-free sales: 1-800-289-2550. The Decisioneering Web site is www.decisioneering.com.

[5]David B. Hertz and Howard Thomas, *Risk Analysis and Its Applications,* Chichester, UK: John Wiley & Sons, Ltd., 1983, p. 1.

[6]Hertz and Thomas, ibid, p. 24.

Thus, risk analysis seeks to examine the impacts of uncertainty in the estimates and their potential interaction with one another on the output variable of interest.

Hertz and Thomas also note that the challenge to risk analysts is to frame the output of risk analysis procedures in a manner that makes sense to the manager and provides clear insight into the problem, suggesting that simulation has many advantages. We have laid the foundation for risk analysis with the technique of Monte-Carlo simulation in previous chapters:

1. Build a model for providing information and insight to evaluate a decision.
2. Recognize and identify the uncertainty associated with the model and its variables.
3. Generate a probability distribution for the outcome variables associated with the decision.
4. Analyze the effects of uncertainty in the outcome variables on the decision. For example, we could answer such questions as What is the probability that we will incur a financial loss? What is the probability that we will run out of inventory? What are the chances that a project will be completed on time?

We will use the financial spreadsheet model we introduced in Figure 1-2 as the basis for much of our discussion of risk analysis and the use of Crystal Ball in this chapter. To set the context behind such a model, think of any retailer that operates many stores throughout the country, such as Old Navy, Hallmark Cards, or Radio Shack, to name just a few. The retailer is often seeking to open new stores and has developed the model to evaluate the profitability of a new site of a certain size that would be leased for 5 years. If you examine the model closely, you will see that the key assumptions in the model might be based on historical data (e.g., cost of merchandise as a percent of sales and operating expenses), current economic forecasts (e.g., inflation rate), or judgmental estimates based on preliminary market research (e.g., first-year sales revenue and annual growth rates). These assumptions represent the "most likely" estimates, and as a deterministic model, the spreadsheet shows that the new store will appear to be quite profitable by the end of 5 years. However, the model does not provide any information about what might happen if these variables do not attain these most likely values, and considerable uncertainty exists about their true values.

To characterize the uncertainty, we first have to ask some questions about the uncertainty of each assumption. For example,

- What are the chances that first-year sales will exceed $900,000?
- Is there any possibility that first-year sales revenue will exceed $1 million?
- What are the chances that first-year sales revenue will be as low as $500,000?

Answers to such questions can help us to select an appropriate probability distribution, as we discussed in the previous chapter. Once we have characterized the uncertainty associated with each model assumption by a probability distribution, we can perform Monte-Carlo simulation to generate a distribution of the cumulative discounted cash flow in each year. Then we can address questions about risk, such as

- What are the chances that the store would not be profitable by the third year?
- How likely is it that cumulative profits over 5 years would not exceed $100,000?
- What profit are we likely to realize with a probability of at least 0.70?

Monte-Carlo Simulation with Crystal Ball

As we have seen in the previous chapters, performing a Monte-Carlo simulation on a spreadsheet can be tedious because of the need to generate large tables for repeated trials and collect output results in a frequency distribution and/or histogram for analysis. We would also want to compute summary statistics for the results and probabilities of various risk scenarios. These require further work.

Fortunately, spreadsheet add-ins exist that relieve the user of the burden of performing all these tasks. The software that we feature in this book is called *Crystal Ball*, which was developed and is published by Decisioneering, Inc. Essentially, Crystal Ball automates the more complex and tedious steps in the Monte-Carlo simulation process, such as sampling from distributions, replicating and aggregating the model results, and computing statistical information, and also provides tools for addressing risk-related questions easily. During installation, Crystal Ball may be set up to run automatically whenever Excel is loaded (however, this causes Excel to take more time to load), or it can be run from the Windows *Start* menu. You may also start Crystal Ball after Excel is running by going to *Tools . . . Add-Ins* and checking the box for Crystal Ball. We discussed the use of Crystal Ball for fitting distributions to data in Chapter 3. If you did not read this section, please see Figure 3-23 for a picture of the Excel screen after Crystal Ball is started. It includes three new menu items—*Cell, Run,* and *CBTools*—and a button bar to provide short-cut commands.

Using Crystal Ball follows the general process that we defined in Chapter 2 for performing Monte-Carlo simulations on spreadsheets. To use Crystal Ball, we must perform the following steps:

1. Develop the spreadsheet model.
2. Define *assumptions* for uncertain variables; that is, the probability distributions that describe the uncertainty.
3. Define the *forecast cells;* that is, the output variables of interest.
4. Set the number of trials and other run preferences.
5. Run the simulation.
6. Interpret the results.

A FINANCIAL ANALYSIS RISK SIMULATION

The financial analysis spreadsheet model is shown in Figure 4-1. In this figure, the **assumption cells** are shaded in the *Model Assumptions* section of the spreadsheet. These represent the uncertain variables for which we must define a probability distribution. In the *Model Outputs* section, the output variables in which we are interested—Crystal Ball calls these **forecast cells**—are also shaded (darker). You may control the shading and formatting of the assumption and forecast cells by selecting *Cell . . . Cell Preferences* in the Crystal Ball menu.

SPECIFYING INPUT INFORMATION

The first step in using Crystal Ball is to define the probability distributions for assumption cells. Suppose that the new-business development manager for the firm has identified the following distributions and parameters for these variables:

First-year sales revenue: normal, mean = $800,000, standard
   deviation = $70,000, minimum = $650,000
Annual growth rate, year 2: lognormal, mean = 20 percent, standard
   deviation = 8 percent

	A	B	C	D	E	F	G
1	**New Store Financial Analysis**						
2							
3	*Model Assumptions*		Year 1	Year 2	Year 3	Year 4	Year 5
4	*Annual Growth Rate*			20%	12%	9%	5%
5	*Sales Revenue*		$ 800,000				
6							
7	*Cost of Merchandise (% of sales)*	30%					
8	*Operating Expenses*						
9	*Labor Cost*	$ 200,000					
10	*Rent Per Square Foot*	$ 28					
11	*Other Expenses*	$ 325,000					
12							
13	*Inflation Rate*	3%					
14	*Store Size (square feet)*	$ 5,000					
15	*Total Fixed Assets*	$ 300,000					
16	*Depreciation period (straight line)*	5					
17	*Discount Rate*	10%					
18	*Tax Rate*	34%					
19							
20	*Model Outputs*		Year 1	Year 2	Year 3	Year 4	Year 5
21	*Sales Revenue*		$ 800,000	$ 960,000	$1,075,200	$1,171,968	$1,230,566
22	*Cost of Merchandise*		$ 240,000	$ 288,000	$ 322,560	$ 351,590	$ 369,170
23	*Operating Expenses*						
24	*Labor Cost*		$ 200,000	$ 205,333	$ 210,809	$ 216,430	$ 222,202
25	*Rent Per Square Foot*		$ 140,000	$ 143,733	$ 147,566	$ 151,501	$ 155,541
26	*Other Expenses*		$ 325,000	$ 333,667	$ 342,564	$ 351,699	$ 361,078
27	*Net Operating Income*		$ (105,000)	$ (10,733)	$ 51,700	$ 100,746	$ 122,575
28	*Depreciation Expense*		$ 60,000	$ 60,000	$ 60,000	$ 60,000	$ 60,000
29	*Net Income Before Tax*		$ (165,000)	$ (70,733)	$ (8,300)	$ 40,746	$ 62,575
30	*Income Tax*		$ (56,100)	$ (24,049)	$ (2,822)	$ 13,854	$ 21,276
31	*Net After Tax Income*		$ (108,900)	$ (46,684)	$ (5,478)	$ 26,893	$ 41,300
32	*Plus Depreciation Expense*		$ 60,000	$ 60,000	$ 60,000	$ 60,000	$ 60,000
33	*Annual Cash Flow*		$ (48,900)	$ 13,316	$ 54,522	$ 86,893	$ 101,300
34	*Discounted Cash Flow*		(44,454.55)	11,004.96	45,059.75	71,812.04	83,718.61
35	*Cumulative Discounted Cash Flow*		(44,454.55)	(33,449.59)	11,610.16	83,422.20	167,140.82

FIGURE 4-1 Financial Analysis Model

Annual growth rate, year 3: lognormal, mean = 12 percent, standard deviation = 4 percent

Annual growth rate, year 4: lognormal, mean = 9 percent, standard deviation = 2 percent

Annual growth rate, year 5: lognormal, mean = 5 percent, standard deviation = 1 percent

Cost of merchandise: uniform between 27 percent and 33 percent

Labor cost: triangular, minimum = $175,000, most likely = $200,000, maximum = $225,000

Rent per square foot: uniform between $26 and $30

Other expenses: triangular, minimum = $310,000, most likely = $325,000, maximum = $350,000

Inflation rate: triangular, minimum = 1 percent, most likely = 2 percent, maximum = 5 percent

To define them in Crystal Ball, first select a cell or range of cells. Assumption cells must contain a value; they cannot be defined for formula, nonnumeric, or blank cells. From the *Cell* menu, select *Define Assumption*. Crystal Ball displays a gallery of probability distributions from which to choose and prompts you for the parameters. For example, let us define the distribution for the first-year sales revenue. First click on cell C5. Then select *Cell . . . Define Assumption*. Crystal Ball displays the distribution gallery, shown in Figure 4-2. Because we assume that this variable has a normal distribution, we click on the normal distribution and then the OK button (or simply double click the distribution). A dialog box is then displayed, prompting you for the parameters associated with this distribution.

We suggest that you first enter a clear, descriptive name for your assumptions in the top box. Crystal Ball will automatically use text in cells next to or above assumption cells, but these may not be the correct ones for your application. For most continuous distributions, you have several options on how to specify the distribution using percentiles. For example, with the normal distribution, the default is to enter the mean and standard deviation; however, you can also define the distribution by its 10th and 90th percentiles, the mean and the 90th percentile, and several other ways. This option is useful when only percentile information is available or when specific parameters such as the mean and standard deviation are unknown. As a practical illustration, suppose that you are interviewing a construction manager to identify the distribution of the time it takes to complete a task. Although a beta distribution is often appropriate in such applications, it would be very difficult to define the beta distribution parameters (see Chapter 3) from judgmental information. However, it would be easy for the manager to estimate the 10th and 90th percentiles for task times. In Crystal Ball, these input options can be found in a pop-up box by clicking on the *Parms* button in the dialog box. We suggest you check this when using a new distribution.

Crystal Ball anticipates the default parameters based on the current value in the spreadsheet model. For example, with a normal distribution, the default mean is the assumption cell value, and the standard deviation is assumed to be 10 percent of the mean. Therefore, in our example, we need to change the standard deviation to $70,000. Clicking on "Enter" fixes these values and rescales the picture to allow you to see what the distribution looks like (this feature is quite useful for flexible families of distributions such as the triangular, gamma, or beta discussed in Chapter 3). The result for the example is shown in Figure 4-3. However, the manager believes that first-year sales will never be less than $650,000. Therefore, we must truncate the distribution to reflect this. This can be done by entering the minimum value in the box above the mean (which currently reads "-Infinity") or by clicking and holding and then moving the small triangular "grabber" under the left tail of the distribution. (When you do this, you will notice

FIGURE 4-2 Crystal Ball Distribution Gallery

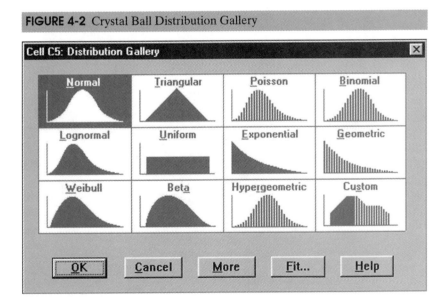

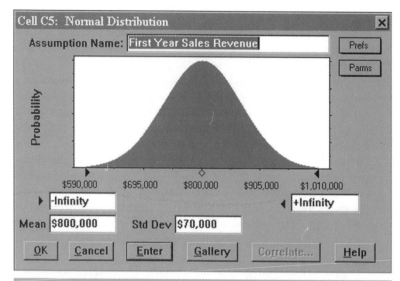

FIGURE 4-3 Normal Distribution Assumption Dialog Box

the value in the minimum box automatically changes.) The result of this change is shown in Figure 4-4. However, as we noted in Chapter 3, truncating distributions changes the true values of the mean and standard deviation. In this case, for example, the mean of the resulting distribution is no longer $800,000. To view the mean value of the assumption, click on *Prefs . . . Show Mean*. Clicking on *OK* accepts your choice and returns to the main screen.

We repeat this process for each of the probabilistic assumptions in the model. Figure 4-5, for example, shows the lognormal distribution for the annual growth rate in year 2. For the lognormal distribution, the default parameters are the mean and standard deviation of the normal random variable; however, you can enter the log mean and log standard deviation and other options from the *Parms* button. Figures 4-6 and 4-7 show the dialog boxes for the uniform distribution assumption for cost of merchandise and triangular assumption for inflation rate, respectively. In Figure 4-7 you may notice three

FIGURE 4-4 Truncated Normal Distribution Assumption

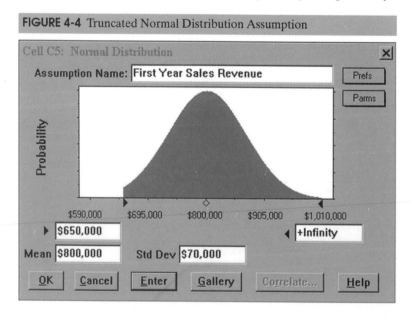

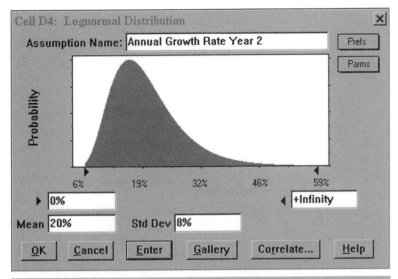

FIGURE 4-5 Lognormal Distribution Assumption Dialog Box

grabbers—two small, black triangles and one open diamond just below the horizontal axis. The end grabbers truncate the distribution; the center grabber allows you to adjust the shape and skewness of the distribution. Many other distributions in the gallery have similar grabbers.

   After all assumptions are defined, you must define one or more forecast cells that define the output variables of interest. In our example, these are cells C35:G35. Instead of choosing them individually, highlight the entire range, and then select *Define Forecast* from the *Cell* menu. The Define Forecast dialog box is shown in Figure 4-8. You may enter a name and unit of measure for each forecast. When you click "OK," Crystal Ball will automatically bring up the dialog box for the next cell in the range. If *Display Window Automatically* is checked, you will see the output distribution being built as Crystal Ball runs each trial of the simulation.

   The Excel cut, copy, and paste commands only address cell values and attributes, but *do not* copy Crystal Ball data. Crystal Ball provides edit commands to let you copy,

FIGURE 4-6 Uniform Distribution Assumption Dialog Box

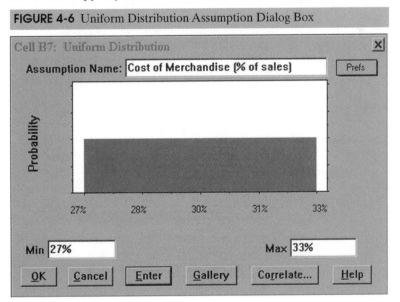

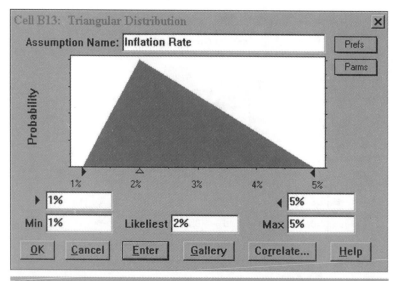

FIGURE 4-7 Triangular Distribution Assumption Dialog Box

paste, or clear assumptions or forecasts from cells. These are useful, for example, when you have several assumptions with the same properties. These commands are found under the *Cell* menu as *Copy Data, Paste Data,* and *Clear Data.* The *Cell* menu options *Select All Assumptions* and *Select All Forecasts* allow you to select all relevant cells to clear.

Prior to running a simulation, you need to define some specifications. To do this, select the *Run Preferences* item from the *Run* menu. The *Trials* dialog box, shown in Figure 4-9, allows you to choose the number of trials. If you check the *Stop if Specified Precision is Reached* box (and set the precision control parameters when you define the forecast by clicking the *More* button in Figure 4-8), the simulation will stop if confidence intervals for all forecast statistics meet the specified level of precision or the maximum number of trials is reached. We recommend you leave this box unchecked until you become an experienced user.

The Sampling Preferences dialog box (see Figure 4-10) allows you to use the same sequence of random numbers for generating random variates; this allows you to repeat the simulation results at other times. This is controlled by the *Initial Seed Value.* Crystal Ball has two types of sampling methods: Monte-Carlo and Latin Hypercube. Monte-Carlo sampling selects random variates independently over the entire range of possible values in the same way as we conducted the examples in Chapters 2 and 3. With Latin Hypercube sampling, Crystal Ball divides each assumptions probability distribution into intervals of equal probability and generates an assumption value randomly within each interval. Latin Hypercube sampling is more precise because it samples the entire

FIGURE 4-8 Define Forecast Dialog Box

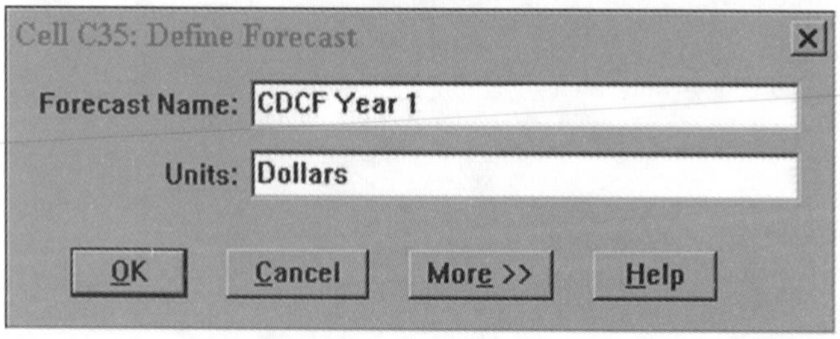

Cell C35: Define Forecast

Forecast Name: CDCF Year 1

Units: Dollars

OK   Cancel   More >>   Help

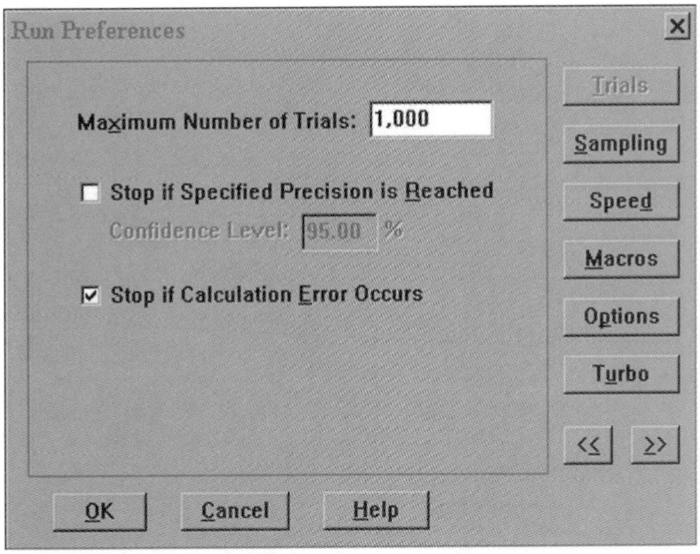

FIGURE 4-9 Run Preferences Dialog Box (Trials)

range of the distribution in a more even and consistent manner, thus achieving the same accuracy as a larger number of Monte-Carlo trials. However, it has additional memory requirements, which typically is not a problem.

The Options dialog box (see Figure 4-11) controls how often Crystal Ball redraws forecast charts during a simulation (we recommend the default value of 50 trials) and allows you to generate sensitivity information and turn off correlations. We will discuss these options later in this chapter. We recommend leaving the *Speed, Macros,* and *Turbo* (which applies only to network servers) options at their default values unless you are an advanced user.

The last step in running a simulation is to select *Run* from the *Run* menu and watch Crystal Ball go to work! The *Run* menu also provides options to stop a simulation in

FIGURE 4-10 Run Preferences Dialog Box (Sampling)

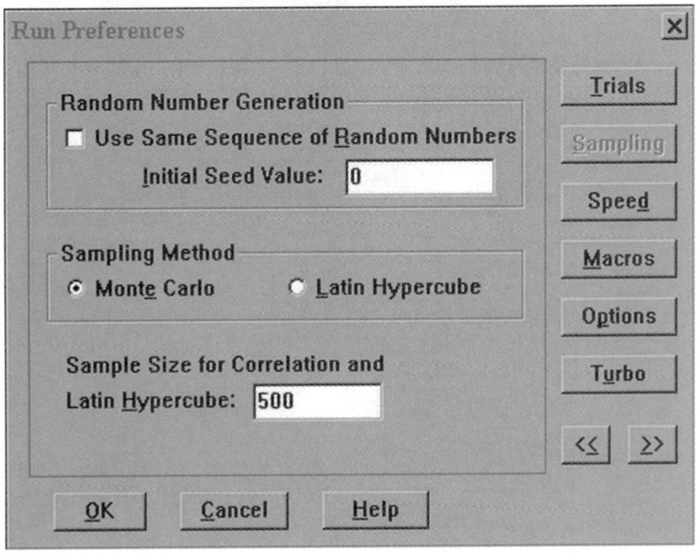

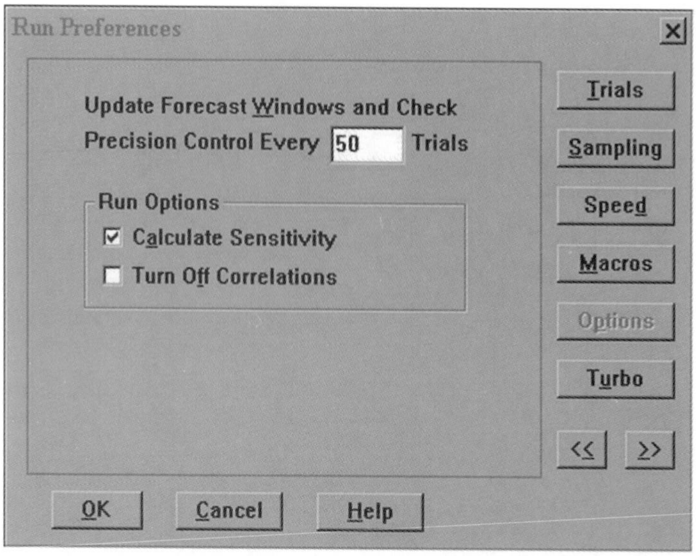

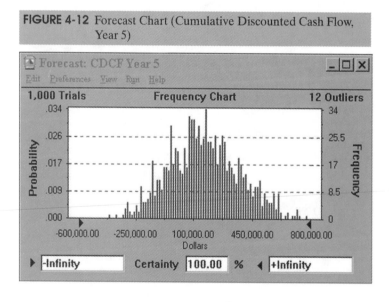

FIGURE 4-11 Run Preferences Dialog Box (Options)

progress, continue, and reset values (that is, clear statistics and results). It is important to clear statistics if you need to rerun a simulation.

CRYSTAL BALL OUTPUT

The principal output reports provided by Crystal Ball are the *forecast chart, percentiles summary,* and *statistics summary.* Figure 4-12 shows the forecast chart for the 5-year cumulative discounted cash flow after 1,000 trials. The forecast chart is simply a histogram of the outcome variable that includes all values within 2.6 standard deviations of the mean, which represents approximately 99 percent of the data. (This may be changed in the *Preferences/Display Range* menu.) The number of outliers is shown in the upper-right corner of the chart. For this example, we have 11 data points outside 2.6 standard deviations of the mean. Just below the horizontal axis at the extremes of the distribution are two endpoint grabbers. The range values of the variables at these positions are

FIGURE 4-12 Forecast Chart (Cumulative Discounted Cash Flow, Year 5)

given in the boxes at the bottom left and right corners of the chart. The percentage of data values between the grabbers is displayed in the *Certainty* box at the lower center of the chart.

Questions involving risk can be answered by manipulating the endpoint grabbers or by changing the range and certainty values in the boxes. Several options exist.

1. *You may move an endpoint grabber by clicking and holding the left mouse button on the grabber and moving it.* As you do, the distribution outside of the middle range changes color, the range value corresponding to the grabber changes to reflect its current position, and the certainty level changes to reflect the new percentage between the grabbers. Figure 4-13 shows the result of moving the left grabber to the value $300,666.67. The dark portion of the histogram represents 26.4 percent of the distribution (as given in the *Certainty* box). This represents the likelihood that the cumulative discounted cash flow will exceed $300,666.67.

2. *You may type in specific values in the range boxes.* When you do, the grabbers automatically move to the appropriate positions, and the certainty level changes to reflect the new percentage of values between the range values. For example, suppose you wanted to determine the percentage of values greater than $0. If you enter this in the left range box, the grabber will automatically move to that position, the portion of the histogram to the right of $0 will lighten, and the certainty level will change to reflect the percentage of the distribution between the grabbers. This is illustrated in Figure 4-14.

3. *You may specify a certainty level.* If the endpoint grabbers are free (as indicated by a black color), the certainty range will be centered around the mean (or median as specified in the *Preferences/Statistics* menu option). For example, Figure 4-15 shows the result of changing the certainty level to 90 percent. The range centered about the mean is from −$135,725 to $615,145. You may anchor an endpoint grabber by clicking on it. When anchored, the grabber will be a lighter color. (To free an anchored grabber, click anywhere in the chart area.) If a grabber is anchored and you specify a certainty level, the free grabber moves to a position corresponding to this level. Finally, you may cross over the grabbers to determine certainty levels for the ends of the distribution.

FIGURE 4-13 Probability That CDCF Exceeds $300,376.89

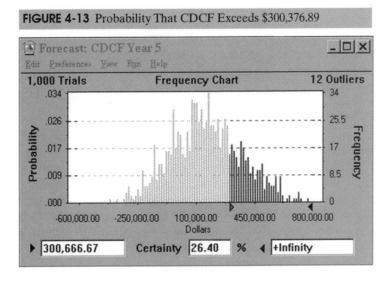

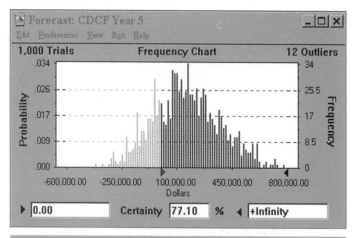

FIGURE 4-14 Probability That CDCF Is Positive

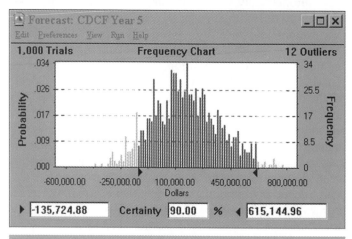

FIGURE 4-15 90 Percent Probability Interval for CDCF

We warn you that a certainty level is *not* a confidence interval! It is simply a probability interval. Confidence intervals, as you may recall, depend on the sample size. We will discuss how to compute them from the Crystal Ball output later.

The forecast chart may be customized to change its appearance through the Preferences/Chart Preferences dialog box. The chart may be displayed as an area, outline, or column (bar) chart and may be displayed as a frequency distribution (the default), cumulative distribution, or reverse cumulative distribution. The number of groups determines the granularity of the chart; a smaller number of groups provides less detail. The number format may be changed in the Preferences/Format Preferences dialog box. You may also improve the appearance of the forecast chart by checking the box *Round Axis Values* from the Preferences/Display Range dialog box.

The percentiles chart can be displayed from the *View* menu. An example is shown in Figure 4-16. For example, we see that the chance that the total profit will be less than $278,723 is about 70 percent. From the *View* menu, you may also select a statistics report. This report, shown in Figure 4-17, provides a summary of key descriptive statistical measures (which we reviewed in Chapter 3). The mean standard error, in particular, defines the standard deviation for the sampling distribution of the mean, based on the run of 1,000 samples, and can be used to construct confidence intervals as we described in Chapter 3.

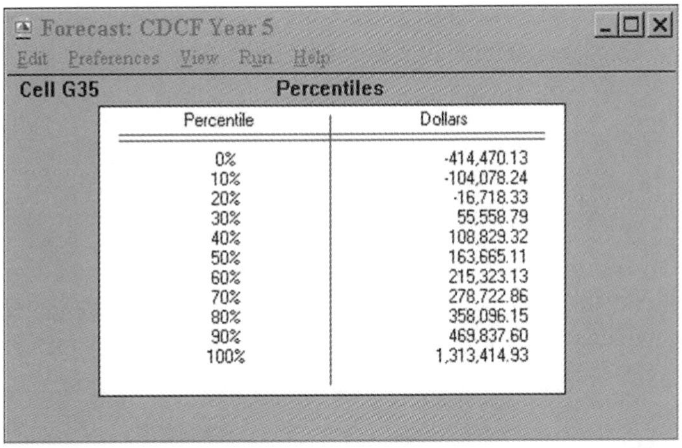

Forecast: CDCF Year 5

Edit Preferences View Run Help

Cell G35 **Percentiles**

Percentile	Dollars
0%	-414,470.13
10%	-104,078.24
20%	-16,718.33
30%	55,558.79
40%	108,829.32
50%	163,665.11
60%	215,323.13
70%	278,722.86
80%	358,096.15
90%	469,837.60
100%	1,313,414.93

FIGURE 4-16 Forecast Chart—Percentiles View

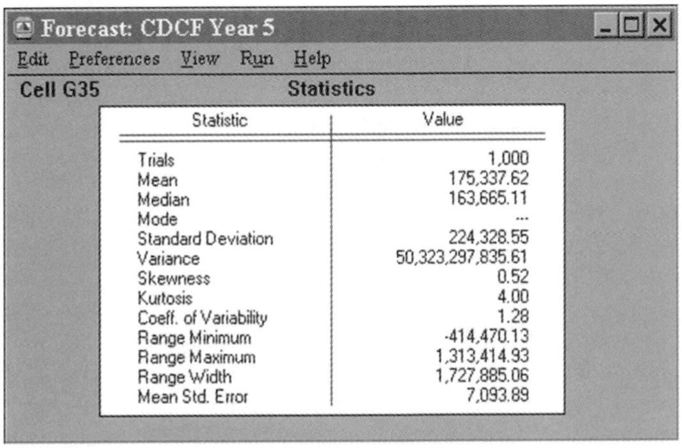

Forecast: CDCF Year 5

Edit Preferences View Run Help

Cell G35 **Statistics**

Statistic	Value
Trials	1,000
Mean	175,337.62
Median	163,665.11
Mode	---
Standard Deviation	224,328.55
Variance	50,323,297,835.61
Skewness	0.52
Kurtosis	4.00
Coeff. of Variability	1.28
Range Minimum	-414,470.13
Range Maximum	1,313,414.93
Range Width	1,727,885.06
Mean Std. Error	7,093.89

FIGURE 4-17 Forecast Chart—Statistics View

To close Crystal Ball without closing Excel, select *Close Crystal Ball* from the *Run* menu. Crystal Ball also closes automatically when you exit Excel. When you save your spreadsheet in Excel, any assumptions and forecasts that you defined for Crystal Ball are also saved. However, this does not save the results of a Crystal Ball simulation. To save a Crystal Ball simulation, select *Run . . . Save Run.* Doing so allows you to save any customized chart settings and other simulation results and recall them without rerunning the simulation. To retrieve a Crystal Ball simulation, choose *Restore* from the *Run* menu. You may also copy a saved simulation to other computers where Crystal Ball is installed.

SKILLBUILDER EXERCISE

Open the file *New Store Financial Model—CB Model.xls.* This file has all the Crystal Ball assumptions and forecasts defined for the example. Make sure Crystal Ball is running within Excel and run your model for 1,000 trials. From the *Run* menu, select *Forecast Windows,* then *Open All Forecasts.* What is the probability that the cumulative discounted cash flow will be positive in each of the first 4 years? What is the mean and range for each year?

CREATING CRYSTAL BALL REPORTS

Customized reports may be created from the *Run* menu by choosing *Create Report*. This option allows you to select a summary of assumptions and output information that we described. These are created in a separate Excel worksheet and may be printed or customized for reporting purposes. In addition, Crystal Ball allows you to extract selected data to a new Excel workbook for further analysis. From the *Run* menu, select *Extract Data*. In the dialog box that appears, you may select various types of data to extract:

- Forecast values—each of the forecast values generated from each simulation trial
- Statistics—the data in the Statistics View of the forecast chart
- Percentiles
- Frequency counts—a tabular frequency distribution of the forecast chart histogram
- Cumulative counts—a cumulative tabular frequency distribution
- Sensitivity data (which will be described later in this chapter)

Finally, you may access Crystal Ball data directly in Excel by using CB functions. We saw the use of CB functions in Chapter 3 when discussing methods for random variate generation. A complete list may be found by invoking the function wizard button [f_x] on the Excel menu bar and going to the Crystal Ball category. Some of the more useful functions for customizing risk analysis reports are

- *CB.GetForeStatFN(forecast_cell_reference, index)*. The first argument is a valid forecast cell; *index* is a number from 1 to 13 that refers to a forecast statistic. These are shown in Figure 4-17. Index = 1 refers to the number of trials; index = 2 refers to the mean, etc. Thus, to get the standard deviation for year 5 CDCF, you would use the function =*CB.GetForeStatFN(G35, 5)*.
- *CB.GetCertaintyFN(forecast_cell_reference, value)*. The first argument is a valid forecast cell; *value* refers to the threshold value for which you want to calculate the certainty level—the probability of achieving a forecast value at or below the threshold. For example, to find the probability that year 5 CDCF is $100,000 or less, use the function =*CB.GetCertaintyFN (G35, 100000)*.
- *CB.GetForePercentFN(forecast_cell_reference, percent)*. Returns a percentile for a forecast cell. The first argument is a valid forecast cell; *percent* is a number from 0 to 100 that specifies the desired percentile. Thus, to find the 75th percentile for year 5 CDCF, use the function =*CB.GetForePercentFN (G35, 75)*.

Using these techniques allows you to create a useful, customized, managerial report for interpreting risk analysis results.

SKILLBUILDER EXERCISE

Open the file *Financial Model Base Case.Run*. This retrieves the Crystal Ball run that we have used in this chapter. Create a Crystal Ball report from the simulation results and extract the statistics for all forecasts to a worksheet. Create a new Excel workbook that you might use as an attachment to a managerial report.

Additional Crystal Ball Modeling and Analysis Options

Crystal Ball contains a variety of additional options that facilitate risk analysis modeling and interpretation of results. In this section we review these capabilities.

CORRELATED ASSUMPTIONS

In Crystal Ball, each random input variable is assumed to be independent of the others by default. In many situations, we might wish to explicitly model dependencies between variables. Although this may be done by using cell formulas directly in Excel, Crystal Ball allows you to specify correlation coefficients to define dependencies between assumptions. Crystal Ball uses the correlation coefficients to rearrange the generated random variates to produce the desired correlations. This can be done only after assumptions have been defined. For example, in the new store financial analysis model in Figure 4-1, suppose that we wish to correlate the annual growth rates. For instance, if the annual growth rate in year 2 is high, then it would make sense that the annual growth rate in year 3 would be high also. You probably would not expect the growth rate in one year to be high if the growth rate in the previous year is low. Thus, we might expect a positive correlation between these variables.

To define correlations in Crystal Ball, select one of these cells—for instance, the annual growth rate in year 2 (D4). Then select *Define Assumption* from the *Cell* menu. When the distribution dialog box appears, click on the *Correlate . . .* button. The Correlation dialog box, shown in Figure 4-18, appears. The *Select Assumption* button provides a list of the assumptions that you have defined. We then select the other assumption to correlate (for instance, the annual growth rate in year 3) and then enter a correlation coefficient.

You may enter a correlation coefficient in one of three ways. First, you can enter a value between −1 and 1 in the *Coefficient Value* box. Second, you may drag the slider control along the *Correlation Coefficient* scale; the specific value you select is displayed in the box to the left of the scale. Third, you may click on *Calc . . .* and enter ranges of cells in the spreadsheet that contain empirical values that should be used to calculate a correlation coefficient. After the correlation coefficient is specified, Crystal Ball displays a sample correlation chart, as shown in Figure 4-19. The solid line indicates where

FIGURE 4-18 Correlation Dialog Box

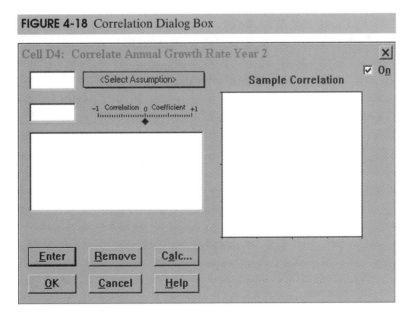

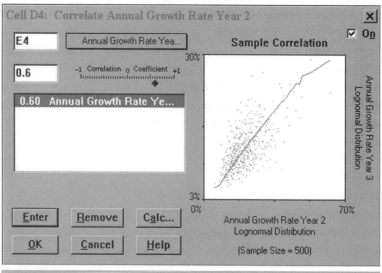

FIGURE 4-19 Correlation Chart

values of a perfect correlation would fall; the points represent the actual pairing of assumption values that would occur during the simulation. When using this option, you must be cautious if you have a large number of correlations, because it is possible that some correlations might conflict with others, preventing Crystal Ball from running. If Crystal Ball detects a conflict, it will try to reset the correlation coefficients by asking the user to update them.

Instead of manually entering the correlations between several variables one at a time, Crystal Ball has a tool that allows you to define a matrix of correlations between assumptions in one step. Under the *CBTools* menu, select *Correlation Matrix*. In the dialog box that appears, include all the assumptions you want to define in the correlation matrix by moving them from the *Available Assumptions* field to the *Selected Assumptions* field by either double-clicking on each assumption to move it or selecting it and clicking on the >> button. When you click "Next," the Specify Options dialog box appears. Select your options (e.g., create a temporary matrix on a new worksheet) and click "Start." You may enter the correlation coefficients into the matrix and click on *Load* to put them into your model (the *Load* button only appears if a temporary matrix has been selected).

SKILLBUILDER EXERCISE

Open the file *New Store Financial Model—CB Model.xls*. Use the Correlation Matrix tool to define the following correlations. Run the Crystal Ball model. How have the results changed?

	Year 2	*Year 3*	*Year 4*	*Year 5*
Year 2	1.00	.60		
Year 3		1.00	.70	
Year 4			1.00	.80
Year 5				1.00

OVERLAY CHARTS

If a simulation has multiple, related forecasts, the overlay chart feature allows you to superimpose the frequency data from selected forecasts on one chart in order to compare differences and similarities that might not be apparent. You may also use the overlay

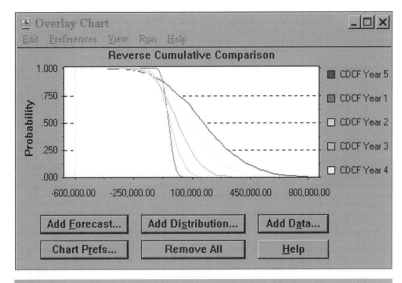

FIGURE 4-20 Overlay Chart Example

chart to fit a standard distribution to any single forecast, similar to the distribution-fitting feature described in Chapter 3. This option is invoked from the *Run* menu by selecting *Open Overlay Chart.* In the Overlay Chart dialog box that appears, click on *Add Forecast* to select the forecasts to include. You may select them individually or click on *Choose All* to include all of them. You may customize the appearance of the chart in many ways by clicking on the *Chart Prefs* button. This allows you to choose the type of chart and the distribution type. Figure 4-20 shows an outline-type overlay chart for the reverse cumulative distribution of each of the 5 years' CDCF. You can easily see that the probability of cumulative discounted cash flow values increases each year.

From the overlay chart, you may also fit a distribution to a forecast. (See Chapter 3 for an explanation of Crystal Ball's distribution-fitting capability. The difference here is that you are fitting the distribution to the forecast values, not historical data for identifying input distributions.) Figure 4-21 shows a fit of the year 5 CDCF forecast to a normal distribution. The p-value of approximately 0.05 suggests a reasonable fit. This would

FIGURE 4-21 Normal Distribution Fit to Forecast Distribution

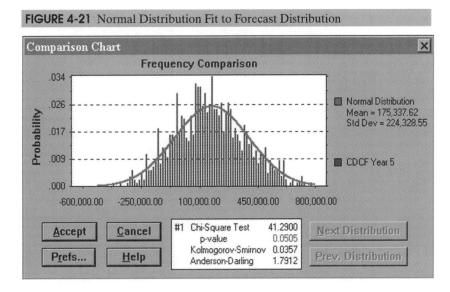

allow you to use analytical calculations with the normal distribution to answer questions about risk.

TREND CHARTS

If a simulation has multiple forecasts that are related to one another (such as over time), you can view the certainty ranges of all forecasts on a single chart, called a *trend chart*. Figure 4-22 shows a trend chart for the 5-year cumulative discounted cash flows in our example. The trend chart displays certainty ranges in a series of patterned bands. For example, the band representing the 90 percent certainty range shows the range of values into which a forecast has a 90 percent chance of falling. From the trend chart in Figure 4-22, we see that although the average cash flow increases over time, so does the variation, indicating that the uncertainty also increases with time. Trend charts are opened from the *Run* menu by selecting *Open Trend Chart*. Clicking on the *Trend Preferences . . .* button displays a dialog box that allows you to customize the trend chart by selecting the number and type of certainty bands as well as the chart type.

SENSITIVITY ANALYSIS

An important reason for using simulation for risk analysis is the ability to conduct sensitivity analyses to understand the impacts of individual variables or their distributional assumptions on forecasts. A somewhat naïve way to investigate the impact of assumptions on forecast cells is to freeze, or hold, certain assumptions constant in the model and compare the results with a base case simulation. The *Freeze Assumptions . . .* option under the *Cell* menu allows you to temporarily exclude certain assumptions from a simulation and conduct this type of sensitivity analysis. Although this might be necessary in certain situations, freezing assumptions is somewhat tedious.

The uncertainty in a forecast is the result of the combined effect of the uncertainties of all assumptions as well as the formulas used in the model. An assumption might have a high degree of uncertainty yet have little effect on the forecast because it is not weighted heavily in the model formulas. For instance, a forecast might be defined as

$$0.9(\text{Assumption 1}) + 0.1(\text{Assumption 2})$$

FIGURE 4-22 Trend Chart for Cumulative Discounted Cash Flows

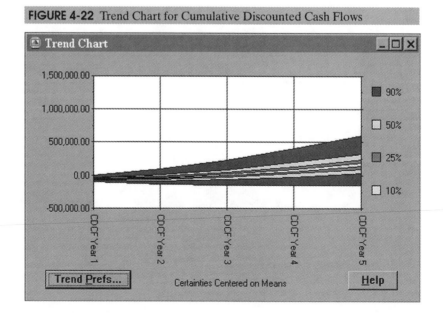

In the model, the forecast is nine times as sensitive to changes in the value of assumption 1 as it is to changes in the value of assumption 2. Thus, even if assumption 2 has a much higher degree of uncertainty as specified by the variance of its probability distribution, it would have a relatively minor effect on the uncertainty of the forecast. The *Sensitivity Chart* feature of Crystal Ball allows you to determine the influence that each assumption cell has individually on a forecast cell. The sensitivity chart displays the rankings of each assumption according to its impact on a forecast cell as a bar chart. This provides three benefits:

1. It tells which assumptions are influencing forecasts the most and which need better estimates.
2. It tells which assumptions are influencing forecasts the least and can be ignored or discarded altogether.
3. By understanding how assumptions affect your model, you can develop more realistic spreadsheet models and improve the accuracy of your results.

To create a sensitivity chart, you must ensure that the *Calculate Sensitivity* box is checked in the Run Preferences Options dialog box before running the simulation. After the simulation is completed, select *Open Sensitivity Chart* from the *Run* menu. The assumptions (and possible other forecasts) are listed on the left, beginning with the assumption having the highest sensitivity. Sensitivity charts for the cash flow example are shown in Figures 4-23 and 4-24. The sensitivities in Figure 4-23 are measured by **rank correlation coefficients.** Correlation coefficients provide a measure of the degree to which assumptions and forecasts change together. Rank correlation is a method whereby Crystal Ball replaces assumption values with their ranking from lowest to highest and uses the rankings to compute a correlation coefficient. Positive coefficients indicate that an increase in the assumption is associated with an increase in the forecast; negative coefficients imply the reverse. The larger the absolute value of the correlation coefficient, the stronger is the relationship.

Figure 4-24 shows the sensitivities as a percent of the contribution to the variance of the forecast. This addresses the question What percentage of the variance or uncertainty in the target forecast is due to a specific assumption? However, it is important to note that this method is only an approximation and not precisely a variance decomposition

FIGURE 4-23 Sensitivity Chart by Rank Correlation

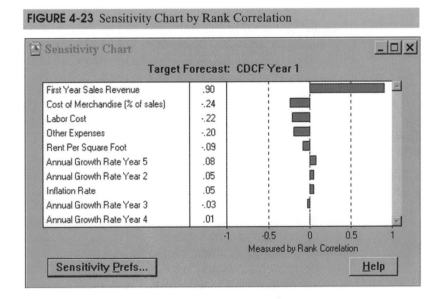

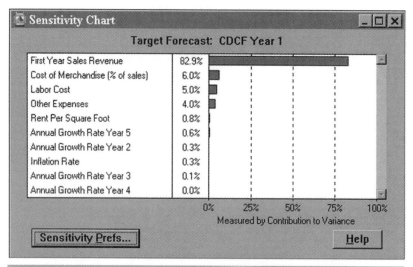

FIGURE 4-24 Sensitivity Chart by Contribution to Variance

in a statistical sense. The option can by changed by clicking on the *Sensitivity Prefs . . .* button on the sensitivity chart. This also allows you to select forecasts to include in the sensitivity chart. However, if assumptions are correlated or assumptions are nonmonotonic with the target forecast (meaning an increase or decrease in the assumption does not necessarily lead to an increase or decrease in the forecast, respectively), the sensitivity results may not be accurate.

Crystal Ball has another means of providing *a priori* sensitivity analysis information—the Tornado Chart tool. This tool tests the range of each variable at specified percentiles and then calculates the value of the forecast at each point. It illustrates the range between the minimum and maximum forecast values for each variable and arranges them from largest to smallest in a "funnel" shape of a tornado, called a **tornado chart.** This differs from the sensitivity chart in that it tests each assumption independently, while freezing the other variables at their base values. Thus, it measures the effect of each variable on the forecast cell while removing the effects of the other variables. Its usefulness lies in quickly prescreening the variables in a model to determine which are good candidates to define as assumptions prior to building a Crystal Ball simulation model. Those variables having little effect on a forecast might not need to be defined as assumptions and can be kept as constants in the model, thus simplifying it. Figure 4-25 shows a tornado chart for year 5 CDCF. This shows that the first-year sales revenue and annual growth rate in year 2 have the most effect on the forecast; the rent per square foot and annual growth rates for years 4 and 5 have comparatively little effect.

The Tornado Chart tool also allows you to create a **spider chart,** as the one shown in Figure 4-26. A spider chart provides similar information as a tornado chart and illustrates the differences between the minimum and maximum forecast values by graphing a curve through all the variable values tested. Curves with steep slopes indicate that those variables have a large effect on the forecast; curves that are almost horizontal have little or no effect on the forecast. The slopes also indicate whether a positive or negative change in the variable has a positive or negative effect on the forecast.

Although tornado and spider charts are useful, they do not consider correlations defined between variables and depend significantly on the base case used for the variables. To confirm the accuracy of the results, you should run the tool multiple times with

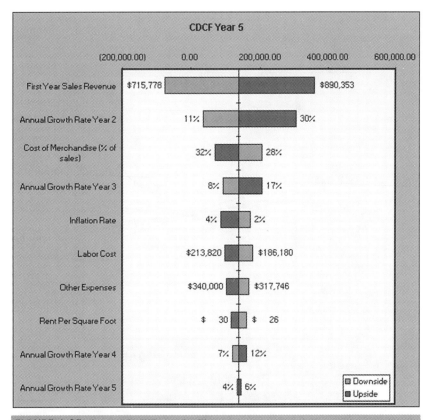

FIGURE 4-25 Crystal Ball Tornado Chart

different base cases. The sensitivity chart is somewhat preferable, because it computes sensitivity by sampling the variables all together while a simulation is running.

STATISTICAL ISSUES IN RISK ANALYSIS

In Chapter 3, we briefly discussed statistical issues associated with interpreting the output of Monte-Carlo simulations. As we have seen, Crystal Ball provides most basic descriptive statistical measures. We also noted that the forecast chart provides probability intervals, not confidence intervals. We may easily construct a confidence interval for the mean using the formula given in Chapter 3:

$$\bar{x} \pm z_{\alpha/2}(\sigma/\sqrt{n})$$

Note that the standard error of the mean, $\sigma/\sqrt{n}$, is reported on the last line of the statistics report in the Crystal Ball output. Because a Crystal Ball simulation will generally have a large number of trials, we may use the standard normal value $z_{\alpha/2}$ instead of the t-distribution. Thus, for the year 5 CDCF results, a 95 percent confidence interval would be

$$\$175,337.62 \pm 1.96(\$7093.89) \text{ or } [\$161,433.60, \$189,241.64]$$

To reduce the size of the confidence interval, we would need to run the simulation for a higher number of trials, as discussed in Chapter 3.

This classical approach for confidence intervals assumes that the sampling distribution of the mean is normal. However, if the sampling distribution is not normally distributed, such a confidence interval is not valid. Also, if we wanted to develop a

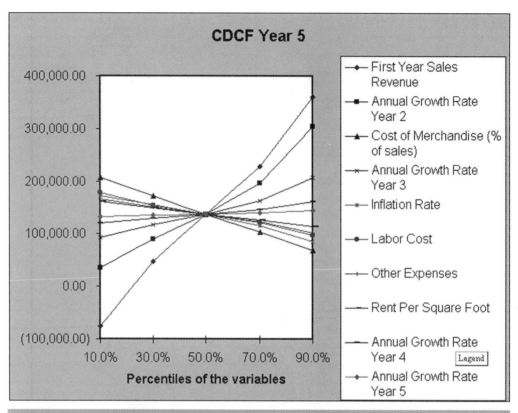

FIGURE 4-26 Crystal Ball Spider Chart

confidence interval for the median, standard deviation, or maximum forecast value, for example, we may not know what the sampling distributions of these statistics are. A statistical technique called **bootstrapping** analyzes sample statistics empirically by repeatedly sampling the data and creating distributions of the statistics. This approach allows you to estimate the sampling distribution of any statistic, even an unconventional one such as the minimum or maximum endpoint of a forecast.

The Crystal Ball Bootstrap tool does this. The tool has two alternative methods:

1. One-simulation method, which simulates the model data once and then repeatedly samples with replacement.
2. Multiple-simulation method, which repeatedly simulates the model and then creates a sampling distribution from each simulation. This method is more accurate, but might take a prohibitive amount of time.

The Bootstrap tool constructs sampling distributions for the following statistics:

- Mean
- Median
- Standard deviation
- Variance
- Skewness
- Kurtosis
- Coefficient of variability
- Range minimum (multiple-simulation method only)
- Range maximum (multiple-simulation method only)
- Range width (multiple-simulation method only)

	A	B	C	D	E	F	G	H	I	J
1		Mean	Median	Standard Deviation	Variance	Skewness	Kurtosis	Minimum	Maximum	Range
2	Number of additional shifts	13.10154	13.005	2.713838678	7.371373026	0.00	2.96	5.06	21.165	16.105
3										
4	Correlations:									
5	Mean	1.000	0.056	0.107	0.107	-0.049	-0.083	0.129	0.107	-0.025
6	Median		1.000	0.039	0.039	0.006	0.069	0.035	0.108	0.062
7	Standard Deviation			1.000	1.000	-0.007	-0.013	-0.297	0.187	0.331
8	Variance				1.000	-0.007	-0.013	-0.297	0.187	0.331
9	Skewness					1.000	-0.007	0.309	0.406	0.067
10	Kurtosis						1.000	-0.481	0.471	0.668
11	Minimum							1.000	-0.150	-0.668
12	Maximum								1.000	0.739
13	Range									1.000

FIGURE 4-27 Bootstrap Tool Results and Correlation Matrix

The forecast chart constructed for each statistic visually conveys the accuracy of the statistic: A narrow and symmetric distribution is better than a wide and skewed distribution. A small standard error of the statistic and coefficient of variability are quantitative indicators of reliable statistical estimates.

To apply the tool, select *Bootstrap* from the *CB Tools* menu and follow the dialog box instructions to select the forecast you wish to analyze; choose the method of analysis; and set the number of samples. Figure 4-27 shows the results for bootstrapping the number of additional shifts (cell G4 in Figure 2-23, which was defined as the forecast) in the Mantel Manufacturing example introduced in Chapter 1. We used 200 bootstrap samples with 350 trials per sample. (We warn you that this tool may take considerable computing time for some simulation models.) Row 2 in the spreadsheet shows the statistics of the sampling distribution of the number of additional shifts; this is displayed in the forecast chart in Figure 4-28. The results spreadsheet also displays a correlation matrix showing the correlations between the various statistics. High correlation between certain statistics, such as between the mean and the standard deviation, usually indicates a highly skewed distribution. Figure 4-29 shows the statistics associated with the forecast chart. Because this is an approximation of the sampling distribution, the standard deviation in this chart represents the standard error of a single 200-trial Crystal Ball simulation. Thus, a confidence interval for the mean is found by adding and subtracting the appropriate z-value times the standard deviation—not the standard error—from the mean.

Figures 4-30 and 4-31 show the sampling distribution and associated statistics for the maximum number of additional shifts. We see that the maximum number of shifts ranges from 19 to 24, and that a 95 percent confidence interval is $21.17 \pm 1.96(.98)$ or [19.25, 23.09]. Thus, if we are concerned with setting a budget to cover the worst case, we might want to plan for as many as 23 additional shifts.

ANALYZING DECISION ALTERNATIVES

Often, the purpose of risk analysis is to help make a choice among several decisions. For example, in the Dave's Candies example introduced in Chapter 1, the choice is the order quantity. One approach to comparing different alternatives using Crystal Ball is to run

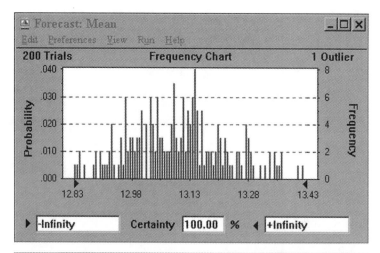

FIGURE 4-28 Sampling Distribution of the Mean Number of Additional Shifts

Statistic	Value
Trials	200
Mean	13.10
Median	13.10
Mode	13.01
Standard Deviation	0.13
Variance	0.02
Skewness	0.24
Kurtosis	2.68
Coeff. of Variability	0.01
Range Minimum	12.83
Range Maximum	13.43
Range Width	0.61
Mean Std. Error	0.01

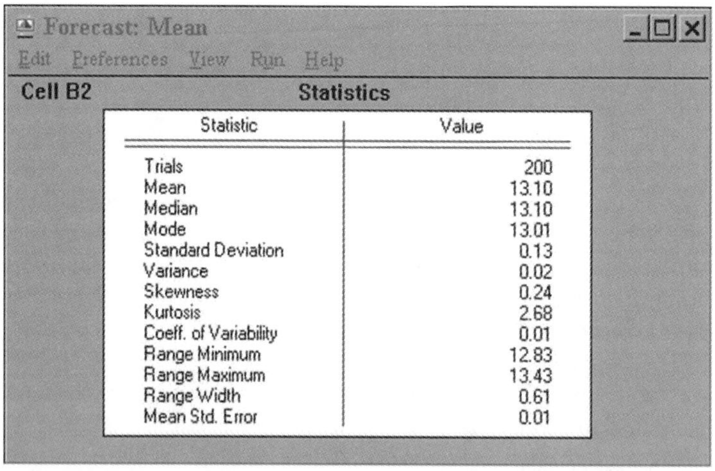

FIGURE 4-29 Sampling Distribution of the Mean Number of Additional Shifts, Statistics View

a new simulation for each decision variable. This can be tedious; fortunately, Crystal Ball includes a Decision Table tool that runs multiple simulations to test different values for one or two *decision variables,* which represent quantities that a decision maker can control in a model. The results are displayed in a table that you can analyze using Crystal Ball forecast, trend, or overlay charts.

To use the tool, you must first define the decision variables in your model. We have modified the Dave's Candies simulation model we developed in Chapter 2 for Crystal Ball, as shown in Figure 4-32. The demand in cell B16 is generated by using the *CB.Custom* function. The formula used is =CB.Custom(A8:B13)—see the discussion of this function in Chapter 3. Alternatively, we could have defined this cell as an assumption cell, but it makes no difference here. (It would make a difference if you wanted to use the sensitivity chart; then you would have to define it as an assumption rather than using the Crystal Ball function.) Cell B15 represents the order quantity. We

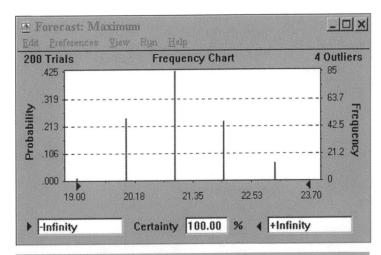

FIGURE 4-30 Sampling Distribution of the Maximum Number of Additional Shifts

Statistic	Value
Trials	200
Mean	21.17
Median	21.00
Mode	21.00
Standard Deviation	0.98
Variance	0.95
Skewness	0.57
Kurtosis	3.16
Coeff. of Variability	0.05
Range Minimum	19.00
Range Maximum	24.00
Range Width	5.00
Mean Std. Error	0.07

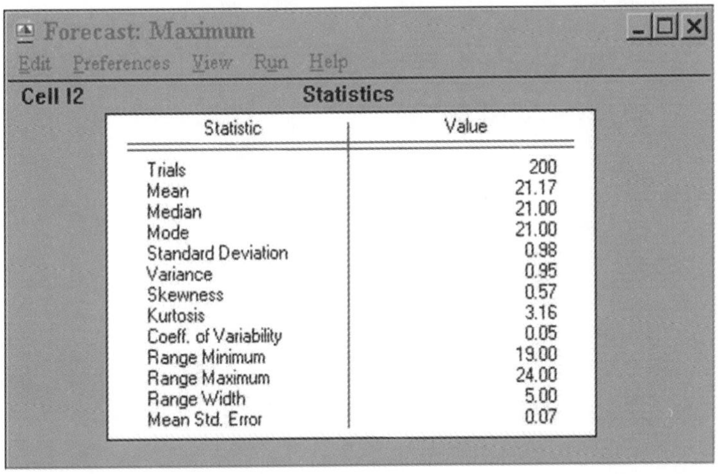

FIGURE 4-31 Sampling Distribution of the Maximum Number of Additional Shifts, Statistics View

must first define this to be a decision variable in Crystal Ball. To do this, click on the cell, then select *Define Decision Variable* from the *Cell* menu. The dialog box is shown in Figure 4-33. We defined the range of decisions to be discrete between 40 and 90 in steps of 10 (corresponding to the levels of demand). Next, set the *Sampling* options in the Run Preferences dialog box to use the same sequence of random numbers. Set an initial seed value (e.g., 999) and select Monte-Carlo simulation. These options make the resulting simulations comparable. Select *Decision Table* from the *CBTools* menu, choose the forecast, the decision variables, and the simulation options (for this example, we used 1,000 trials).

Figure 4-34 shows the results. Crystal Ball constructs a new Excel workbook showing the simulation results for each level of the decision variable. If you have two decision variables, the results would be displayed in a matrix. By clicking on one or more forecast cells (shaded), you can display a trend chart, develop an overlay chart, or show the forecast charts by clicking on the buttons in cell A1. Figure 4-35 shows the trend chart after selecting the range B2:G2.

	A	B
1	**Dave's Candies Simulation**	
2		
3	*Selling price*	$ 12.00
4	*Cost*	$ 7.50
5	*Discount price*	$ 6.00
6		
7	*Demand*	*Probability*
8	40	1/6
9	50	1/6
10	60	1/6
11	70	1/6
12	80	1/6
13	90	1/6
14		
15	*Order Quantity*	65
16	*Demand*	65
17	*Profit*	$ 292.50

FIGURE 4-32 Dave's Candies Crystal Ball Simulation Model

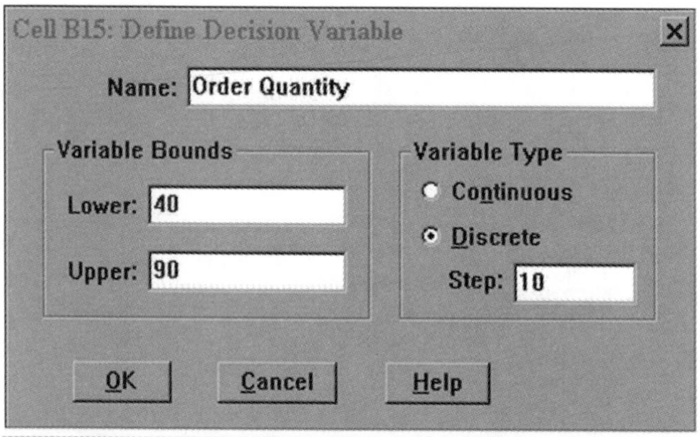

Cell B15: Define Decision Variable

Name: Order Quantity

Variable Bounds

Lower: 40

Upper: 90

Variable Type

○ Continuous
● Discrete

Step: 10

OK   Cancel   Help

FIGURE 4-33 Define Decision Variable Dialog Box

SKILLBUILDER EXERCISE

Open the file *New Store Financial Model—CB Model.xls.* Suppose that the store size (cell B14) is an important decision variable. Apply the Decision Table tool to evaluate the year 5 CDCF over store sizes from 4,000 to 6,000 square feet in increments of 500.

OTHER CRYSTAL BALL TOOLS

Crystal Ball has two other tools in the *CB Tools* menu that we will not formally illustrate but briefly discuss: *Batch Fit* and *Two-Dimensional Simulation.* The Batch Fit tool fits probability distributions to multiple data series in the same fashion that we discussed in Chapter 3 for fitting distributions to data. The advantage of this tool is that it eliminates the necessity to fit each distribution individually. The only requirement is that the data must be in adjacent rows or columns.

The Two-Dimensional Simulation tool allows you to distinguish between uncertainty in assumptions due to limited information or data, and variability—assumptions that change because they describe a population with different values. For instance, in the Man-

	A	B	C	D	E	F	G
1	Trend Chart Overlay Chart Forecast Charts	Order Quantity (40)	Order Quantity (50)	Order Quantity (60)	Order Quantity (70)	Order Quantity (80)	Order Quantity (90)
2		$180.00	$215.10	$239.52	$254.28	$259.32	$254.40
3		1	2	3	4	5	6

FIGURE 4-34 Decision Table Results

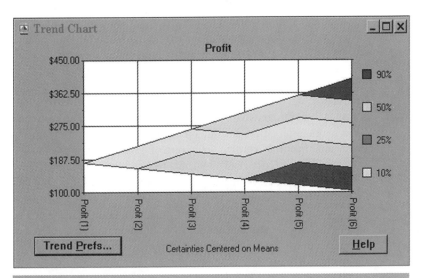

FIGURE 4-35 Trend Chart for Profit Forecasts

tel Manufacturing example, the demand assumption is uncertain because we do not know the population from which it is determined. Theoretically, you can eliminate uncertainty by gathering more information; practically, it is usually impossible or cost-prohibitive. However, suppose that the production rate is variable, perhaps due to absenteeism, late shipments of parts, and so on. Variability is inherent in the system, and you cannot eliminate it by gathering more information. Separating these two types of assumptions lets you more accurately characterize risk. The Two-Dimensional Simulation tool runs an outer loop to simulate the uncertainty values and then freezes them while it runs an inner loop to simulate variability. The process repeats for some small number of outer simulations, providing a portrait of how the forecast distribution varies due to the uncertainty. Further information may be found in the Crystal Ball manual or Help options.

Simulation in Practice

APPLICATIONS OF RISK ANALYSIS USING CRYSTAL BALL[7]

Crystal Ball has been applied by numerous companies. Several applications are described here, and others are continually being added to Decisioneering's Web site.

---

[7]Adapted from case studies described at Decisioneering, Inc. Web site, www.decisioneering.com/.

Optimizing Portfolio Profit   ProVise Management Group, a SEC-registered investment advisor that specializes in asset management and financial planning, understands that proper portfolio management requires a rigorous analysis of risk. When ProVise begins to construct a client's portfolio, the two critical elements that must be thoroughly identified are the client's risk tolerance and a corresponding asset allocation. ProVise establishes the client's risk tolerance through extensive interviews and a risk tolerance questionnaire program developed by the American College in Bryn Mawr, Pennsylvania. Once ProVise is satisfied that they understand the types and amount of risk a client can accept, they perform a Crystal Ball analysis of the client's current portfolio. In an Excel spreadsheet, they examine the potential of each asset. Key asset characteristics include its category (aggressive growth, global/foreign, high-yield bond, etc.), its dollar value, the percentage of the total portfolio that it represents, its measure of risk, its target return (using historical asset class returns) and weighted target return, and its 3- or 5-year historical average annualized return. Crystal Ball assumptions include each asset's weighted target return and weighted beta.

After the portfolio simulation is complete, ProVise helps their client to understand the current risk level by showing them the certainty level for the overall portfolio's target return. Very often, the client will see that their portfolio has a substantial potential for negative returns, sometimes 30 percent or more. ProVise then constructs a new portfolio spreadsheet that contains a combination of current and new assets. They import the assets' monthly returns from one of several available services, such as Morningstar, and use Excel's Data Analysis tools to correlate the data. The correlated data are then entered into the Crystal Ball Correlation Matrix. Uncertainty within the model is reflected in the measure of risk (beta) and target return for each asset. ProVise defines these variables as Crystal Ball assumption cells, either as a normal distribution, using an asset's historical mean return and standard deviation, or as a uniform or lognormal distribution, in the case of T-bills and money market assets. ProVise typically runs 8,000 trials using Latin Hypercube sampling.

The proposed portfolio is then optimized for the client's risk tolerance and target return. In the new model, the value of each asset selected for the new portfolio is defined as a decision cell. A constraint is applied for each asset class, such as aggressive growth, where the asset class's value will equal a certain amount. The OptQuest program (a part of the Crystal Ball Professional Edition) then optimizes the portfolio by maximizing target return, while risk (beta or standard deviation) is set to a selected maximum limit. Then, a final Crystal Ball simulation is run on the optimized portfolio to demonstrate the portfolio's reduced risk and/or enhanced return to the client. Using the graphic "confidence bands" in the trend chart for projected accumulation values or cash flow, clients can understand that they have an *X* percent chance of not running out of money over their life expectancy, and they are able to actually see the risk that is in the proposed portfolio.

Marketing Applications at Hewlett-Packard   Michael Hart, a marketing product specialist within the market intelligence area of Hewlett-Packard, has been using Crystal Ball for over 7 years to aid in difficult marketing decisions surrounding the forecasting of demand, new product launches, and product line extensions for Hewlett-Packard's printers. When evaluating the launching of a new product or extending a current product line, significant market research and testing are conducted to obtain a general idea as to the viability of the product in question. The research explores areas such as potential market share, possible sales levels, desirable product attributes, market segmentation, and product usage patterns and is used to forecast future product performance.

Once data is obtained from this research, Crystal Ball's distribution-fitting capabilities are used to fit distributions around the unknown variables in the raw data. In connection with this process, parameters are selected to define lower- and upper-end boundaries. The next step is to perform a Monte-Carlo simulation that provides an associated probability level to potential areas of concern such as possible cannibalization of existing products and optimum pricing points for market entry. Additionally, Crystal Ball is used to produce forecasts at a 95 percent confidence level around important financial information such as expected profit and ROI.

Hart uses other features of Crystal Ball. For example, he employs the use of Tornado Charts and the Correlation Matrix to give him a better indication of the variables that are having the greatest effect on his models. The Tornado Charts allow him to visualize as well as quantify the impact of each variable within his forecast, and the Correlation Matrix easily allows variables that depend on each other in any way to be correlated. Using these tools, he can isolate "problem" variables and conduct additional market research to confirm the validity of the data; or, he can adjust the assumptions and variables of the model, rerun the simulation, and compare the results.

Assessing Risk from Radioactive Sources   The Environmental Evaluation Group (EEG) is an independent technical oversight group for the Department of Energy's (DOE) Waste Isolation Pilot Plant (WIPP) project. The WIPP, a deep geologic repository for defense-generated transuranic waste, is located in southeastern New Mexico. The EEG performs independent technical analyses covering various aspects of the WIPP, including reviewing the safety analysis report. The safety analysis report is a DOE-required report for nuclear facilities that documents all aspects of safety related to the handling of radioactive materials.

As part of its responsibilities, the DOE performs safety analysis studies to identify the potential for accidents such as an airborne release of transuranic radioactive particles. This type of accident would pose a significant inhalation hazard, so the DOE calculates the maximum radiological dose from an accidentally released airborne contaminant. The DOE analyses are conducted deterministically, and the values are chosen based on standard DOE assumptions from various handbooks and guides. For example, the calculated dose to a receptor is simply the product of values representing worst case and typical site conditions. The DOE models do not take into consideration actual measurements of meteorological conditions or expected radioactivity in the waste containers.

After the DOE safety studies have been completed, the EEG reviews the results and assesses the potential radiological release consequences. To ensure that no potential risk factors have been overlooked, the EEG performs a complementary probabilistic analysis, a procedure mandated by the International Council of Radiation Protection (ICRP). Unlike the deterministic approach, a probabilistic assessment of the inhalation dose calculations can consider a multitude of independent scenarios. Single-value input parameters are replaced by probability distributions that represent meaningful variations in the range of conditions. These parameters are then randomly sampled to simulate realistic accident scenarios.

For the probabilistic model, Crystal Ball assumptions included site-specific meteorological conditions and varying quantities of radioactive sources within the waste containers. Using 10,000 trials, the forecasted doses were plotted to show 95 percent, 50 percent, and 5 percent dose likelihood. The 95 percent dose likelihood showed that the DOE deterministic doses were generally higher, and, therefore more conservative than the probabilistic doses. This result was attributed to the effect of smearing the high- and low-end values from the probability distribution functions. The more conservative DOE assessment showed that the release consequences may be overrepresented by DOE, and that the high doses attributed to accident scenarios are less likely to occur.

The Crystal Ball analysis also identified a potential weakness of the deterministic approach for accidents involving a large number of waste containers. The standard method used to assess the radioactive source in an accident assumes that one container was loaded to its administrative limit of 80 Ci, with the remaining containers at 8 Ci. Although some accidents were assessed as having only one container, with the average source of 80 Ci, other accidents involved a larger number of containers. In these multi-container cases, the deterministic assessment showed a lower average source from each container than was reflected in the probabilistic analysis. Based on this finding, the EEG suggested to the DOE that a reassessment of the source term values should be conducted in the evaluation of deterministic doses.

SIMULATION IN NEW-PRODUCT SCREENING AT CINERGY CORPORATION[8]

Cinergy Corporation was created in October 1994 by a merger of The Cincinnati Gas & Electric Company and PSI Energy. Cinergy is the nation's thirteenth-largest electric utility company, with approximately 11,000 megawatts of generating capacity, annual revenues of approximately $3 billion, and 8,000 employees. The company spans 25,000 square miles in north central, central, and southern Indiana, southwestern Ohio, and northern Kentucky, and serves the energy needs of 1.4 million electric customers and 435,000 gas customers.

A common situation faced by Cinergy and many other companies is the screening of new-product ideas, which clearly involves much uncertainty. At Cinergy, the new-product screening process starts with a product idea assessment, including market potential, annual sales, and profit estimates. Crystal Ball is used to help identify the risks involved with new-product development.

In one case, a new energy service product called the Uninterruptible PC Power Service was considered for introduction. This product and service would provide several benefits for a home computer, including an uninterruptible power supply, protection against power line surges and sags, and electric sinusoidal wave enhancement (i.e., improved power quality). The first step in the assessment process is to estimate the market potential for the product. This was accomplished through primary and secondary market research, leading to the spreadsheet in Figure 4-36.

Growth rates; customer forecasts; the saturation of computers, home offices, and home businesses; and the purchase intentions among the different subgroupings of the target market are all uncertain and defined as model assumptions. For many of these, no historical information was available to assign probability distributions. The saturation values and growth rates for computers, home offices, and home businesses were obtained from secondary sources, and subjective estimates of the uncertainty related to these values were obtained by reading additional material and industry reviews, and talking with experts in the field. The percent of households with a PC, PC owners who have a home office, and home office PC owners with a home business were modeled by uniform distributions; growth-rate assumptions were triangular.

For the household random error term, historical data was available that suggested errors are random and within 0.5 percent of the actual values. Therefore, a normally-distributed assumption was made for these cells. The standard deviation of this random-error term increases throughout the forecast horizon because there is more uncertainty related to the household forecast for 2001 than for 1997. Purchase intent values were obtained from a primary research study that was designed to estimate these values within ±3 percent. Normal distributions were used.

---

[8]We are indebted to Bruce Sailers and Cinergy Corporation for providing this case study.

	A	B	C	D	E	F	G
1	**Uninterrupted PC Power Supply Service**						
2	**Estimated Customer Potential**						
3	**Market Research Information:**						
4							
5	Percent of Households with a PC:						34%
6	Compound 5-year Growth Rate:						7.6%
7	Percent of PC Owners Who Have a Home Office:						55%
8	Compound 5-year Growth Rate:						12%
9	Percent of Home Office PC Owners Who Have a Home Business:						29%
10	Compound 5-year Growth Rate:						9%
11							
12	**Target Market Size**						
13	Year	Estimated Cinergy Households	Household Random Error	Cinergy Households	Household ds with a PC	PC Owners with a Home Office	PC Owners & Home Business
14	1997	1,228,000	0	1,228,000	417,520	229,636	66,594
15	1998	1,244,000	0	1,244,000	449,112	257,192	72,587
16	1999	1,261,000	0	1,261,000	483,095	288,055	79,120
17	2000	1,278,000	0	1,278,000	519,649	322,622	86,241
18	2001	1,293,000	0	1,293,000	558,969	361,337	94,003
19							
20	**Purchase Intent for UPS System & Service**						
21		Very Serious	Somewhat Serious	Purchase Intent			
22	PC only	9.0%	48.0%	15.3%			
23	PC & Home Office	16.0%	47.0%	20.1%			
24	PC & Home Business	13.0%	47.0%	17.9%			
25							
26	Assume 80% of very serious purchase:			72%			
27	Assume 20% of somewhat serious purchase:			18%			
28							
29	**Estimated Customer Potential for UPS System & Service**						
30							
31	Year	PC Only Customers	Home Office Customers	Business Customers	Market Potential		
32	1997	28,652	32,744	11,943	73,339		
33	1998	29,268	37,075	13,017	79,360		
34	1999	29,744	41,961	14,189	85,894		
35	2000	30,047	47,473	15,466	92,986		
36	2001	30,139	53,690	16,858	100,687		

FIGURE 4-36 Spreadsheet for New-Product Analysis

Finally, the market potential cells are Crystal Ball forecast cells. Cinergy was interested in making inferences regarding the market potential for this product and, therefore, information needed to be gathered on them. However, these same cells are also used as input into another spreadsheet that projects the number of customers purchasing the product each year (i.e., adoptions). This is shown in Figure 4-37. We will not discuss the intricacies of the Bass diffusion model here. However, the parameters that are included in the model are uncertain. These values can be estimated from secondary sources, such as other utility programs that involve this same type of technology, and were modeled using triangular distributions.

Once sales projections are made, financial estimates can be developed. At the early stages of a product's assessment, many uncertainties exist. Warranty costs, advertising expenditures, administrative costs, billing and initial investments are all uncertain until just before the decision is made to introduce a new product. Often, they are uncertain even after this decision. The financial spreadsheet is shown in Figure 4-38 on page 144.

Most of the assumptions were modeled using triangular or normal distributions. A truncated normal was used for the inflation factor, having a mean of 3 percent, a standard deviation of 1 percent, and a minimum value of 2 percent. The main variable of interest is the project's net present value, which was a Crystal Ball forecast cell.

The last piece of information needed was the number of iterations to use. Cinergy management wished to estimate the NPV to within $10,000 with 95 percent confidence. An estimate of the standard deviation was obtained from running a preliminary analysis

	A	B	C	D	E	F	G
1	**Sales Forecast**						
2			**Bass Diffusion Model**				
3							
4		$$\text{Adopt}_t = \left(p + q \left(\frac{\text{CumAdopt}_{t-1}}{\text{Population}_t} \right) + bX \right) * (\text{Population}_t - \text{CumAdopt}_{t-1})$$					
5							
6							
7							
8		p		0.1	(Coefficient of Innovation)		
9		q		0.6	(Coefficient of Imitation)		
10		b	0.0000				
11							

FIGURE 4-37 Sales Forecast Model

FIGURE 4-38 Financial Spreadsheet

	A	B	C	D	E	F	G
1	**Financial Analysis**						
2							
3	Assumptions:						
4		Price of UPS system:				$	300.00
5		Price of Installation:				$	100.00
6		Price of Service Contract:				$	150.00
7		Percentage of Customers Electing Service Contract:					15.0%
8		Percentage of Customers Electing Installation Service:					5%
9							
10		Cost of UPS System to Cinergy:				$	200.00
11		Installation and Fulfillment Costs per Unit:					
12		Order Processing				$	1.25
13		Billing				$	1.00
14		Product Installation					
15			Hours of technician time				2
16			Cost per hour (fully loaded or contract)			$	40.00
17							
18		Service Contract Costs:					
19		Hours of technician time					2
20		Materials (batteries, etc.)				$	25.00
21		Systems serviced each year					30%
22							
23		*Administrative Costs - 3 FTE fully loaded (Manager, Asst, Clerk)*				$	200,000
24		Inflation Factor:					3%
25		Taxes:					
26		Federal					35%
27		State					8.9%
28		Acquisition Costs Per Customer:					
29		Year 3 - 5				$	50.00
30							
31		*Initial Investment*				$	300,000

(a)

	A	B	C	D	E	F
33	**Summary of Analysis:**					
34	**Year**	**Customers**	**Average Revenue per Cust**	**Average Cost per Cust**	**Average Profits per Cust**	**Total Profit**
35	1	7,334	$ 327.50	$ 306.42	$ 21.08	$ 91,543
36	2	11,196	$ 327.50	$ 275.37	$ 52.13	$ 345,587
37	3	15,456	$ 327.50	$ 274.70	$ 52.80	$ 483,212
38	4	18,839	$ 327.50	$ 272.58	$ 54.92	$ 612,709
39	5	19,853	$ 327.50	$ 272.31	$ 55.19	$ 648,772
40						
41	**Initial Investment:**			$300,000		
42	**NPV @ 14%:**			1,072,098		
43	**Internal Rate of Return:**			87%		

(b)

	A	B	C	D	E	F	G
46	**Financial Analysis:**						
47							
48	Year	Customers	Product Revenue	Service Contract Revenue	**Total Revenue**		
49	1997	7,334	$ 2,236,870	$165,015	$ 2,401,885		
50	1998	11,196	$ 3,414,780	251,910	3,666,690		
51	1999	15,456	$ 4,714,080	347,760	5,061,840		
52	2000	18,839	$ 5,745,895	423,878	6,169,773		
53	2001	19,853	$ 6,055,165	446,693	6,501,858		
54							
55	Year	UPS System Cost	Installation and Fulfillment Costs	Acquisition Costs	Maintenance Service Costs	Administrative Costs	**Total Costs**
56	1997	$ 1,466,800	$ 45,838	$500,000	$ 34,653	$ 200,000	$ 2,247,291
57	1998	$ 2,239,200	$ 69,975	515,000	$ 52,901	206,000	3,083,076
58	1999	$ 3,091,200	$ 96,600	772,800	$ 73,030	212,180	4,245,810
59	2000	$ 3,767,800	$ 117,744	941,950	$ 89,014	218,545	5,135,053
60	2001	$ 3,970,600	$ 124,081	992,650	$ 93,805	225,101	5,406,237
61							
62	Year	Profits Before Taxes	State Tax	Federal Tax	**Total Annual Profit**		
63	1997	$ 154,594	$ 13,759	$ 49,292	$ 91,543		
64	1998	583,614	51,942	186,085	345,587		
65	1999	816,030	72,627	260,191	483,212		
66	2000	1,034,720	92,090	329,921	612,709		
67	2001	1,095,621	97,510	349,339	648,772		

(c)

FIGURE 4-38 *(continued)*

with 1,000 iterations. Calculations showed that 5,000 replications were necessary to achieve this precision.

The Crystal Ball output is shown in Figure 4-39. The sensitivity analysis shows that the coefficients of the Bass model are important. This makes sense, because these values directly determine the number of customers each year. As with most products, the financial projections are sensitive to the units sold. This suggests that if there is time to perform further research, any effort expended to estimate the Bass model coefficients as accurately as possible will be time well spent.

The mean of the NPV is $400,748 with a 95 percent confidence interval around the mean value of $386,969 to $414,527. This interval provides a high degree of confidence

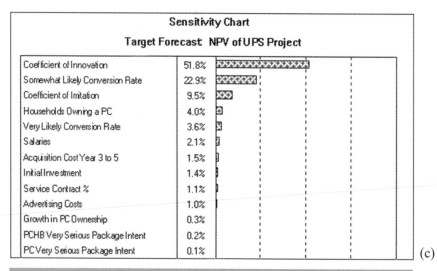

Forecast: NPV of UPS Project

Summary:
    Certainty Level is 78.90%
    Certainty Range is from 0 to +Infinity
    Display Range is from -1,000,000 to 1,750,000
    Entire Range is from -1,039,145 to 2,150,522
    After 5,000 Trials, the Std. Error of the Mean is 7,030

Statistics:	Value
Trials	5000
Mean	400,748
Median	391,489
Mode	---
Standard Deviation	497,098
Variance	2E+11
Skewness	0.10
Kurtosis	2.83
Coeff. of Variability	1.24
Range Minimum	-1,039,145
Range Maximum	2,150,522
Range Width	3,189,667
Mean Std. Error	7,030.03

(a)

Forecast: NPV of UPS Project

5,000 Trials   **Frequency Chart**   21 Outliers

Mean = 400,748

Certainty is 78.90% from 0 to +Infinity

(b)

Sensitivity Chart

Target Forecast NPV of UPS Project

Coefficient of Innovation	51.8%
Somewhat Likely Conversion Rate	22.9%
Coefficient of Imitation	9.5%
Households Owning a PC	4.0%
Very Likely Conversion Rate	3.6%
Salaries	2.1%
Acquisition Cost Year 3 to 5	1.5%
Initial Investment	1.4%
Service Contract %	1.1%
Advertising Costs	1.0%
Growth in PC Ownership	0.3%
PCHB Very Serious Package Intent	0.2%
PC Very Serious Package Intent	0.1%

(c)

FIGURE 4-39 Crystal Ball Output

that the mean NPV value is positive. However, this does not say that no risk exists for this project. For any single trial, there is a 21.1 percent chance of the NPV being negative. In fact, a 95 percent confidence interval around a single trial is −$573,565 to 1,375,061. Every manager has a different degree of aversion to risk. However, at this stage in the product's development, an almost 80 percent chance of returning a positive NPV is encouraging. It was recommended that this product move forward to the next phase in the product development process.

Questions and Problems

1. Modify the simulation model spreadsheet for roulette that we developed in Chapter 2 to use Crystal Ball. What advantages does Crystal Ball have compared to the original model? Simulate the model for 5,000 trials using the same inputs as in Figure 2-27. What conclusions can you reach?

2. Apply Crystal Ball to the Excel model for the game of craps developed in Chapter 2 (Figure 2-28), setting cell I2 as a forecast cell (no assumptions need be defined). Simulate for 5,000 trials. What are your conclusions? Additionally, replace the lookup tables with Crystal Ball Custom distributions and repeat the simulation. Compare your results.

3. Modify the blackjack simulation spreadsheet in Chapter 2 (Figure 2-29) to replace the lookup tables by custom distributions. Define appropriate forecast cells and simulate for 5,000 trials. How much more accurate are your conclusions?

4. A garage band wants to hold a concert. The expected crowd is normally distributed, with mean of 3,000 and standard deviation 500 (minimum of 0). The average expenditure on concessions is also normally distributed, with mean $5, standard deviation $2, and minimum 0. Tickets sell for $10 each, and the band's profit is 80 percent of the gate, along with concession sales, minus a fixed cost of $10,000. Develop a spreadsheet model and simulate 100 trials using Crystal Ball to identify the mean profit, the minimum observed profit, maximum observed profit, and the probability of achieving a positive profit. Repeat for 1,000 trials, and compute and compare confidence intervals for the mean profit for both cases.

5. A warehouse manager currently has one truck for local deliveries and is considering the purchase of an additional one to handle occasional high volumes. Currently, she rents additional trucks as needed for $300 per day. The truck that the company currently owns is charged off at a rate of $200 per day whether it is used or not. A new one would have a charge-off rate of $250 per day. Historical records show that the number of trucks needed each day has the distribution:

Number Needed	Probability
0	0.20
1	0.30
2	0.40
3	0.10

Develop a spreadsheet model to evaluate both policies (using the same demand for each) and simulate it for 2,000 trials with Crystal Ball, using the custom distribution for the number of trucks needed each day. Identify the average, minimum, and maximum profit for both policies. In your spreadsheet, identify when the new policy is inferior to the current one, and use the simulation to estimate the percentage of trials for which the current policy is better. Create an overlay chart to compare the mean profits.

6. A firm produces guava juice in distinctive 1-gallon jugs. The profit function is

$$\text{Profit} = (\$10 - \text{variable cost})*\text{gallons sold} - \text{fixed cost}$$

The current plant capacity is 5 million gallons per year. The firm can expand the plant by an additional 1 million or 2 million gallons per year, with the following cost impacts:

	Plant Capacity (gallons/year)	Fixed Cost	Variable Cost
Current	5 million	$5 million	$5/gallon
Add 1 million	6 million	$6 million	$4.90/gallon
Add 2 million	7 million	$7 million	$4.80/gallon

Annual demand follows the normal distribution with a mean of 5 million and a standard deviation of 1 million. Build a Crystal Ball model and, based on 500 simulation trials, identify the average profit for each option, as well as the percentile report for profit for each option.

7. Develop a Crystal Ball model for a 3-year financial analysis of total profit based on the following data and information. Sales volume in the first year is estimated to be 100,000 units and is projected to grow at a rate that is normally distributed with a mean of 7 percent per year, and a standard deviation of 4 percent. The selling price is $10 and the amount of sales increase is normally distributed with a mean increase of $0.50 (and standard deviation of $0.05) each year. Per-unit variable costs are $3, and annual fixed costs are $200,000. Per-unit costs are expected to increase by an amount normally distributed with a mean of 5 percent per year (standard deviation 2 percent). Fixed costs are expected to increase following a normal distribution with a mean of 10 percent per year (standard deviation 3 percent). Based on 1,000 simulation trials, identify the average 3-year cumulative profit, as well as the percentile report for profit. Identify the number of times a negative cumulative profit was experienced, and generate a trend chart showing net profit by year.

8. Most firms maintain a minimum amount of cash on hand to meet their daily obligations or as a requirement from the firms' banks. A maximum amount may also be specified to reflect the trade-off between the transaction cost of investing in liquid assets and the cost of lost interest if the cash is not invested. The Miller-Orr model for cash management computes the spread between the minimum and maximum cash balance limits as:

$$\text{Spread} = 3*(.75*\text{transaction cost}*\text{variance of daily cash flows/daily interest rate})^{.333}$$

The maximum cash balance is the spread plus the minimum cash balance, which is assumed to be known. The "return point" is defined as the minimum cash balance plus spread/3. Whenever the cash balance hits the maximum, the firm should invest the difference between the maximum and the return point; if the minimum is reached, sufficient securities should be sold to bring it up to the return point. Develop a Crystal Ball model for computing these decision points if the minimum cash balance is $100,000, the interest rate is 9 percent annually, and the transaction cost is $20. Model 250 banking days per year, with net cash flow per day normally distributed with a mean of $500 and a standard deviation of $2,500 (variance of $6,250,000). Use a beginning balance of $110,000, and identify the average number and dollars of security sales as well as investments.

9. A hotel wants to analyze its room-pricing structure prior to a major remodeling effort. Currently, rates and average number of room sold are as follows:

Room Type	Rate	Average Sold/Day
Standard	$ 85	250
Gold	$ 98	100
Platinum	$139	50

Each market segment has its own elasticity of demand to price. The elasticity values are standard, −1.5; gold, −2; platinum, −1. These mean, for example, if the price of a standard room is reduced by 1 percent, the number of rooms sold each day is expected to increase by 1.5 percent. The projected number of rooms sold can be determined using the following formula:

Projected number sold = elasticity*(price change in $)*(average daily sales/room rate)

where the average daily sales and room rates are those in the table in this problem. Develop a Crystal Ball model to compute the projected revenue for each market segment and the total projected revenue for any set of price changes. Develop a 90 percent confidence interval for revenues for the policy of room rates of $78, $90, and $145 for standard, gold, and platinum rooms, respectively.

10. An auto parts store stocks special halogen bulbs. These bulbs are high quality, and the demand averages 12 per day, which the manager's son (who is getting an MBA) tells him is Poisson distributed. The manager currently has a stocking policy of reordering 12 bulbs whenever current stock drops to 12 or below. Bulbs are delivered overnight at a cost of $1 per delivery. Holding one bulb one day costs the store $0.01. Build a Crystal Ball model to simulate 14 days of operations using 1,000 trials, and calculate holding costs, ordering costs, and shortage costs. Compute a 95 percent confidence interval for the total of these three costs for policies of ordering 12, 24, or 36 bulbs whenever an order is placed. Examine the skewness and kurtosis of the distribution of the holding, order, and shortage costs. What can you conclude?

11. Apply Crystal Ball to a 3-year spreadsheet model with the following assumptions. First-year sales is normally distributed with mean 250,000 and standard deviation 15,000. Sales growth rate is triangular with minimum = 2 percent, most likely = 5 percent, and maximum = 7%. The inflation factor for per-unit costs is uniform between 2 and 6 percent. The inflation factor for fixed costs is uniform between 2 and 4 percent. Run the simulation for 1,000 trials. Construct trend and sensitivity charts, and explain the results in a memo to company management.

12. Repeat the simulation in problem 11 by freezing the assumption of sales growth rate at 2 percent, 5 percent, and 7 percent. What differences do you observe?

13. A neighborhood grocery store orders a weekly entertainment magazine. Demand varies each week, but historical records show the following distribution:

Number of Magazines	Probability
12	3/36
13	12/36
14	11/36
15	7/36
16	3/36

Each magazine costs $1.50 and sells for $2.50. Any magazines left over at the end of the week are donated to a local retirement home. Use simulation to identify the optimal ordering policy (based upon average profit for each policy). Model each of the five policies for the same demand, and calculate the profit for each.

Simulate 52 weeks (52 trials) using random-number stream 1234, and compare results of Monte-Carlo simulation with using Latin Hypercube sampling for both 500 and 2,500 trials.

14. The manager of an apartment complex has observed that the number of units rented during any given month varies between 30 and 40. (Use a triangular distribution with minimum 30, most likely 34, maximum 40.) Rent is $500 per month. Operating costs average $15,000 per month, but vary following a normal distribution with mean $15,000. Operating costs are assumed to be normal with a standard deviation of $300. Use Crystal Ball to estimate the 80 percent, 90 percent, and 95 percent confidence intervals for profitability of this business.
    a. What is the probability that monthly profit will be positive?
    b. What is the probability that monthly profit will exceed $4,000?
    c. Compare the 80 percent, 90 percent, and 95 percent certainty ranges.
    d. What is the probability that profit will be between $1,000 and $3,000?

15. For problem 14, identify the distribution of the forecasted profit among the options of normal, triangular, and uniform distributions. After simulating the 1,000 trials, select *Extract Data,* and select *Forecast Values* from the *Run* menu. This should provide you with 1,000 rows of forecasts. Then from the *Tools* menu, select *Batch Fit,* where you provide the data range. You can unselect all of the distributions available except for normal, triangular, and uniform. Use the Chi-Square test as a selection technique. Which distribution provides the best fit?

16. A trucking company deals with many flat tires every year. The number of tires that needed repairing last year was 20,000. This growth rate has been normally distributed with a mean of 10 percent and a standard deviation of 3 percent. Currently, the company has truckers get their flats fixed on the open market, which last year cost an average of $50 per tire. Inflation in tire repair cost has been $5 per year (normally distributed, mean $5, standard deviation $2). The company is considering an opportunity to sign a contract with a tire firm. Under this contract, the tire company would provide patrolling vehicles that would repair tires in about the same time as the current system at a fixed cost of $60 per tire. This price would be constant over the 3-year contract period. Management wants a comparison of net present costs for repairing tires under both options. Use 1,000 trials and Monte-Carlo sampling.
    a. Obtain the mean and maximum net present cost for each option.
    b. Identify the 90 percent confidence interval for each option.
    c. Identify the probability for each system of having a net present cost greater than $3.5 million; greater than $3.7 million; greater than $3.9 million.
    d. Compare the decile reports (from the *Percentile* view of the Forecast charts).
    e. Develop a tornado and spider chart for both net present costs.

17. Repeat parts a, b, and c from problem 16 using Latin Hypercube sampling. Are there any significant differences?

18. Apply the Crystal Ball Bootstrap tool to see the dispersion of each option's net present costs. Compare mean estimates and their 90 percent confidence intervals. (From the *CBTools* menu, select *Bootstrap,* and the cell with the net present cost for the market option. Use the option based on running one simulation and resampling.)

19. A lawn mower supplier has to bid on an annual basis with a large retail chain. The demand for this particular model of lawn mower is lognormally distributed with a mean of 10,000 and standard deviation of 1,000. The bid is competitive, and the low competitor bid is lognormally distributed with a mean of 100 and a standard deviation of 10. This supplier has a cost of production of $90 per mower, which establishes his bottom bid. He wants to simulate bids from $90 to $110 (which he feels someone is sure to bid under). Use a Crystal Ball decision table to compare expected profits from different bids.

20. An entrepreneur has agreed to provide software to a distributor on demand. Over the next year, the demand is highly variable, following the exponential distribution with a mean of 1,000 sales per year. The contract price is $500 per copy, and variable cost to produce this product (including packaging and manuals) is $400. There is a fixed cost per year of $50,000. The entrepreneur wants to estimate expected profit for this operation. Compare the following statistics for Monte-Carlo and Latin Hypercube sampling for 1,000 trials:
    a. Average expected profit
    b. The decile report
    c. The distribution of the output (based on visual inspection).

CHAPTER **5**

Applications of Risk Analysis

Chapter Outline

- Operations Management Applications
  Machine Reliability and Maintenance
  Project Management
  Hotel Overbooking
  Call Center Capacity Analysis
- Finance Applications
  Rate of Return Analysis
  Retirement Planning
  Cash Budgeting
  New-Product Development
  Recycling Decision Model
- Marketing Applications
  Sales Projection
  Marketing Analysis for Pharmaceuticals
  Analysis of Distribution Strategies
- Engineering Applications
  Reliability of a Spring
  Risk Assessment of Toxic Waste
  Electrical Circuit Analysis
- Simulation in Practice
  Risk Analysis at Getty Oil Company
  Assessing Tritium Supply Alternatives by the U.S. Department of Energy
- Questions and Problems

Monte-Carlo simulation can be used to analyze risk in a variety of business disciplines, including operations management, finance, and marketing, as well as other fields such as engineering. For example, *Fortune* reported on May 30, 1994, that a Monte-Carlo simulation model played a key role in determining whether Merck Corporation should pay $6.6 billion to acquire Medco, a mail-order pharmaceutical company.[1] The model included estimates of possible futures of the U.S. health care system, possible future changes in mix of generic and brand-name drugs, probability distributions of profit margins for each product, and assumptions of competitors' behavior. The model helped Merck management decide that the Medco acquisition made

---

[1]"A New Tool to Help Managers," *Fortune,* Vol. 129, no. 11, May 30, 1994, pp. 135–140.

sense, whether or not health care reform materializes. Merck, which has operations in more than 140 companies, also uses Monte-Carlo simulation to manage foreign-exchange risks.

The purpose of this chapter is to illustrate a variety of applications of risk analysis using Crystal Ball. Specifically, we present applications in four major areas: operations management, finance, marketing, and engineering. These examples also illustrate different principles of simulation modeling and features of Crystal Ball. To conserve space, we have not shown the cell formulas in the figures; however, all models are available on the CD-ROM accompanying this book if you wish to study the Excel formulas. In addition, the Crystal Ball run files for these examples are also on the CD. To access them, choose *Restore Run* from the Crystal Ball *Run* menu in Excel and open the appropriate file. (If the *Restore Run* command is not visible, first select *Reset* to close out any previous Crystal Ball runs.)

Operations Management Applications

Simulation has a wide variety of applications in operations, ranging from manufacturing to service. Although the majority of applications generally involve complex systems simulation, many OM issues can be addressed by using Monte-Carlo techniques. In this section, we present examples of machine reliability and maintenance, project management, hotel overbooking, and capacity analysis of a call center.

MACHINE RELIABILITY AND MAINTENANCE[2]

A large milling machine has three different bearings that fail in service. The probability distribution of the life of each bearing is identical. When a bearing fails, the mill stops, a repairperson is called, and a new bearing is installed. The delay time of the repairperson's arriving at the milling machine is also a random variable. Downtime for the mill is estimated at $10 per minute, and the direct on-site cost of the repairperson is $24 per hour. It takes 20 minutes to change one bearing, 30 minutes to change two bearings, and 40 minutes to change all three. Each bearing costs $30. A proposal has been made to replace all three bearings whenever a bearing fails instead of replacing them individually. Is this proposal worthwhile?

Figure 5-1 shows a spreadsheet designed to simulate the current approach of replacing each bearing when it fails. We will assume that historical data on bearing life has been collected and rounded to the nearest 100 hours, but that the detailed data have been discarded. In addition, the delay time has been estimated judgmentally as either 5, 10, or 15 minutes based on conversations with workers. The form of the data forces us to make some assumptions about the simulation model. For example, because bearing lives are rounded to increments of 100 hours, there will be instances when the model will generate multiple bearing failures at the same time. This is unlikely to occur in practice, so we can reasonably assume that no more than one bearing will be replaced at any breakdown time. To make this more realistic, we might consider transforming the frequency distribution to a continuous distribution by, for example, assuming that a life of 1,000 hours represents the midpoint of a uniform interval between 950 and 1,050 and using the custom distribution with continuous ranges in Crystal Ball to generate random variates. However, realize that the purpose of the simulation is to compare decision alternatives; thus, even though the approximation is a bit crude, it should provide

---

[2]Adapted from Jerry Banks and John S. Carson, *Discrete Event Simulation,* Upper Saddle River, NJ: Prentice Hall, 1984.

	A	B	C	D	E	F	G	H	I	J	K	L
1	**Bearing Reliability Model**											
2												
3	Bearing unit cost	$30										
4	Delay unit cost	$10										
5	Repairperson wage	$24										
6												
7	Bearing Life	Probability			Delay Time	Probability		*Simulation Results (Current Method)*				
8		1000	0.10		5	0.6						
9		1100	0.13		10	0.3		Total bearings replaced		45		
10		1200	0.25		15	0.1		Total delay time		337.5		
11		1300	0.13									
12		1400	0.09		Max Time	20000		Cost of bearings		$1,350		
13		1500	0.12					Cost of delay time		$3,375		
14		1600	0.02					Cost of downtime during repair		$9,000		
15		1700	0.06					Cost of repairpersons		$360		
16		1800	0.05					Total cost		$14,085		
17		1900	0.05									
18												
19	*Current Method of Bearing Replacement*											
20		Bearing 1	Failures		Total Delay		Bearing 2	Failures		Total Delay	Bearing 3	Failures	Total Delay
21			15		112.5			15		112.5		15	112.5
22		Life	Cumulative	Delay			Life	Cumulative	Delay		Life	Cumulative	Delay
23		1337	1337	7.5			1337	1337	7.5		1337	1337	7.5
24		1337	2674	7.5			1337	2674	7.5		1337	2674	7.5
25		1337	4011	7.5			1337	4011	7.5		1337	4011	7.5
26		1337	5348	7.5			1337	5348	7.5		1337	5348	7.5
27		1337	6685	7.5			1337	6685	7.5		1337	6685	7.5
28		1337	8022	7.5			1337	8022	7.5		1337	8022	7.5
29		1337	9359	7.5			1337	9359	7.5		1337	9359	7.5
30		1337	10696	7.5			1337	10696	7.5		1337	10696	7.5
31		1337	12033	7.5			1337	12033	7.5		1337	12033	7.5
32		1337	13370	7.5			1337	13370	7.5		1337	13370	7.5
33		1337	14707	7.5			1337	14707	7.5		1337	14707	7.5
34		1337	16044	7.5			1337	16044	7.5		1337	16044	7.5
35		1337	17381	7.5			1337	17381	7.5		1337	17381	7.5
36		1337	18718	7.5			1337	18718	7.5		1337	18718	7.5
37		1337	20055	7.5			1337	20055	7.5		1337	20055	7.5

FIGURE 5-1 Simulation Model for Bearing Replacement (current method)

comparable results. The spreadsheet is designed to simulate the process for a fixed time specified in cell E12; in this case, 20,000 hours.

For each bearing, we simulate the life until failure and delay time before repair by using the Crystal Ball *CB.Custom* function for the frequency distributions. This model is similar to the craps game simulation we presented in Chapter 2 in the sense that we cannot predict exactly how many failures will occur during the 20,000 hours of simulated time. To get around this, we extend the worksheet formulas well beyond the number of failures we might expect (in this case, down to row 100 for a total of 78 possible failures) and maintain a record of the cumulative time. We use an *IF* function to determine when the cumulative time exceeds 20,000 hours and then replace the cell values with blanks. This allows us to count the number of failures (by subtracting the value of the *COUNTBLANK* function from 78 over the entire range) and the total delay time. We define the total bearings replaced and total cost as forecast cells.

A useful means of debugging a Crystal Ball model is to use the *Single Step* command under the *Run* menu. This allows you to generate one trial at a time to verify that the model is doing what is intended. An example is shown in Figure 5-2. By clicking on the *Single Step* command repeatedly, it becomes easier to identify errors and provide data by which you may check your calculations manually to verify the accuracy of your model formulas.

Figure 5-3 shows the model for the proposed system. The life of each bearing is simulated independently, and the minimum time representing the time to first failure is recorded in column R. Cell S21 tracks the number of sets of bearings replaced during the time of the simulation. Figure 5-4 shows the results of simulating both situations.

	A	B	C	D	E	F	G	H	I	J	K	L
19	*Current Method of Bearing Replacement*											
20		Bearing 1	Failures	Total Delay		Bearing 2	Failures	Total Delay		Bearing 3	Failures	Total Delay
21			15	110			16	135			14	110
22		Life	Cumulative	Delay		Life	Cumulative	Delay		Life	Cumulative	Delay
23		1100	1100	5		1200	1200	5		1700	1700	5
24		1400	2500	5		1200	2400	10		1100	2800	5
25		1400	3900	5		1200	3600	5		1900	4700	15
26		1000	4900	5		1100	4700	5		1200	5900	10
27		1500	6400	5		1400	6100	15		1500	7400	10
28		1900	8300	15		1900	8000	10		1200	8600	5
29		1900	10200	10		1200	9200	10		1200	9800	10
30		1900	12100	15		1200	10400	5		1200	11000	10
31		1800	13900	5		1500	11900	5		1400	12400	5
32		1300	15200	10		1000	12900	10		1500	13900	10
33		1100	16300	5		1500	14400	15		1500	15400	10
34		1100	17400	5		1500	15900	5		1800	17200	5
35		1100	18500	5		1100	17000	5		1600	18800	5
36		1300	19800	10		1200	18200	10		1700	20500	5
37		1200	21000	5		1200	19400	10				
38						1900	21300	10				
39												

FIGURE 5-2 Example of Single Step Command in Crystal Ball

FIGURE 5-3 Simulation Results for Bearing Replacement (proposed method)

	N	O	P	Q	R	S	T
7		*Simulation Results (Proposed Method)*					
8							
9		Total bearings replaced			57		
10		Total delay time			140		
11							
12		Cost of bearings			$1,710		
13		Cost of delay time			$1,400		
14		Cost of downtime during repair			$7,600		
15		Cost of repairpersons			$304		
16		Total cost			$11,014		
17							
18							
19		*Proposed Method of Bearing Replacement*					
20		Bearing 1	Bearing 2	Bearing 3		Failures	Total Delay
21					First	19	140
22		Life	Life	Life	Failure	Cumulative	Delay
23		1500	1300	1500	1300	1300	5
24		1000	1200	1300	1000	2300	5
25		1600	1000	1500	1000	3300	15
26		1000	1500	1200	1000	4300	10
27		1400	1500	1100	1100	5400	15
28		1700	1900	1900	1700	7100	5
29		1900	1100	1000	1000	8100	10
30		1500	1200	1400	1200	9300	5
31		1200	1000	1300	1000	10300	10
32		1600	1400	1000	1000	11300	10
33		1400	1400	1100	1100	12400	10
34		1800	1000	1300	1000	13400	5
35		1000	1200	1600	1000	14400	5
36		1000	1200	1200	1000	15400	5
37		1800	1000	1200	1000	16400	5
38		1900	1200	1100	1100	17500	5
39		1800	1100	1000	1000	18500	5
40		1300	1300	1500	1300	19800	5
41		1400	1700	1800	1400	21200	5
42							

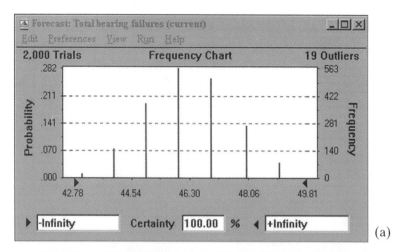

(a)

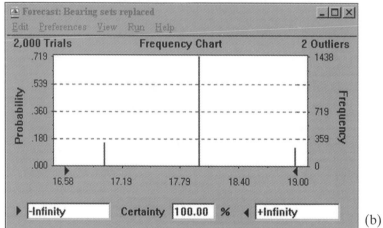

(b)

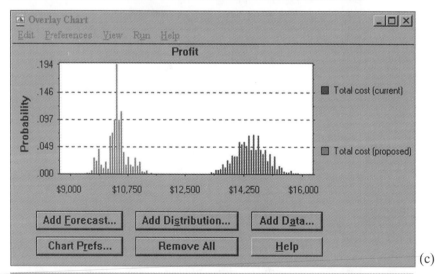

(c)

FIGURE 5-4 Simulation Results for Bearing Replacement Alternatives

Under the current system, between 42 and 52 individual bearings are replaced during the 20,000 hour time; on average, 18 sets (54 bearings) are replaced under the proposed approach. However, because replacing all bearings saves changing time and reduces delay time, the cost savings offset the cost of the additional bearings. The overlay chart of the cost forecasts clearly shows the superiority of the proposal.

SKILLBUILDER EXERCISE

Open the file *Bearing Reliability.xls.* Run the simulation for 40,000 simulated hours. Compare the results with the example. On a per-hour basis, is there any significant difference?

PROJECT MANAGEMENT

Project management is concerned with scheduling the activities of a project involving interrelated activities.[3] An important aspect of project management is identifying the expected completion time of the project. Activity times can be deterministic or probabilistic. We often assume that probabilistic activity times have a beta or triangular distribution, especially when times are estimated judgmentally. Analytical methods such as PERT allow us to determine probabilities of project completion times by assuming that the expected activity times define the critical path and invoking the Central Limit Theorem to make an assumption of normality of the distribution of project completion time. However, this assumption may not always be valid, and we will explore this later. Simulation can provide a more realistic characterization of the project completion time and the associated risks. We will illustrate risk analysis in project management through the following example.

A consulting firm has been hired to assist in the evaluation of new software. The manager of the IS department is responsible for coordinating all of the activities involving consultants and the company's resources. The activities shown in Table 5-1 have been defined for this project, which is depicted graphically in Figure 5-5. The target project completion date is 150 working days. Because this is a new application, no historical data on activity times are available; they must be estimated judgmentally. The IS manager has determined the most likely time for each activity but, recognizing the uncertainty in the times to complete each task, has estimated the 10th and 90th percentiles. Setting the parameters in this way is a common approach for estimating the distribution, because managers typically cannot estimate the absolute minimum or maximum times but can reasonably determine a time that might be met or exceeded 10 percent of the time. With only these estimates, a triangular distribution is an appropriate assumption.

Figure 5-6 shows a spreadsheet designed to simulate the project completion time. For those activity times that are not constant, we define the cell for the activity time as a Crystal Ball assumption. After selecting the triangular distribution in the Crystal Ball gallery, click on the *Parms* button in the top right of the dialog box. This provides a list of alternative ways to input the data. We select the 10th percentile, most likely, and 90th percentile option. To facilitate data input, we may use cell references in the input boxes

---

[3]Complete discussions of project management can be found in various operations management textbooks, such as James R. Evans, *Production/Operations Mangement: Quality, Performance, and Value,* Fifth Edition, St. Paul: West, 1996.

TABLE 5-1 Activity and Time Estimate List

Activity		Predecessors	10th %ile	Most Likely	90th %ile
A	Select steering committee	—	15	15	15
B	Develop requirements list	—	40	45	60
C	Develop system size estimates	—	10	14	30
D	Determine prospective vendors	—	2	3	5
E	Form evaluation team	A	5	7	9
F	Issue request for proposal	B,C,D,E	4	5	8
G	Bidders' conference	F	1	1	1
H	Review submissions	G	25	30	50
I	Select vendor short list	H	3	5	10
J	Check vendor references	I	3	7	10
K	Vendor demonstrations	I	20	30	45
L	User site visit	I	3	4	5
M	Select vendor	J,K,L	3	3	3
N	Volume sensitive test	M	10	13	20
O	Negotiate contracts	M	10	14	28
P	Cost-benefit analysis	N,O	2	2	2
Q	Obtain board of directors approval	P	5	5	5

FIGURE 5-5 Project Network Corresponding to Table 5-1

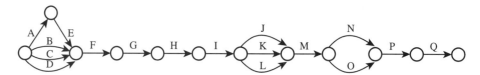

FIGURE 5-6 Project Management Simulation Model

	A	B	C	D	E	F	G	H	I	J	K
1	**Project Management Model**										
2											
3		**10th**	**Most**	**90th**	**Activity**	**Early**	**Early**	**Latest**	**Latest**		**On Critical**
4	**Activity**	**Percentile**	**Likely**	**Percentile**	**Time**	**Start**	**Finish**	**Start**	**Finish**	**Slack**	**Path?**
5	A	15	15	15	15.00	0.00	15.00	27.32	42.32	27.32	0
6	B	40	45	60	49.32	0.00	49.32	0.00	49.32	0.00	1
7	C	10	14	30	19.19	0.00	19.19	30.12	49.32	30.12	0
8	D	2	3	5	3.43	0.00	3.43	45.88	49.32	45.88	0
9	E	5	7	9	7.00	15.00	22.00	42.32	49.32	27.32	0
10	F	4	5	8	5.86	49.32	55.18	49.32	55.18	0.00	1
11	G	1	1	1	1.00	55.18	56.18	55.18	56.18	0.00	1
12	H	25	30	50	36.49	56.18	92.67	56.18	92.67	0.00	1
13	I	3	5	10	6.29	92.67	98.96	92.67	98.96	0.00	1
14	J	3	7	10	6.57	98.96	105.53	124.54	131.11	25.58	0
15	K	20	30	45	32.15	98.96	131.11	98.96	131.11	0.00	1
16	L	3	4	5	4.00	98.96	102.96	127.11	131.11	28.15	0
17	M	3	3	3	3.00	131.11	134.11	131.11	134.11	0.00	1
18	N	10	13	20	14.72	134.11	148.83	137.71	152.43	3.60	0
19	O	10	14	28	18.32	134.11	152.43	134.11	152.43	0.00	1
20	P	2	2	2	2.00	152.43	154.43	152.43	154.43	0.00	1
21	Q	5	5	5	5.00	154.43	159.43	154.43	159.43	0.00	1
22											
23				**Project completion time**		159.43					

as shown in Figure 5-7 for activity B (cell E6 in Figure 5-6). Then, we may use the *Copy* and *Paste Data* commands under the *Cell* menu to copy the Crystal Ball assumptions to the other appropriate cells. (Remember not to use the Excel copy and paste commands.) Note that Crystal Ball determines the appropriate minimum (*a*) and maximum (*b*) values for the triangular distribution based on the percentile information.

The project completion time depends on the specific time for each activity. To find this, we compute the activity schedule and slack for each activity. Activities A, B, C, and D have no immediate predecessors and, therefore, have early start times of 0. The early start time for each other activity is the maximum of the early finish times for the activity's immediate predecessor. Early finish times are computed as the early start time plus the activity time. The early finish time for the last activity, Q, represents the earliest time the project can be completed; that is, the minimum project completion time. This is defined as the forecast cell for the Crystal Ball simulation.

To compute late start and late finish times, we set the late finish time of the terminal activity equal to the project completion time. The late start time is computed by subtracting the activity time from the late finish time. The late finish time for any other activity, say X, is defined as the minimum late start of all activities to which activity X is an immediate predecessor. Slack is computed as the difference between the late finish and early finish. The critical path consists of activities with zero slack. Based on the expected activity times, the critical path consists of activities B-F-G-H-I-K-M-O-P-Q and has an expected duration of 159.43 days.

In the analytical approach found in most textbooks, probabilities of completing the project within a certain time are computed assuming:

1. The distribution of project completion times is normal (by appealing to the Central Limit Theorem).
2. The expected project completion time is the sum of the expected activity times along the critical path, which is found using the expected activity times.
3. The variance of the distribution is the sum of the variances of those activities along the critical path, which is found by using the expected activity times. If more than one critical path exists, use the path with the largest variance.

FIGURE 5-7 Triangular Distribution

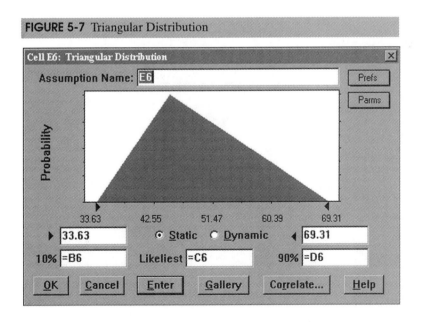

Using the minimum and maximum values for the triangular distribution computed by Crystal Ball from the percentile data input, we may use the formula presented in Chapter 3 to compute the variance for each activity as shown in Figure 5-8. Thus, for this example, the variance of the critical path is 281.88 (found by adding the variances of those activities with zero slack). With the normality assumption, the probability that the project will be completed within 150 days is found by computing the z-value:

$$z = \frac{150 - 159.43}{16.789} = -.562$$

From Appendix A, this corresponds to a probability of approximately 0.29.

Variations in actual activity times may yield different critical paths than the one resulting from expected times. This may change both the mean and variance of the actual project completion time, resulting in an inaccurate assessment of risk. Simulation can easily address these issues.

Figure 5-9 shows the Crystal Ball forecast and statistics charts for 2,000 trials. The mean and variance are quite close to that predicted; in fact, fitting the normal distribution to the forecast in an overlay chart results in a very good fit. Whereas the analytical approach computed the probability of completing the project within 150 days as 0.29, analysis of the forecast chart shows this to be smaller, 22 percent. Note, however, that the structure of the project network in Figure 5-5 is quite "linear," resulting in few options for critical paths. For projects that have many more parallel paths, the results between the simulation and analytical models may differ significantly more.

SKILLBUILDER EXERCISE

Open the file *Project Management.xls*. Replace the assumption cells with triangular distributions using the parameters *a*, *m*, and *b*, shown in Figure 5-8, which were determined based on the original estimates of percentiles. Rerun the simulation. Would you expect to see nearly identical results?

	M	N	O	P	Q
	a	m	b	mean	variance
4					
5	15	15	15	15	0
6	33.63	45	69.31	49.3133	55.369872
7	4.07	14	39.51	19.1933	55.704406
8	0.95	3	6.34	3.43	1.2336167
9	3.38	7	10.62	7	2.1840667
10	2.73	5	9.89	5.87333	2.2314056
11	1	1	1	1	0
12	17.58	30	61.89	36.49	87.07235
13	0.67	5	13.21	6.29333	6.7612389
14	0.03	7	12.68	6.57	6.6907167
15	10.66	30	55.79	32.15	85.441017
16	2.19	4	5.81	4	0.5460167
17	3	3	3	3	0
18	6.62	13	24.55	14.7233	13.766439
19	4.48	14	36.48	18.32	44.999467
20	2	2	2	2	0
21	5	5	5	5	0
22	On critical path:			159.44	281.87535
23			standard deviation		16.789144

FIGURE 5-8 Triangular Distribution Parameters and Activity Means and Variances

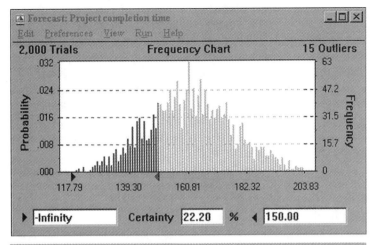

FIGURE 5-9 Probability of Project Completion Within 150 Days

HOTEL OVERBOOKING

An important operations decision for service businesses such as hotels, airlines, and car rental companies is the number of reservations to accept to effectively utilize capacity with the knowledge that some customers may not use their reservations or tell the business. If a hotel, for example, holds rooms for customers that do not show up, they lose revenue opportunities. (Even if they charge a night's lodging as a guarantee, rooms held for additional days may go unused.) A common practice in these industries is to overbook reservations. When more customers arrive than can be handled, the business usually incurs some cost to satisfy them (by putting them up at another hotel or, for most airlines, not only providing a free flight but an additional ticket bonus, as well). Therefore, the decision becomes how much to overbook to balance the costs of overbooking against the lost revenue for underutilization.

We will illustrate how a simulation model can help in making this decision. Figure 5-10 shows a spreadsheet simulation model for a hypothetical hotel that has 300 rooms. The hotel charges $120 per room. Cell B6 represents the decision variable of how many reservations to accept. In cell B7, we estimate the probability of a no-show. If we assume this to be constant for all customers, then the number of no-shows (cell B8) can be modeled using a binomial distribution with n = number of reservations accepted and p = probability of a no-show. If the actual number of customer arrivals exceeds the room capacity, the hotel incurs a cost of $100 to put them up at a nearby facility.

FIGURE 5-10 Hotel Overbooking Simulation Model

	A	B	C	D	E	F	G	H	I	J	K
1	**Hotel Overbooking Model**										
2					Net Revenue	Overbookings	Mean	Standard	Return to	Mean	Standard
3					$34,200	18	revenue	Deviation	Risk	Overbookings	Deviation
4	Rooms available	300		300	$ 32,160	0	$ 33,468	$ 545	61.45566	-	-
5	Price	$120		310	$ 33,360	0	$ 34,666	$ 540	64.17382	0.01	0.13
6	Reservations accepted	350		320	$ 34,560	0	$ 35,590	$ 344	103.3132	1.27	2.12
7	Probability of No-Show	0.06		330	$ 35,760	0	$ 35,094	$ 423	82.96858	8.97	4.37
8	No shows	32		340	$ 35,200	8	$ 34,110	$ 454	75.16193	18.90	4.54
9	Customer arrivals	318		350	$ 34,200	18	$ 33,110	$ 454	72.95842	28.90	4.54
10	Overbooked customers	18		360	$ 33,200	28	$ 32,110	$ 454	70.75491	38.90	4.54
11	Overbooking cost	$100									
12	Net revenue	$34,200									

Because we are interested in simulating results for different values of the decision variable, we use a data table similar to the examples in Chapter 2. We define both the net revenue and the number of overbookings as forecast cells in the data table. To provide a convenient summary of the Crystal Ball results, we extract the mean and standard deviation for the forecasts using the function *CB.GetForeStatFN,* as described in Chapter 4. Assuming a no-show probability of 0.06, which might be determined from historical data, the results suggest that accepting 320 reservations maximizes the net revenue over the range of possible decision values. Figure 5-11 shows the forecast charts for this scenario. Although this decision might result in up to 13 overbookings, over 60 percent of the time there are none.

As we noted in Chapter 1, a measure of risk is the standard deviation, because it describes how the forecast varies from its mean. The coefficient of variability (C.V.), which is the ratio of the standard deviation to the mean, provides a relative measure of risk. The smaller the C.V., the smaller the relative risk for the return provided. The reciprocal of the C.V., called **return-to-risk,** is often used because it is easier to interpret. This is similar to the Sharpe ratio in finance, which is the ratio of a fund's excess returns (annualized total returns minus Treasury bill returns) to its standard deviation. If several investment opportunities have the same mean but different variances, a risk-averse in-

FIGURE 5-11 Forecast Charts for Decision to Accept 320 Reservations

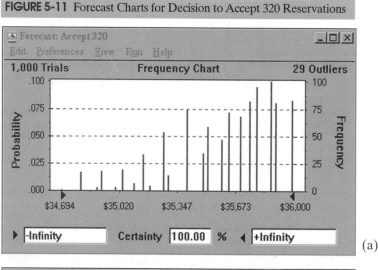

(a)

(b)

vestor will select the one that has the smallest variance. If maximization of return is the objective, a large return-to-risk is equivalent to a relatively large mean in proportion to the standard deviation. Interestingly in this example, the return-to-risk is also maximized at the reservation level corresponding to the maximum net revenue.

SKILLBUILDER EXERCISE

Open the file *Overbooking.xls*. Use the Crystal Ball Decision Table tool to run multiple simulations for values of the decision variable between 315 and 325 and find a more accurate solution.

CALL CENTER CAPACITY ANALYSIS[4]

Many companies depend on call centers to handle customer orders, technical support, and other service activities. One of the challenging problems associated with call centers is providing the right amount of staff at various times of the week and day to match demand. Too little staff results in long hold times or lost customers; too many staff results in excess costs. This example presents a spreadsheet model to calculate the predicted overcapacity and undercapacity of a call center. By calculating the percentiles during days of the week and time blocks during the day, the model shows where the number of calls exceeds the capacity of the staff to service the calls or where the call center is overstaffed.

The model relies on calling frequencies and staffing/productivity profiles by time of day and day of the week. The calling frequencies are aggregated by blocks representing key times of the day. The first block occurs between midnight and 7:00 A.M. (preworkday), the second between 8:00 A.M. and 11:00 A.M. (morning office hours), the third between 12:00 and 1:00 P.M. (lunch), the fourth between 2:00 P.M. and 5:00 P.M. (afternoon of workday), and the fifth between 6:00 P.M. and 12:00 A.M. (after work hours). These blocks were chosen to reflect the traffic that occurs at the call center during any time of the day.

A Monte-Carlo simulation using Crystal Ball is used to vary the call frequencies during any 1-hour period. The model is illustrated in Figure 5-12. Columns C through I show the historical call volumes; columns J through P show the staffing levels by time of day and day of week. Column B represents the Crystal Ball assumptions by which the call frequencies are varied randomly. Each is a triangular distribution with parameters 0, 1, and 2. In addition, call frequencies during the same calling block are moderately correlated to each other ($r = 0.5$) and are loosely correlated to frequencies in adjacent calling blocks ($r = 0.2$) using Crystal Ball's Correlation Matrix tool. The correlation between hours in the same block and hours in adjacent time blocks allows the model to simulate the fluctuation of the calling frequency during the day. The model outputs are shown in the third section (labeled "Over/Under Capacity"), which give the percent of calls over ($+$) or under ($-$) the staff's capacity to handle these calls. The marginal values show the averages for each time period and day of week, as well as for the aggregated time blocks. These are the Crystal Ball forecast cells.

At first glance, the static representation of the problem shows that the call center is appropriately staffed during Tuesday, Wednesday, and Thursday. During Mondays and

---

[4]This model was provided by Raymond Covert of Fairfax, VA (RPCovert.com). We gratefully acknowledge his contribution.

	A	B	C	D	E	F	G	H	I	J	K	L	M	N	O	P	Q	R	S	T	U	V	W	X	Y	Z
1			Average Calls/min							Staff Capability (calls/min)							Over / Under Capacity							Time	Over/Under
2	Time		M	I	W	R	F	Sa	Su	M	I	W	R	F	Sa	Su	M	I	W	R	F	Sa	Su	Avg.	Block	Capacity
3	0:00	1	2	1	2	2	2	2	1	2	2	2	2	2	2	2	0%	-50%	0%	0%	0%	0%	-50%	-14%	0-7	-6%
4	1:00	1	2	2	2	2	2	2	1	2	2	2	2	2	2	2	0%	0%	0%	0%	0%	0%	-50%	-7%		
5	2:00	1	2	2	2	2	2	2	2	2	2	2	2	2	2	2	0%	0%	0%	0%	0%	0%	0%	0%		
6	3:00	1	4	4	4	4	3	3	2	4	4	4	4	4	3	3	0%	0%	0%	0%	-25%	0%	-33%	-8%		
7	4:00	1	4	4	4	4	3	3	2	4	4	4	4	4	3	3	0%	0%	0%	0%	-25%	0%	-33%	-8%		
8	5:00	1	8	7	7	7	5	5	2	7	7	7	7	7	4	4	14%	0%	0%	0%	-29%	25%	-50%	-6%		
9	6:00	1	12	10	10	10	8	8	3	10	10	10	10	10	6	6	20%	0%	0%	0%	-20%	33%	-50%	-2%		
10	7:00	1	18	15	15	15	12	12	5	15	15	15	15	15	9	9	20%	0%	0%	0%	-20%	33%	-44%	-2%		
11	8:00	1	25	20	20	20	15	15	5	20	20	20	20	20	10	10	25%	0%	0%	0%	-25%	50%	-50%	0%	8-11	-1%
12	9:00	1	30	25	25	25	19	19	8	25	25	25	25	25	14	14	20%	0%	0%	0%	-24%	36%	-43%	-2%		
13	10:00	1	25	21	21	21	17	17	8	21	21	21	21	21	13	13	19%	0%	0%	0%	-19%	31%	-38%	-1%		
14	11:00	1	20	17	17	17	14	14	8	17	17	17	17	17	11	11	18%	0%	0%	0%	-18%	27%	-27%	0%		
15	12:00	1	35	29	20	20	20	18	8	25	25	25	25	25	13	13	40%	16%	-20%	-20%	-20%	38%	-38%	-1%	12-13	-1%
16	13:00	1	20	17	17	17	13	13	6	17	17	17	17	17	10	10	18%	0%	0%	0%	-24%	30%	-40%	-2%		
17	14:00	1	20	16	16	16	12	12	4	16	16	16	16	16	8	8	25%	0%	0%	0%	-25%	50%	-50%	0%	14-17	-1%
18	15:00	1	20	16	16	16	12	12	4	16	16	16	16	16	8	8	25%	0%	0%	0%	-25%	50%	-50%	0%		
19	16:00	1	18	15	15	15	11	11	4	15	15	15	15	15	8	8	20%	0%	0%	0%	-27%	38%	-50%	-3%		
20	17:00	1	18	14	14	14	10	10	2	14	14	14	14	14	6	6	29%	0%	0%	0%	-29%	67%	-67%	0%		
21	18:00	1	15	12	12	12	9	9	2	12	12	12	12	12	6	6	25%	0%	0%	0%	-25%	50%	-67%	-2%	18-24	-4%
22	19:00	1	10	8	8	8	6	6	1	8	8	8	8	8	4	4	25%	0%	0%	0%	-25%	50%	-75%	-4%		
23	20:00	1	6	5	5	5	4	4	1	5	5	5	5	5	3	3	20%	0%	0%	0%	-20%	33%	-67%	-5%		
24	21:00	1	5	4	4	4	3	3	1	4	4	4	4	4	2	2	25%	0%	0%	0%	-25%	50%	-50%	0%		
25	22:00	1	4	4	4	4	3	3	1	4	4	4	4	4	2	2	0%	0%	0%	0%	-25%	50%	-50%	-4%		
26	23:00	1	2	2	2	2	2	2	1	2	2	2	2	2	2	2	0%	0%	0%	0%	0%	0%	-50%	-7%		
27																Average	16%	-1%	-1%	-1%	-20%	31%	-47%			

FIGURE 5-12 Call Center Simulation Model

Saturdays, the call center is understaffed, and during Fridays and Sundays, the call center is overstaffed. The static output shows that the call center is appropriately staffed during the average of the calling periods (Monday through Sunday). The problem with the static results is that it does not provide a level of confidence or reliability that the staffing level is appropriate. The simulation results show when staffing is appropriate during these days and time blocks and with an associated level of confidence or reliability. Figure 5-13 shows the percentiles for each forecast along with the charts by day of week and time block. The percentiles were obtained using the *CB.GetForePercentFN* function. By examining the charts, we get an idea of the confidence of the staffing profile. The Monte-Carlo simulation confirms the static case findings that the call center is understaffed during Mondays and Saturdays, but can handle the incoming calls 10 percent of the time on Saturdays and 30 percent of the time on Mondays. The call center is nominally staffed from Tuesday through Thursday with 50 percent confidence. The call center is overstaffed 100 percent of the time on Sunday and 90 percent of the time on Friday. If the staffing profile were adjusted to reach call capacity 50 percent of the time, the staffing profile could be reduced on Sundays and Fridays and increased on Mondays and Saturdays. To check these results, we could rerun the model to reflect these staff changes (see the following SkillBuilder Exercise).

FIGURE 5-13 Percentile Results and Charts for Call Center Simulation

	AB	AC	AD	AE	AF	AG	AH	AI	AJ	AK	AL	AM	AN
1	Percentiles	Monday	Tuesday	Wednesday	Thursday	Friday	Saturday	Sunday	0-7	8-11	12-13	14-17	18-24
2	0%	-25%	-36%	-35%	-35%	-47%	-15%	-65%	-54%	-67%	-83%	-76%	-60%
3	10%	3%	-12%	-12%	-12%	-28%	17%	-53%	-23%	-27%	-38%	-31%	-24%
4	20%	8%	-8%	-8%	-8%	-25%	22%	-51%	-17%	-18%	-26%	-20%	-17%
5	30%	11%	-6%	-5%	-5%	-23%	25%	-49%	-13%	-12%	-17%	-14%	-12%
6	40%	14%	-4%	-3%	-3%	-21%	28%	-48%	-9%	-6%	-9%	-8%	-8%
7	50%	16%	-2%	-1%	-1%	-20%	31%	-47%	-6%	-1%	-1%	-2%	-4%
8	60%	18%	1%	1%	1%	-18%	33%	-46%	-2%	4%	5%	5%	1%
9	70%	21%	3%	3%	3%	-16%	36%	-44%	1%	10%	13%	11%	5%
10	80%	24%	5%	6%	6%	-14%	40%	-43%	5%	16%	23%	18%	9%
11	90%	28%	9%	9%	9%	-11%	45%	-41%	11%	25%	35%	29%	17%
12	100%	54%	30%	31%	31%	7%	73%	-29%	39%	66%	84%	72%	54%

(a)

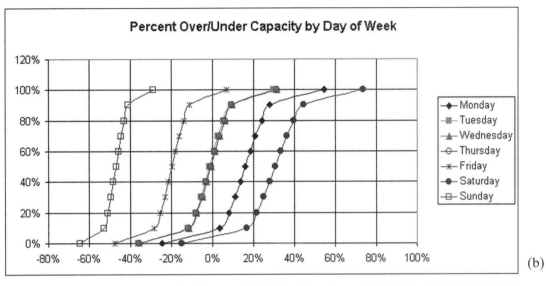

(b)

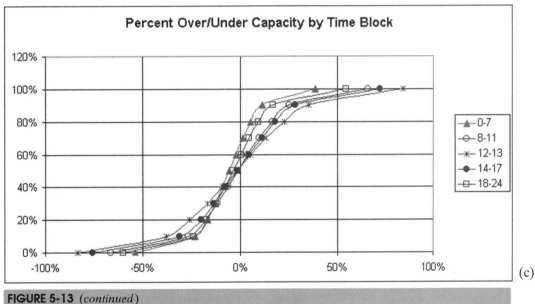

(c)

FIGURE 5-13 *(continued)*

Examination of the chart of over- and under-capacity by percentile for each time block shows that there is an approximately 50 percent confidence that the staffing is appropriate for the call volumes during these periods. The slope of the over/undercapacity percentile curve shows that there is a large variance in call volume and that, for confidence levels between 20 and 80 percent, the marginal gain in confidence that a call will be serviced has approximately a linear relation to the number of staff.

SKILLBUILDER EXERCISE

Open the file *Call Volume.xls*. Adjust the staffing volume to reach call capacity 50 percent of the time and rerun the model. How do your results compare with the base case?

Finance Applications

Financial problems are among the most appropriate for simulation because they lend themselves naturally to spreadsheet models and in reality contain a great deal of uncertainty because they extend many years into the future. Characterizing the uncertainty of financial and economic assumptions provides a better picture of the potential impacts of a capital project, strategic plan, or acquisition. We will illustrate applications to evaluating internal rate of return, retirement planning, cash budgeting, new-product development, and a recycling decision model.

RATE OF RETURN ANALYSIS[5]

Figure 5-14 shows a spreadsheet model for computing the internal rate of return of some financial investment, such as building a plant, acquiring a company, or adopting a strategic plan. Costs and revenues are assumed to start at an uncertain level and grow at uncertain rates for 5 years. Each of these assumptions is modeled by a triangular distribution with parameters given in the range B5:D8. The initial cash outflow in year 0, cell B19, is also uncertain and defined by a triangular distribution with the parameters in row 9. Using these assumptions, revenues and operating costs for the next 5 years are forecasted. The pre-tax cash flow is computed by subtracting costs from revenues. The going concern residual value (GCRV) is computed by dividing the cash flow in year 5 by the discount rate. GCRV represents the value of an annuity that throws off the year 5 cash flow ad infinitum. Because the productive life of the project, acquisition, or strategic plan is not over in year 5, and because we may want to use a 20-year spreadsheet model, we have assumed that cash flow levels off in year 5 and continues from there.

FIGURE 5-14 Internal Rate of Return Simulation Model

	A	B	C	D	E	F	G
1	**Internal Rate of Return**						
2							
3	*Assumptions*						
4		Baseline	Low	High			
5	Starting Revenue ($)	$ 18,367	$ 17,500	$ 19,100			
6	Revenue Growth (%)	10.0%	9.5%	10.50%			
7	Starting Op Cost ($)	$ 15,067	$ 14,700	$ 15,500			
8	Op Cost Growth (%)	10.0%	9.5%	10.50%			
9	Total Project Cost	$ 26,578	$ 22,500	$ 31,000			
10	Discount Rate	15%					
11							
12	*Model*						
13							
14		Year 0	Year 1	Year 2	Year 3	Year 4	Year 5
15	Revenues		$18,366.67	$20,203.33	$22,223.67	$24,446.03	$26,890.64
16	Operating Cost		$15,066.67	$16,573.33	$18,230.67	$20,053.73	$22,059.11
17	Cash Flow		$ 3,300.00	$ 3,630.00	$ 3,993.00	$ 4,392.30	$ 4,831.53
18	Going concern Residual Value						$32,210.20
19	Project Cash Flow	$(26,577.67)	$ 3,300.00	$ 3,630.00	$ 3,993.00	$ 4,392.30	$37,041.73
20							
21	Internal Rate of Return	17.7%					

[5]We thank Dr. David Hulett, Senior Vice President, Project Risk Management, International Institute for Learning (www.iil.com) for providing this example. Copyright © 2000 International Institute for Learning, Inc. Used with Permission.

The measure of success is the internal rate of return (IRR). The IRR is calculated with the Excel function from the stream of negative and positive cash flows in row 19. If the IRR is above the discount rate, here assumed to be 15 percent, the project exceeds the organization's "hurdle rate" and can go forward. If there are several projects, the one with the greatest IRR is preferred. In the baseline case, the IRR is calculated at 19.8 percent, well above the discount rate. In a risky environment, the confidence that the IRR is above 15 percent should be explored. Given the uncertainties, how likely is it that the IRR will achieve the 15 percent necessary for consideration? If the organization is a fairly conservative company, it may wish to choose between projects based on some IRR that is at least 80 percent likely to occur, or one that is only 20 percent likely to be underrun. The Crystal Ball forecast chart is shown in Figure 5-15. Although the mean IRR is 16.8 percent, the probability that the IRR will fall below the hurdle rate of 15 percent is 31.7 percent. A company that requires an 80 percent certainty will not accept this project. Of course, there is some chance that the IRR will be greater than 20 percent, which may appeal to some aggressive companies.

SKILLBUILDER EXERCISE

Open the file *Internal Rate of Return.xls*. Set the cash flows for years 1 through 5 (without the GCRV) as Crystal Ball forecasts. Run the simulation, and show and interpret your results on a trend chart.

RETIREMENT PLANNING

Forecasting income streams and rates of return of investments forms the basis for retirement planning. We will demonstrate an example of using simulation to evaluate a retirement plan for a 30-year-old individual. We will assume that the employee's current income is $30,000 per year, and that 7 percent is contributed into a retirement fund that is matched equally by the employer. Two key inputs in determining a future retirement fund balance are the individual's future income as determined by annual raises

FIGURE 5-15 IRR Forecast Chart

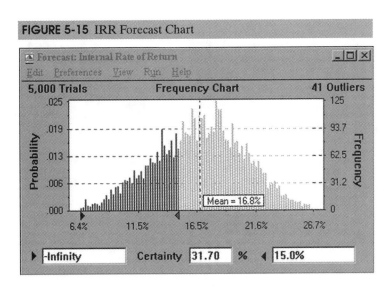

and the rate of return of the investment portfolio. We will assume that the individual's annual raise is estimated to be 4 percent and that investment returns are indexed to changes in the Standard & Poor's 500.

Figure 5-16 shows a histogram for the historical annual returns of the S&P 500 over the time period 1960–1999. If we smooth these 40 data points to form a continuous distribution by assuming a uniform distribution within each cell interval, we find an average annual return of 9.13 percent. (We note that the actual average annual return for the 40 data points is 9.3 percent; the difference is due to the fact the actual data are not truly uniformly distributed within each cell interval.) As an alternative, we may sample from the actual distribution or fit a theoretical distribution to the data, as discussed in Chapter 3. We leave it as an exercise for you to investigate the sensitivity of the results to the distributional assumptions in problem 10. Assuming an annual return of 9.13 percent, we may construct a spreadsheet and estimate a retirement fund of about $1.6 million at age 65, as shown in Figure 5-17. However, with the high volatility in market returns as seen over the past 40 years, the timing of the actual returns and influence of compounding can result in a significantly different result.

To simulate uncertainty in the annual returns, we define each year's annual return as a Crystal Ball assumption using the frequency distribution in the range F3:H14 and the Crystal Ball custom distribution. To facilitate data input for the custom distribution, you may click the *Data* button in the custom distribution dialog box and simply enter the cell reference of the data range. (See Chapter 3 for a discussion of the input options for the custom distribution.) In this case, the range is F3:H14, as shown in Figure 5-16. The result is shown in Figure 5-18. You need only do this once for the first year return (cell D8), and then use the *Copy* and *Paste Data* commands under the *Cell* menu to copy these assumptions to the other cells. (Do not use the Excel copy command as it does not copy Crystal Ball data!) The fund balance at age 65, cell E43, is defined as the forecast cell.

Figure 5-19 shows some of the Crystal Ball results for this example. Although the mean fund balance is $1.565 million, the percentiles chart shows that there is about a 30 percent chance that it will be less than $1 million. Knowing this risk might help the individual to consider other investment strategies.

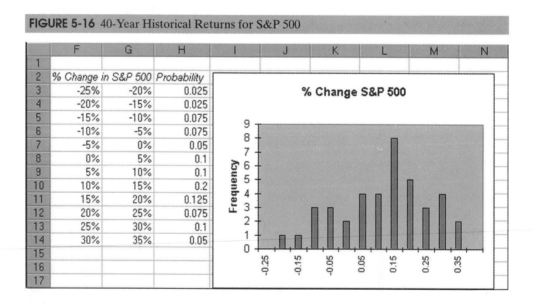

FIGURE 5-16 40-Year Historical Returns for S&P 500

	F	G	H
2	% Change in S&P 500		Probability
3	-25%	-20%	0.025
4	-20%	-15%	0.025
5	-15%	-10%	0.075
6	-10%	-5%	0.075
7	-5%	0%	0.05
8	0%	5%	0.1
9	5%	10%	0.1
10	10%	15%	0.2
11	15%	20%	0.125
12	20%	25%	0.075
13	25%	30%	0.1
14	30%	35%	0.05

	A	B	C	D	E
1	**Retirement Planning**				
2					
3	*Contribution rate (indiv. & matching)*			7%	
4	*Expected annual raise*			4%	
5					
6				Annual	
7	Age	Income	Contribution	Return	Fund Value
8	30	$30,000	$2,100	9.13%	$4,583
9	31	$31,200	$2,184	9.13%	$9,369
10	32	$32,448	$2,271	9.13%	$14,767
11	33	$33,746	$2,362	9.13%	$20,839
12	34	$35,096	$2,457	9.13%	$27,654
13	35	$36,500	$2,555	9.13%	$35,287
14	36	$37,960	$2,657	9.13%	$43,822
15	37	$39,478	$2,763	9.13%	$53,347
16	38	$41,057	$2,874	9.13%	$63,963
17	39	$42,699	$2,989	9.13%	$75,778
18	40	$44,407	$3,109	9.13%	$88,910
19	41	$46,184	$3,233	9.13%	$103,488
20	42	$48,031	$3,362	9.13%	$119,656
21	43	$49,952	$3,497	9.13%	$137,568
22	44	$51,950	$3,637	9.13%	$157,394
23	45	$54,028	$3,782	9.13%	$179,320
24	46	$56,189	$3,933	9.13%	$203,550
25	47	$58,437	$4,091	9.13%	$230,305
26	48	$60,774	$4,254	9.13%	$259,829
27	49	$63,205	$4,424	9.13%	$292,387
28	50	$65,734	$4,601	9.13%	$328,270
29	51	$68,363	$4,785	9.13%	$367,795
30	52	$71,098	$4,977	9.13%	$411,310
31	53	$73,941	$5,176	9.13%	$459,194
32	54	$76,899	$5,383	9.13%	$511,861
33	55	$79,975	$5,598	9.13%	$569,765
34	56	$83,174	$5,822	9.13%	$633,401
35	57	$86,501	$6,055	9.13%	$703,309
36	58	$89,961	$6,297	9.13%	$780,080
37	59	$93,560	$6,549	9.13%	$864,361
38	60	$97,302	$6,811	9.13%	$956,856
39	61	$101,194	$7,084	9.13%	$1,058,336
40	62	$105,242	$7,367	9.13%	$1,169,643
41	63	$109,451	$7,662	9.13%	$1,291,696
42	64	$113,829	$7,968	9.13%	$1,425,500
43	65	$118,383	$8,287	9.13%	$1,572,150

FIGURE 5-17 Retirement Planning Spreadsheet Model

FIGURE 5-18 Crystal Ball Custom Distribution

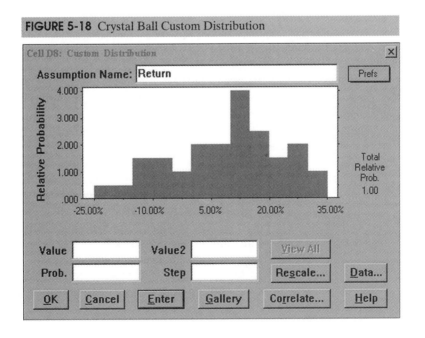

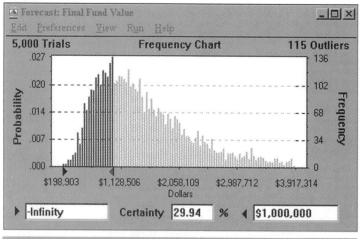

FIGURE 5-19 Retirement Planning Simulation Results

SKILLBUILDER EXERCISE

Open the file *Retirement Planning.xls*. Modify the spreadsheet to include the assumption that the annual raise is uncertain. Assume a triangular distribution with a minimum of 1 percent, most likely value of 4 percent, and maximum value of 5 percent, and use Crystal Ball to find the distribution of the ending retirement fund balance under this assumption. How do the results compare with the base case?

CASH BUDGETING[6]

Cash budgeting is the process of projecting and summarizing a company's cash inflows and outflows expected during a planning horizon, usually 6 to 12 months. The cash budget also shows the monthly cash balances and any short-term borrowing used to cover cash shortfalls. Positive cash flows can increase cash, reduce outstanding loans, or be used elsewhere in the business; negative cash flows can reduce cash available or be offset with additional borrowing. Most cash budgets are based on sales forecasts. With the inherent uncertainty in such forecasts, Monte-Carlo simulation is an appropriate tool to analyze cash budgets.

Figure 5-20 shows an example of a cash budget spreadsheet. The budget begins in April (thus, sales for April and subsequent months are uncertain). These are assumed to be normally distributed with a standard deviation of 10 percent of the mean. In addition, we assume that sales in adjacent months are correlated with one another, with a correlation coefficient of 0.6. On average, 20 percent of sales is collected in the month of sale, 50 percent in the month following the sale, and 30 percent in the second month following the sale. However, these figures are uncertain, so a uniform distribution is used to model the first two values, (15 percent–20 percent and 40 percent–50 percent, respectively) with the assumption that all remaining revenues are collected in the sec-

---

[6]Adapted from Douglas R. Emery, John D. Finnerty, and John D. Stowe, *Principles of Financial Management,* Upper Saddle River, NJ: Prentice Hall, 1998, pp. 652–654.

	A	B	C	D	E	F	G	H	I	J	K
1	**Cash Budgeting**										
2	Desired Minimum Balance	$ 100,000									
3			February	March	April	May	June	July	August	September	October
4		Sales	$400,000	$500,000	$600,000	$700,000	$800,000	$800,000	$700,000	$600,000	$500,000
5	*Cash Receipts*										
6	Collections (current)	20%			$120,000	$140,000	$160,000	$160,000	$140,000	$120,000	
7	Collections (previous month)	50%			$250,000	$300,000	$350,000	$400,000	$400,000	$350,000	
8	Collections (2nd month previous)	30%			$120,000	$150,000	$180,000	$210,000	$240,000	$240,000	
9	Total Cash Receipts				$490,000	$590,000	$690,000	$770,000	$780,000	$710,000	
10											
11	*Cash Disbursements*										
12	Purchases				$420,000	$480,000	$480,000	$420,000	$360,000	$300,000	
13	Wages and Salaries				$ 72,000	$ 84,000	$ 96,000	$ 96,000	$ 84,000	$ 72,000	
14	Rent				$ 10,000	$ 10,000	$ 10,000	$ 10,000	$ 10,000	$ 10,000	
15	Cash Operating Expenses				$ 30,000	$ 30,000	$ 30,000	$ 30,000	$ 25,000	$ 25,000	
16	Tax Installments				$ 20,000			$ 30,000			
17	Capital Expenditure						$150,000				
18	Mortgage Payment					$ 60,000					
19	Total Cash Disbursements				$552,000	$664,000	$766,000	$586,000	$479,000	$407,000	
20											
21	*Ending Cash Balance*										
22	Net Cash Flow				$ (62,000)	$ (74,000)	$ (76,000)	$184,000	$301,000	$303,000	
23	Beginning Cash Balance				$150,000	$100,000	$100,000	$100,000	$122,000	$423,000	
24	Available Balance				$ 88,000	$ 26,000	$ 24,000	$284,000	$423,000	$726,000	
25	Monthly Borrowing				$ 12,000	$ 74,000	$ 76,000	$ -	$ -	$ -	
26	Monthly Repayment				$ -	$ -	$ -	$162,000	$ -	$ -	
27	Ending Cash Balance			$150,000	$100,000	$100,000	$100,000	$122,000	$423,000	$726,000	
28	Cumulative Loan Balance			$ -	$ 12,000	$ 86,000	$162,000	$ -	$ -	$ -	
29											

FIGURE 5-20 Cash Budgeting Model

ond month following the sale. Purchases are 60 percent of sales and are paid for 1 month prior to the sale. Wages and salaries are 12 percent of sales and are paid in the same month as the sale. Rent of $10,000 is paid each month. Additional cash operating expenses of $30,000 per month will be incurred for April through July, decreasing to $25,000 for August and September. Tax payments of $20,000 and $30,000 are expected in April and July, respectively. A capital expenditure of $150,000 will occur in June, and the company has a mortgage payment of $60,000 in May. The cash balance at the end of March is $150,000, and managers want to maintain a minimum balance of $100,000 at all times. The company will borrow the amounts necessary to ensure that the minimum balance is achieved. Any cash above the minimum will be used to pay off any loan balance until it is eliminated. The available cash balances in row 24 of the spreadsheet are the Crystal Ball forecast cells.

The Crystal Ball Correlation Matrix tool is used to specify the correlations between sales assumptions. Selecting this tool from the *CBTools* menu brings up a series of dialog boxes that allow you to specify what assumptions to correlate and how to input the data. For this example, we chose to include a correlation matrix in the spreadsheet that is linked to the assumptions. This is shown in Figure 5-21, and it allows you to change the correlations to investigate different scenarios. Figure 5-22 shows the trend chart for the monthly cash balances. We see that there is a high likelihood that cash balances will be negative for the first 3 months before increasing. Viewing the forecast chart and statistics for individual months will provide the details on the distribution of cash balances and the likelihood of requiring loans. For example, the forecast chart shows that there is only a probability of 0.11 that the balance will exceed the minimum of $100,000 and not require an additional loan. (If you want to check this, open the file *Cash Budget.run*, which will restore these results.)

	Sales April	Sales May	Sales June	Sales July	Sales August	Sales September	October
31 Correlation Matrix							
32 Sales April	1.000	0.600					
33 Sales May		1.000	0.600				
34 Sales June			1.000	0.600			
35 Sales July				1.000	0.600		
36 Sales August					1.000	0.600	
37 Sales September						1.000	0.600
38 October							1.000

FIGURE 5-21 Correlation Matrix for Cash Budgeting Model

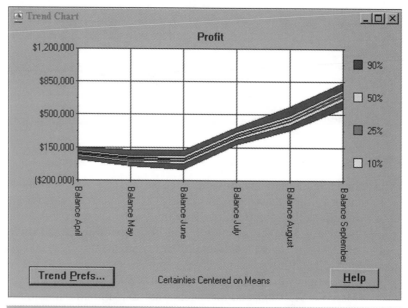

FIGURE 5-22 Trend Chart for Monthly Cash Balances

SKILLBUILDER EXERCISE

Open the file *Cash Budget.xls*. In the Run Preferences Options dialog box, check the box to turn off correlations, and rerun the model. What effect does not correlating the sales data have on the forecasts?

NEW-PRODUCT DEVELOPMENT[7]

Vision Research has completed preliminary development of a new drug, code-named *ClearView*, that corrects nearsightedness. This revolutionary new product could be completely developed and tested in time for release next year if the FDA approves the product.

---

[7]Adapted from *Crystal Ball User Manual*, Copyright 1988–2000, Decisioneering, Inc.

Although the drug works well for some patients, the overall success rate is marginal, and Vision Research is uncertain whether the FDA will approve the product. With millions of dollars at risk, the company has to assess the decision of continuing to proceed to develop and market this drug versus scrapping the project altogether.

So far, Vision Research has spent $10 million developing ClearView and expects to spend an additional $3 to $5 million to test it based on the costs of previous tests. The exact amount is uncertain, and the company cannot estimate a single most likely value, so a uniform distribution is appropriate. The company also plans on spending a sizable amount marketing ClearView if the FDA approves it. They expect to hire a large sales force and kick off an extensive advertising campaign to educate the public about this product. Including sales commissions and advertising costs, Vision Research expects to spend between $12 and $18 million, although the marketing executives believe that the most likely case will be $16 million.

Given these values, a triangular distribution would seem appropriate to model these costs. However, before the FDA will approve ClearView, Vision Research must conduct a controlled test on a sample of 100 patients for 1 year. The company expects that the FDA will grant an approval if ClearView completely corrects the nearsightedness of 20 or more of these patients without any significant side effects. Preliminary testing has showed a success rate of about 25 percent, which the company feels is very encouraging.

Vision Research has determined that nearsightedness afflicts nearly 40 million people in the United States, and an additional 0 to 5 percent of these people will develop this condition during the year in which ClearView is tested. The marketing department has recently learned that there is a 25 percent chance that a competing product will be released on the market soon. This product would decrease ClearView's potential market by 5 to 15 percent. The marketing department also estimates that Vision Research's eventual share of the total market for the product will be about 8 percent, but given the interest shown in preliminary testing, they do not believe the market will be below 5 percent. Analysts have determined that a truncated normal distribution with a mean of 8 percent and standard deviation of 2 percent would suffice to model this assumption. The estimated profit per customer is $12.00.

Figure 5-23 shows a spreadsheet designed to model the factors that Vision Research has identified that affect the profitability of this product development effort. Most of the distributions to use in a Crystal Ball simulation model have been described in the problem statement; however, how to model the test results and market growth rate requires further discussion. For the test results, we note that preliminary testing has shown a cure rate of 25 percent. Therefore, the number of patients cured in a sample of 100 would have a binomial distribution with $n = 100$ and $p = 0.25$. Figure 5-24 shows the Binomial Distribution dialog box from the Crystal Ball Distribution Gallery for cell B10. The growth rate of nearsightedness, cell B15, cannot be described by any standard probability distribution. The distribution can be described as:

Uniform between 0 percent and 5 percent with probability 0.75
Uniform between −15 percent and −5 percent with probability 0.25

However, it can easily be modeled using Crystal Ball's custom distribution. In the dialog box, the input boxes *Value* and *Value2* refer to the continuous range of the random variable having a probability specified in the *Prob.* input box. For example, you would input 0 percent in the *Value* field, 5 percent in the *Value2* field, and 0.75 in the *Prob.* field. After inputting these values, click "Enter" and then input the second uniform range and its associated probability. The result is shown in Figure 5-25.

	A	B
1	**Vision Research**	
2		
3	*Costs (in millions):*	
4	Development Cost of ClearView to Date	$10.0
5	Testing Costs	$4.0
6	Marketing Costs	$15.3
7	Total Costs	$29.3
8		
9	*Drug Test (sample of 100 patients):*	
10	Patients Cured	25
11	FDA Approved if 20 or More Patients Cured	TRUE
12		
13	*Market Study (in millions):*	
14	Persons in U.S. with Nearsightedness Today	40.0
15	Growth Rate of Nearsightedness	-0.63%
16	Persons with Nearsightedness After One Year	39.8
17		
18	*Gross Profit on Dosages Sold:*	
19	Market Penetration	8.28%
20	Profit Per Customer in Dollars	$12.00
21	Gross Profit if Approved (MM)	$39.5
22		
23	*Net Profit (MM)*	$10.1

FIGURE 5-23 Spreadsheet Model for Vision Research

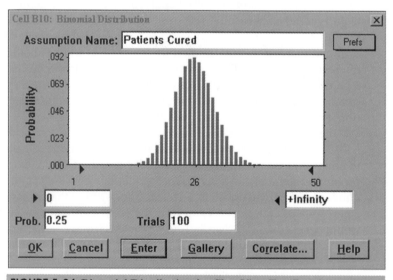

FIGURE 5-24 Binomial Distribution for ClearView Test Results

The forecast charts for this model are shown in Figure 5-26. The odd shape of the net profit distribution results from the way in which the net profit is calculated. If you examine the formula in cell B23, you see that if the FDA approves the drug (cell B11 is *TRUE*), the net profit is calculated by subtracting total costs from the gross profit. This characterizes the large portion of the distribution. However, if the FDA does not approve the drug, then the net profit is simply the development and testing costs incurred to date and results in the peaked portion of the distribution on the left. From the net profit forecast, we can determine that there is a 79 percent chance of a positive net profit

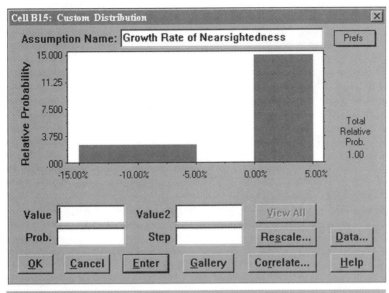

FIGURE 5-25 Custom Distribution for Growth Rate of Nearsightedness

or, correspondingly, a 21 percent chance of a net loss. In addition, the forecast charts show that a two-thirds chance exists of realizing a net profit of at least $4 million. We would expect that Vision Research management would decide to proceed with the project based on these analyses.

SKILLBUILDER EXERCISE

Open the file *Vision Research.xls*. Choose the *Sensitivity* option in the *Run Preferences* box. Rerun the model to study the sensitivity of the assumptions in the model. Also, fit an overlay chart to the gross profit if approved forecast.

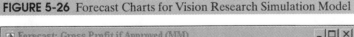

FIGURE 5-26 Forecast Charts for Vision Research Simulation Model

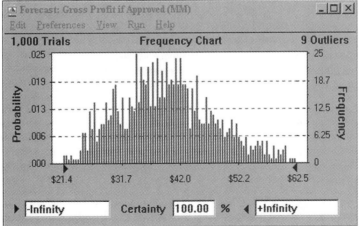

(a)

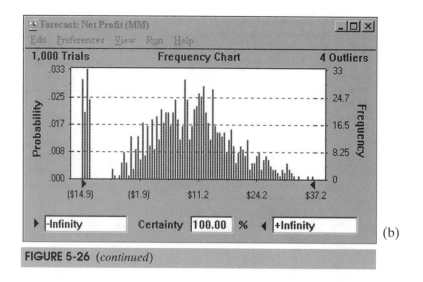

FIGURE 5-26 (*continued*)

RECYCLING DECISION MODEL

Demand for capacity at a municipal garbage recycling center has grown tremendously, requiring a new incinerating system. Three options to fuel this system are natural gas (NG), bunker oil (BO), and wood (W). Each of the three fuels involves different investment cost, operating expense rates, and material cost rates. Operating expenses are functions of the quantity of garbage processed, which is expected to be 1 million tons during the first year and grow at a rate of 10 percent per year. Material cost is also calculated on the basis of a ton of garbage. Material cost for BO and NG depends on the price of a barrel of oil, P. Specific cost parameters (all in millions of dollars) are

		NG	*BO*	*W*
Investment	Year 1	$6	$6	$4
	Year 2	0	0	$2
Operating expense rate/ton		$0.50	$0.60	$1.00
Material expense/ton		$(0.30 + 0.01P)	$0.02P	$0.10

A new system is expected to last for 20 years, and the objective is to identify the system with the lowest net present cost.

There are two major sources of uncertainty. First is demand growth. The expected rate of growth is 10 percent, but it could easily range from 5 to 15 percent. Management estimates that the growth rate is lognormally distributed with a mean of 10 percent and a standard deviation of 2 percent. Second, the future price of a barrel of oil (P) is highly uncertain. In 1982, it was about $3 per barrel. In the early 1980s, the price rose to about $35 per barrel. Throughout the 1990s, it fluctuated in the $15 to $20 range before rising again in the year 2000. Management estimates future prices to be normally distributed with a mean of $20 per barrel and a standard deviation of $4. A third source of potential uncertainty is the cost of capital, which we assume to be normally distributed with a mean of 12 percent and a standard deviation of 1 percent.

This is a typical cash flow decision problem, usually analyzed with ordinary financial spreadsheet models. An Excel model for this problem is given in Figure 5-27. Yearly

	A	B	C	D	E	F	G	H	I
1	**Garbage Recycling Model**								
2									
3	*Demand growth rate*	10.00%							
4	*Oil price ($/barrel)*	$20.00							
5	*Cost of capital*	12.00%							
6									
7	**Year**	0	1	2	3	4	5		
8	Demand (millions of tons)		1.000	1.100	1.210	1.331	1.464		
9									
10	*NATURAL GAS*								Best
11	Investment ($millions)	6							Option
12	Operating cost ($millions)		0.500	0.550	0.605	0.666	0.732	Net Present	
13	Material cost ($millions)		0.500	0.550	0.605	0.666	0.732	Value	
14	Total cost ($millions)	6.000	1.000	1.100	1.210	1.331	1.464	$10.308	1
15									
16	*BUNKER OIL*								
17	Investment ($millions)	6							
18	Operating cost ($millions)		0.600	0.660	0.726	0.799	0.878	Net Present	
19	Material cost ($millions)		0.400	0.440	0.484	0.532	0.586	Value	
20	Total cost ($millions)	6.000	1.000	1.100	1.210	1.331	1.464	$10.308	1
21									
22	*WOOD*								
23	Investment ($millions)	4	2						
24	Operating cost ($millions)		1.000	1.100	1.210	1.331	1.464	Net Present	
25	Material cost ($millions)		0.100	0.110	0.121	0.133	0.146	Value	
26	Total cost ($millions)	4.000	3.100	1.210	1.331	1.464	1.611	$10.524	0

FIGURE 5-27 Recycling Model Spreadsheet

demands are given in row 8, increasing by the growth rate in cell B3. For each alternative, we compute the total cost for each year. The net present value (NPV) of the annual cash flows is computed using the *NPV* function after adding the initial investment in year 0. In column H, we use logical *IF* and *AND* functions to determine the lowest cost option, designated by a 1, by checking if the discounted cost is less than or equal to the other two alternatives. At a growth rate of 10 percent per year and an expected price of $20 per barrel, there is a virtual tie between natural gas and bunker oil. The wood system is a bit more expensive.

In a traditional analysis, further insight might be gained by developing data tables to examine the impact of changes in the price and growth rate. The results of such an analysis, in millions of dollars of net present cost, are shown in Figure 5-28, with the best cost in bold. From this figure, it is clear that for any growth rate scenario, the bunker oil option is best at low prices of oil; the natural gas option is best at a price over $20 per barrel; and for prices over $25 per barrel, the wood option is best. We also observe that wood is relatively better at lower demand growth rates. However, the uncertainty in the assumptions might help to provide a clearer decision.

Figure 5-29 shows the key results for a Crystal Ball simulation of 1,000 trials using Latin Hypercube sampling. All net present value and best option cells are defined as forecast cells. The bunker oil system was the best option 50 percent of the time; natural gas, 39 percent of the time; and wood, only 11 percent of the time. However, although it appears that the bunker oil option might be the best choice, the forecast charts and statistics show that the standard deviation of net present cost is also larger than the other alternatives, indicating a higher level of risk. Therefore, the center managers might wish to seek better information about the future prices of oil and growth rate of demand and perform further analyses.

	K	L	M	N	O
1	Growth rate				
2	5%	Price	Nat. Gas	Bunker Oil	Wood
3		$ 5	$ 9.35	$ 8.76	$ 10.12
4		$ 10	$ 9.55	$ 9.15	$ 10.12
5		$ 15	$ 9.74	$ 9.55	$ 10.12
6		$ 20	$ 9.94	$ 9.94	$ 10.12
7		$ 25	$ 10.14	$ 10.33	$ 10.12
8		$ 30	$ 10.33	$ 10.73	$ 10.12
9		$ 35	$ 10.53	$ 11.12	$ 10.12
10	Growth rate				
11	10%	Price	Nat. Gas	Bunker Oil	Wood
12		$ 5	$ 9.66	$ 9.02	$ 10.52
13		$ 10	$ 9.88	$ 9.45	$ 10.52
14		$ 15	$ 10.09	$ 9.88	$ 10.52
15		$ 20	$ 10.31	$ 10.31	$ 10.52
16		$ 25	$ 10.52	$ 10.74	$ 10.52
17		$ 30	$ 10.74	$ 11.17	$ 10.52
18		$ 35	$ 10.95	$ 11.60	$ 10.52
19	Growth rate				
20	15%	Price	Nat. Gas	Bunker Oil	Wood
21		$ 5	$ 10.00	$ 9.30	$ 10.97
22		$ 10	$ 10.24	$ 9.77	$ 10.97
23		$ 15	$ 10.47	$ 10.24	$ 10.97
24		$ 20	$ 10.71	$ 10.71	$ 10.97
25		$ 25	$ 10.95	$ 11.18	$ 10.97
26		$ 30	$ 11.18	$ 11.65	$ 10.97
27		$ 35	$ 11.42	$ 12.12	$ 10.97

FIGURE 5-28 Data Table Analysis of Growth Rate and Price Scenarios

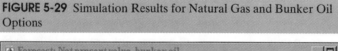

SKILLBUILDER EXERCISE

Open the file *Garbage Recycling.xls*. Suppose that the near-term forecasts of the price of oil stabilize so that the standard deviation is decreased to $2 per barrel. Make this change to the model, and modify the spreadsheet to extract the percent of time that each option is best using the *CB.GetForeStatFN* function. What impact does the change in standard deviation have on the results?

FIGURE 5-29 Simulation Results for Natural Gas and Bunker Oil Options

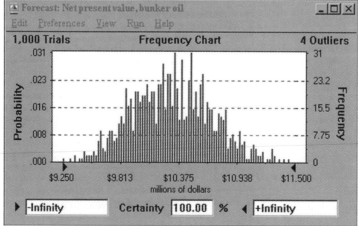

(a)

Forecast: Net present value, bunker oil

Edit   Preferences   View   Run   Help

Cell H20   **Statistics**

Statistic	Value
Trials	1,000
Mean	$10.312
Median	$10.303
Mode	---
Standard Deviation	$0.393
Variance	$0.154
Skewness	0.10
Kurtosis	2.82
Coeff. of Variability	0.04
Range Minimum	$9.130
Range Maximum	$11.632
Range Width	$2.502
Mean Std. Error	$0.012

(b)

Forecast: Net present value, natural gas

Edit   Preferences   View   Run   Help

1,000 Trials   **Frequency Chart**   **7 Outliers**

-Infinity   Certainty 100.00 %   +Infinity

(c)

Forecast: Net present value, natural gas

Edit   Preferences   View   Run   Help

Cell H14   **Statistics**

Statistic	Value
Trials	1,000
Mean	$10.313
Median	$10.298
Mode	---
Standard Deviation	$0.257
Variance	$0.066
Skewness	0.22
Kurtosis	2.94
Coeff. of Variability	0.02
Range Minimum	$9.503
Range Maximum	$11.134
Range Width	$1.631
Mean Std. Error	$0.008

(d)

FIGURE 5-29 (*continued*)

Marketing Applications

Risk analysis has numerous applications in marketing, because consumer data by its very nature exhibits considerable uncertainty and variability. In this section, we present three examples dealing with sales projection, evaluation of new-product introduction, and distribution strategies.

SALES PROJECTION

Many marketing models and analyses require projections of future sales. Generally, the rates of growth of sales are uncertain, although estimates for new products might increase over time and be accompanied by increasing uncertainty the further into the future one goes. For planning purposes, Monte-Carlo simulation may be used to project market demand for some future time period.

A simple model that can be used is shown in Figure 5-30. The mean sales growth projections might be estimated by using a deterministic forecasting model, and this is depicted in the Excel chart. (See Chapter 10 for a discussion of simulation and forecasting.) Each growth value is defined as a Crystal Ball assumption that is normally distributed with the mean and standard deviation shown in the worksheet. Because all these assumptions have the same distributional form, cell references can be used in the input fields to facilitate data entry, as shown in Figure 5-31. The ending sales for each quarter are defined as forecasts. Figure 5-32 shows the results for the sales projection for the last quarter. The trend chart shown in Figure 5-33 illustrates the expected variability in sales over time.

SKILLBUILDER EXERCISE

Open the file *Sales Projection.xls.* Use the Crystal Ball tools to correlate successive assumptions using a correlation coefficient of 0.7. Does this make a significant difference in the results?

MARKETING ANALYSIS FOR PHARMACEUTICALS

This model is a simplified version of several marketing models developed for different pharmaceutical companies by Hervé Thiriez, a management professor at Groupe HEC and CEO of Logma SA (a consulting company). This model measures the net present value, over 10 years, of the introduction of a new pharmaceutical product, based on key assumptions of when the product is introduced and whether any other competing products have beaten it to the market.

The model is shown in Figure 5-34. A Crystal Ball custom distribution defines the first year of production (cell B4), either 1, 2, or 3, with probabilities 0.2, 0.7, and 0.1, respectively. Another custom distribution indicates the position of the product when it is introduced (that is, whether it was the first, second or third product of its kind on the market, with probabilities 0.5, 0.3, and 0.2, respectively). These two assumptions are correlated (using a correlation coefficient of 0.8); the earlier the product comes out, the better the chance it is first on the market.

The table in rows 8 through 10 shows the expected market share when the product is first, second, or third on the market, depending on whether it is its first, second, or tenth year of production. The remainder of the spreadsheet calculates the expected number of sales based on the market size and growth rate, the yearly profit, and the net

	A	B	C	D	E	F	G
1	**Quarterly Sales Projections**						
2							
3	*Quarter*	*Starting Sales*	*Growth*	*Std Dev*	*Ending Sales*		
4	First Qtr 2000	$12,200,000	3.00%	2.00%	$12,566,000		
5	Second Qtr 2000	$12,566,000	2.50%	2.00%	$12,880,150		
6	Third Qtr 2000	$12,880,150	3.50%	2.00%	$13,330,955		
7	Fourth Qtr 2000	$13,330,955	3.50%	2.00%	$13,797,539		
8	First Qtr 2001	$13,797,539	3.00%	2.00%	$14,211,465		
9	Second Qtr 2001	$14,211,465	4.00%	2.00%	$14,779,923		
10	Third Qtr 2001	$14,779,923	4.50%	2.00%	$15,445,020		
11	Fourth Qtr 2001	$15,445,020	5.50%	3.00%	$16,294,496		
12	First Qtr 2002	$16,294,496	4.50%	3.00%	$17,027,748		
13	Second Qtr 2002	$17,027,748	5.00%	3.00%	$17,879,136		
14	Third Qtr 2002	$17,879,136	6.00%	3.00%	$18,951,884		
15	Fourth Qtr 2002	$18,951,884	5.50%	4.00%	$19,994,238		
16							

Sales Growth

(line chart: Growth % by Quarter, ranging from 0.00% to 7.00%, plotting values: First Qtr 2000 3.00%, Second Qtr 2000 2.50%, Third Qtr 2000 3.50%, Fourth Qtr 2000 3.50%, First Qtr 2001 3.00%, Second Qtr 2001 4.00%, Third Qtr 2001 4.50%, Fourth Qtr 2001 5.50%, First Qtr 2002 4.50%, Second Qtr 2002 5.00%, Third Qtr 2002 6.00%, Fourth Qtr 2002 5.50%)

Quarter

FIGURE 5-30 Sales Projection Model

FIGURE 5-31 Defining Cell Reference Inputs for the Normal Distribution

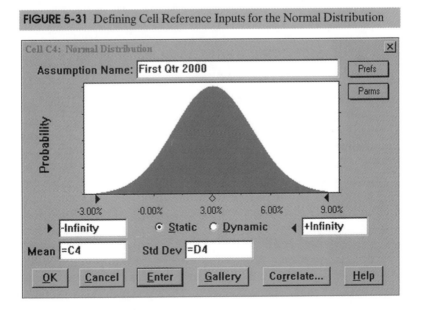

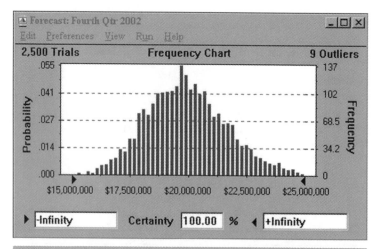

FIGURE 5-32 Forecast Chart for Final Sales Projection

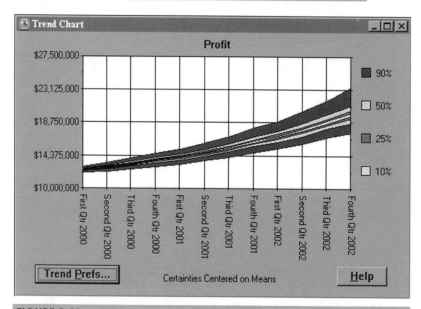

FIGURE 5-33 Trend Chart of Sales Projections

present value. The market growth rates follow a normal distribution with an average of 2 percent and a standard deviation of 0.5 percent; quite logically, they are all positively correlated with a correlation coefficient of 0.7. The market share values in row 13 change depending on the position in the market assumption during the simulation. The forecast chart (see Figure 5-35) shows three distinct distributions that depend on the initial position. These results show that the product's positioning in the market has a dramatic effect on profits. The sensitivity chart (see Figure 5-36) also shows that initial position has the greatest impact on the NPV as measured by rank correlation.

ANALYSIS OF DISTRIBUTION STRATEGIES

Izzy Rizzy's Novelties manufacturers a variety of novelties such as magic tricks and Halloween items. Most items are distributed directly through wholesalers. However, one of the senior managers believes that a new product has the potential to become a national

	A	B	C	D	E	F	G	H	I	J	K
1	**Pharmaceutical Marketing Model**										
2											
3											
4	First production year	2		Initial market size	10,000,000						
5	Position in market	1		Unit profit	$ 5.00						
6											
7	Year of production	1	2	3	4					9	10
8	Market share if product is out 1st	5.0%	12.0%	15.0%	18.0%					21.0%	21.0%
9	Market share if product is out 2nd	3.0%	5.0%	7.0%	9.0%					9.0%	9.0%
10	Market share if product is out 3rd	2.0%	2.0%	2.0%	2.0%					2.0%	2.0%
11											
12		Year 1	Year 2	Year 3	Year 4					Year 9	Year 10
13	Market share		5.0 %	12.0 %	15.0 %					21.0 %	21.0 %
14	Market growth rate	2.0 %	2.0 %	2.0 %	2.0 %					2.0 %	2.0 %
15	Market size	10,000,000	10,200,000	10,404,000	10,612,080					11,716,594	11,950,926
16	Share of market		510,000	1,248,480	1,591,812					2,460,485	2,509,694
17	Profit		$ 2,550,000	$ 6,242,400	$ 7,959,060					$ 12,302,424	$ 12,548,472
18											
19	Interest rate	10%									
20	Net present value of profit	$ 50,891,253									

FIGURE 5-34 Pharmaceutical Marketing Simulation Model

fad and that it would be wise to market it directly to retail stores. The accounting department conducted an analysis of each strategy to determine the estimated production and marketing costs, anticipated prices, and estimated demand. Because the company can directly control distribution to retail stores under the alternative strategy, it feels that the sales potential is significantly greater than distributing to wholesalers. Specifically, the mean demand for the wholesale strategy is estimated to be 400,000 units; that

FIGURE 5-35 Forecast Chart for Net Present Value of Profit

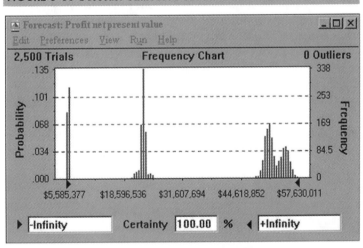

SKILLBUILDER EXERCISE

Open the file *Pharmaceutical Marketing Model.xls.* Use the Crystal Ball *Freeze Assumption* feature to fix the first year of production at 1, 2, and 3 (i.e., you will run three separate simulations). What do the results tell you about the market potential as measured by net present value when you know the first year of production?

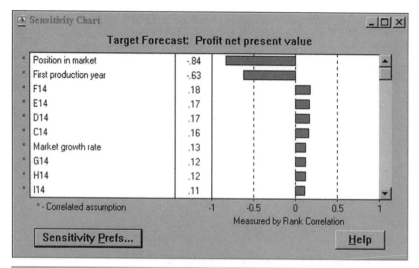

FIGURE 5-36 Sensitivity Chart for Marketing Model Assumptions

for the retail strategy is 600,000 units. Both demand distributions are normally distributed with standard deviations equal to 10 percent of the mean. A spreadsheet analysis of each strategy using the estimated mean demands is shown in Figure 5-37.

Although the retail strategy yields a higher net profit for the average estimated demand, simple what-if analysis shows that if the demand varies by as much as 20 percent of the mean, the retail strategy has a significant downside risk (as well as upside potential). For example, if the demand is lower than the mean by 20 percent, the wholesale strategy still nets a $9,200 profit while the retail strategy would incur a loss of $20,400. Thus, Monte-Carlo simulation would help to assess this risk more accurately.

Figure 5-38 shows the results of 3,000 trials using Crystal Ball. Specifically, we see that the retail strategy has about an 8.5 percent chance of incurring a net loss while this is virtually zero for the wholesale strategy. However, the retail strategy has about a 50 percent chance of realizing a net profit that exceeds 42,000; for the wholesale strategy, this probability is only about 25 percent. We can make similar comparisons by analyzing the percentile report. What decision do you think the company should make?

FIGURE 5-37 Distribution Strategy Spreadsheet Model

	A	B	C	D	E
1	**Izzy Rizzy's Novelties**				
2					
3	*Market through wholesalers*			*Market directly to retailers*	
4	Fixed production costs	$ 50,000		Fixed production costs	$ 50,000
5	Per unit variable production costs	$ 0.20		Per unit variable production costs	$ 0.20
6	Fixed marketing costs	$ 40,000		Fixed marketing costs	$ 220,000
7	Per unit variable marketing costs	$ 0.04		Per unit variable marketing costs	$ 0.08
8	Price charged to wholesaler	$ 0.55		Manufacturer's selling price	$ 0.80
9					
10	Total fixed costs	$ 90,000		Total fixed costs	$ 270,000
11	Total variable cost/unit	$ 0.24		Total variable cost/unit	$ 0.28
12	Sales quantity	400,000		Sales quantity	600,000
13	Net profit	$ 34,000.00		Net profit	$ 42,000.00

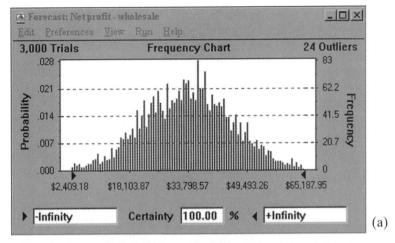

(a)

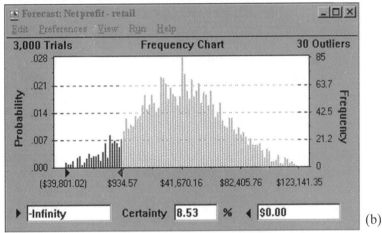

(b)

FIGURE 5-38 Forecast Charts for Alternative Distribution Strategies

---

SKILLBUILDER EXERCISE

Restore the Crystal Ball run file *Distribution Stategies.run.* This will recover the run from which the figures in the previous example were taken. Use the Crystal Ball *Run* menu option *Create Report* to create a Crystal Ball report for this problem. Edit the Excel workbook to put the report in language understandable to the marketing manager as if you were going to use it as an attachment to a formal memo. You may wish, for example, to copy and paste other forecast chart scenarios that show relevant analyses.

Engineering Applications

Risk is a critical factor in many engineering decisions, and many types of engineering problems can be modeled and analyzed using Monte-Carlo simulation. Understanding the applications of risk analysis to engineering problems is important to students of business, as many engineering analyses have economic, legal, or other business consequences.

Likewise, engineering students should not only know how to apply simulation and risk analysis to the problems they face, but they also need to understand the business implications of their design decisions. In this section, we present three applications of Crystal Ball to engineering problems.

RELIABILITY OF A SPRING

In this example, a design engineer is given the task of choosing the best material to use for a helical spring. The best material is the one that will give the spring the highest likelihood of withstanding its peak requirements. This is measured by the spring's reliability factor. Newer practices in design have led to the use of probabilistic methods for computing reliability factors. That is, the variables that affect a component's characteristics (e.g., strength and stress in this case) are assigned probability distributions. The use of probability distributions allows the engineer to more accurately reflect the properties of the materials. Because mathematical operations with these distributions are usually difficult, Monte-Carlo simulation is frequently used to compute the range of possible values for the reliability factor.

Figure 5-39 shows a spreadsheet model for this problem. In the forecast cells B18:B20, a reliability factor is computed for three different materials under consideration. These formulas are simply the ratio of material strengths to the peak shearing stress of the spring, which is a function of the stress parameters in cells B4:B8. Values greater than 1 indicate reliability; values less than 1 indicate unreliability. Because of the variability in the component characteristics, the reliability factors are uncertain.

Figure 5-40 displays the forecast results in an overlay chart. Although there is considerable variation in each individual forecast, the standard deviations are similar. The overlay chart shows that material 3 has the highest reliability compared with the other two materials.

FIGURE 5-39 Spring Reliability Simulation Model

	A	B	C	D	E
1	**Reliability of a Helical Spring**				
2					
3	**Stress Parameters:**				
4	Wire Diameter, in.	0.1500			
5	Coil Diameter, in.	0.8000			
6	Shear Modulus of Elasticity, psi	1.15E+07			
7	Number of Coils	20			
8	Deflection, in.	0.95			
9					
10	**Peak Shearing Stress, psi**	44,573			
11					
12	**Steel Strength (from Vendor)**				
13	Material 1 Strength	50,000			
14	Material 2 Strength	50,333			
15	Material 3 Strength	51,783			
16					
17	**Reliability**				
18	Material 1 Reliability	1.12			
19	Material 2 Reliability	1.13			
20	Material 3 Reliability	1.16			

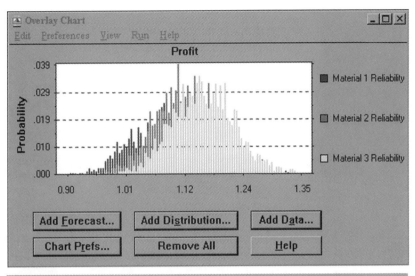

FIGURE 5-40 Overlay Chart for Spring Reliability Results

SKILLBUILDER EXERCISE

Open the file *Spring Reliability.xls*. Use the Crystal Ball Tornado Chart tool to illustrate the impact of the assumptions in the model.

RISK ASSESSMENT OF TOXIC WASTE

A small community gets its water from wells that tap into an old, large aquifer. Recently, an environmental impact study found toxic contamination in the groundwater due to improperly disposed chemicals from a nearby manufacturing plant. Estimates are available of the contamination levels of the various chemicals. Each contaminant's concentration in the water is measured in micrograms per liter. The cancer potency factor (CPF) for each chemical is uncertain. The CPF is the magnitude of the impact the chemical exhibits on humans; the higher the cancer potency factor, the more harmful the chemical is. The population risk assessment must account for the variability of body weights and volume of water consumed by the individuals in the community per day. All these factors lead to the following equation for population risk:

Population risk = CPF*contaminant concentrations*water consumed per day/(body weight*conversion factor)

Figure 5-41 shows a Crystal Ball model for this situation. Body weight is assumed to be normally distributed with a mean of 70 kg and standard deviation of 10 kg. The volume of water in liters/day consumed per day is also normally distributed with a mean of 2 and standard deviation of 1, but it is truncated below zero. The contaminant concentrations are assumed to be triangular with $a = 80$, $m = 110$, and $b = 120$ micrograms/liter. Finally, the CPF is assumed to be lognormal, with a mean of 0.03 and standard deviation of 0.02. The population risk is defined as the forecast. The forecast chart shown in Figure 5-42 shows that the population risk is highly skewed with a mean of 0.0000931. The mean can be displayed on the forecast chart from the *Preferences . . . Chart* option in the forecast chart menu bar.

	A	B	C
1	Risk Assessment at a Toxic Waste Site		
2			
3			
4	Body Weight	70	kilograms
5	Volume of Water per Day	2.05	liters/day
6	Concentration of Contaminant	103.33	micrograms/liter
7	Cancer Potency Factor	0.030	inverse mg/kg/d
8			
9	Risk Assessment	9.10E-05	

FIGURE 5-41 Simulation Model for Toxic Waste Risk Assessment

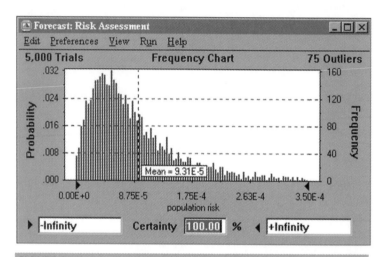

FIGURE 5-42 Distribution of Population Risk

ELECTRICAL CIRCUIT ANALYSIS[8]

The current of a simple electrical circuit is computed as

$$I = V/R$$

where V = voltage and R = resistance. A spreadsheet model is shown in Figure 5-43. Due to random fluctuations in performance, the voltage and resistance may not be constant, resulting in uncertainty of the current. For example, suppose that the voltage is normally distributed with a mean of 12 and a standard deviation of 2.5, and the resistance is also

---

[8]Bryan Dodson, "Reliability Modeling with Spreadsheets," American Society for Quality: ASQ's 53rd Annual Quality Congress Proceedings, Anaheim, CA, 1999, pp. 575–585.

	A	B	C
1	**Electrical Circuit Analysis**		
2			
3			Standard
4		Mean	Deviation
5	Voltage	12	2.5
6	Resistance	3	0.8
7			
8	Current	4	

FIGURE 5-43 Spreadsheet Model for a Simple Electrical Circuit

normal with a mean of 3.0 and standard deviation of 0.8. One might think that the mean current is $12/3 = 4$. However, the ratio of two normal distributions is not normally distributed, as the forecast chart in Figure 5-44 illustrates. When the resistance is low, the current increases significantly more than when the resistance is high. Thus, the distribution is skewed to the right, and the mean is larger than 4.

Simulation can be used to help in engineering design problems, for example, finding the parameter values that reduce the variance of the output to acceptable levels. Suppose an engineer needs to design a circuit where the current must be between 3.0 and 5.0. One solution is to use voltage and resistance components with tighter tolerances; that is, smaller standard deviations. Another option is to change the nominal levels (means) of the voltage and resistance (while maintaining the same standard deviation) in order to reduce the variance of the distribution of the current. The following SkillBuilder Exercise asks you to do this.

FIGURE 5-44 Forecast Chart for the Distribution of Electrical Current

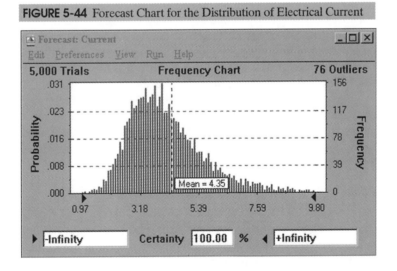

SKILLBUILDER EXERCISE

Open the file *Electrical Circuit.xls*. Find a combination of voltage and resistance that provides a current that meets the specifications of 4.0 ± 1. Should you decide to use the Crystal Ball Decision Table tool to assist you, you will need to define new cells for the means as decision variables and redefine the assumption cells, referencing the decision variable cells as the means for the distributions.

Simulation in Practice

RISK ANALYSIS AT GETTY OIL COMPANY[9]

Oil companies make billions of dollars of investment decisions each year in a very high-risk environment. Poor choices can lead to squandering of precious resources and large losses, as a typical onshore exploration program will cost over $10 million and have only about a 10 percent chance of being successful. The monetary risk is even greater for off-shore and foreign exploration. Even if oil is discovered, considerable uncertainties exist regarding production performance, costs, product prices, and government actions.

In the late 1960s, Getty Oil Company developed two Monte-Carlo simulation programs for risk analysis to help make decisions about investment in oil exploration and well development, and also in setting bids for competitive lease sales held by the federal government. Subsequent efforts focused on developing other simple risk analysis methods that did not rely on simulation. One of the outputs of this effort was the development of a company-wide system for describing uncertainty in projected income and cash flow, called the *Plan Uncertainty Analysis System*. The structure of this system included histograms of net income based on the model assumptions, and 80 percent confidence intervals on gains and losses.

The system provided numerous benefits. First, better information about the variability of projected net income and cash was provided. The data were useful for evaluating the significant deviations of actual income from plan projections. Second, the results from using the system over several years helped managers make better subjective assessments of risks and also improved the quality of both written and oral presentation of results and management feedback. Finally, applications of risk analysis to other investment situations had increased substantially, not from top management edicts, but because of strong interest and support by managers at different levels of the company.

One important application of risk analysis was for property acquisitions. Uncertainty assessments, in conjunction with discounted cash flow models, were used to obtain probability distributions of rates of return for each of several proposed purchase prices, leading to graphs such as the one shown in Figure 5-45. This information is used to gauge the compatibility of a possible bid with management's risk preferences and show the extent to which bids should be changed to optimally balance potential risk and reward. Other uses of risk analysis indicated that Getty should retain a particular operation when an outside consultant suggested divestiture, and showed that a proposed $400 million new processing plant would be too risky.

ASSESSING TRITIUM SUPPLY ALTERNATIVES BY THE U.S. DEPARTMENT OF ENERGY[10]

The U.S. Department of Energy (DOE) is responsible for designing, producing, and maintaining the nuclear weapons required for the defense of the United States. Tritium is a radioactive isotope of hydrogen used in nuclear weapons that decays at a rate of 5.5 percent per year, and it must be periodically replenished. No tritium has been produced since 1988, and the reserve inventory will be depleted below the required reserve level by 2011; in 1994, the DOE was faced with developing a tritium-supply source. The DOE assessed 10 different tritium supply alternatives—heavy-water reactors, advanced light-

---

[9]Adapted from Lynn B. Davidson and Dale O. Cooper, "Implementing Effective Risk Analysis at Getty Oil Company," *Interfaces,* Vol. 10, no. 6, December 1980, pp. 62–75.

[10]Adapted from D. Von Winterfeldt and E. Schweitzer, "An Assessment of Tritium Supply Alternatives in Support of the U.S. Nuclear Weapons Stockpile," *Interfaces* Vol. 28, no. 1, 1998, pp. 92–112.

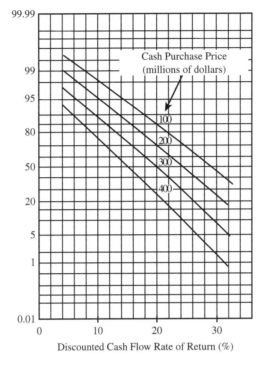

FIGURE 5-45 Example of Rate of Return Probability Distributions for Different Cash Purchase Prices

Source: Reprinted with permission, L. B. Davidson and D. O. Cooper. Copyright 1980, The Institute of Management Sciences and the Operations Research Society of America (currently INFORMS), 2 Charles Street, Suite 300, Providence, RI 02904 USA.

water reactors, modular high-temperature gas-cooled reactors, and commercial reactors—against criteria of production reliability, cost, and environmental impact.

To assess how well the alternatives met these objectives, a model that simulated production patterns over 40 years was developed. The model had three uncertain inputs: the schedule, the production capacity, and the annual availability. To quantify these uncertainties, expert opinions were elicited. For example, schedule uncertainties were obtained from three panels of experts who had knowledge of tritium operations. The experts were presented with the base-case schedules and asked to provide a median (50th percentile) estimate, optimimistic (5th percentile) estimate, and pessimistic (95th percentile) estimate. The participants made their estimates independently and discussed them; however, they were not required to reach consensus. The Weibull distribution was used to approximate the distribution for each individual's estimates because of its flexibility in fitting a variety of shapes (see Chapter 3). Monte-Carlo simulation was then used to create the overall probability distribution over the schedule for each individual; distributions were then averaged to give each participant equal weight. Participants were also asked to identify technical and institutional problems that might lead to major delays, including probabilities for catastrophic events and their expected impact on the schedule. Similar approaches were used to estimate uncertainties in production capacity, availability, and reliability.

Analysts incorporated this information into a dynamic simulation of the production behavior of each alternative by using the following steps:

1. Sample the start year of tritium production once from the schedule distribution.
2. Sample the maximum capacity once from the capacity distribution.
3. Sample the maximum availability once every 40 years after the start date.
4. Calculate the amount of tritium produced in any given year using the data from steps 1 to 3.
5. Add this amount to the amount remaining in the tritium inventory after 1 year of decay.

Each iteration of these steps (i.e., trial) provided an output that shows the tritium inventory over time. Repeating this process several thousand times provided a probability distribution over the available tritium inventory at any point in time. The small advanced light-water reactor, for instance, had a 0.98 probability of providing tritium by 2011 if no delays were encountered, and a 0.78 probability with both types of delays. A table of probabilities was generated, allowing DOE to see the trade-offs involved in the decision and identify alternatives with technical and regulatory uncertainties. The results showed that all alternatives were able to replenish the 5-year decay in less than 5 years. In addition, Monte-Carlo simulations of total life-cycle costs helped to identify the least expensive options. A sensitivity analysis of the effect of discounting found that it had no significant impact on the decision.

In December 1995, the Secretary of Energy published the record of decision. It was decided to pursue commercial reactor options (which had cost advantages, but involved uncertainties of availability) as well as accelerater production of tritium for 3 years. After the 3-year period, the DOE would select one of the alternatives to serve as the primary source of tritium and use the other as a backup alternative. This was the first time that a significant, formal effort was conducted, published, and extensively reviewed on all technical, financial, and environmental aspects of a major programmatic decision by the DOE. The use of Monte-Carlo simulation to understand the risks involved was instrumental in the final decision.

Questions and Problems

1. Use the *Bearing Reliability.xls* model to compare 95 percent confidence intervals for each system's number of bearing failures and each system's total cost. Use Monte-Carlo simulation with 1,000 trials. What conclusions do you reach?
2. A tire company is designing a radial tire that they wish to guarantee for 50,000 miles. The tire components most likely to fail and their expected mileage until failure are shown here. Clearly, if any one fails, the tire fails.

Sidewall failure	Exponential
Tread wearout	Normal (standard deviation = 5,000 miles)
Steel shear of tread	Exponential

Engineers believe they can design the tires so that the expected mileage until failure for each of the three components has a mean of 50,000, 100,000, or 200,000 miles, respectively. Some tires fail for reasons not covered under warranty (such as puncture which leads to tire failure) and, therefore, the company is not liable. The mean time to such failures is 30,000 miles (exponentially distributed). Furthermore, some owners sell their cars with tires before failure (mean 70,000, exponentially distributed).

Identify a 95 percent confidence interval for mean tire life from puncture or sale, as well as for each of the three alternative designs. Also, identify the probabilities of puncture or sale prior to failure, of failure after 50,000 miles, and probability of tire replacement under warranty.
3. Use the Crystal Ball model *Project Management.xls* and identify the probability that paths AD, ABEF, ACEF, and AD are critical. Find the mean and 90 percent certainty range for each path.
4. A software development project consists of six activities, each with time estimates expressed in terms of minimum, most likely, and maximum durations. Assume a triangular distribution for the durations of activities A, B, E, and F. Predecessor relationships are given in the following table. Assume that activity start time is zero if there are no predecessor activities.

	Activity	Min Duration	Likely Duration	Max Duration	Predecessors
A	Requirements analysis	2 months	3 months	4 months	None
B	Programming	6 months	7 months	9 months	A
C	Hardware	Constant	3 months		A
D	User training	Constant	12 months		A
E	Implementation	3 months	5 months	7 months	B,C
F	Testing	1 month	1 month	2 months	E

Find:
a. Mean project completion time
b. Minimum and maximum project completion times
c. Skewness of the completion time distribution
d. Probability of completing the project in 140, 150, 160, or 170 months

5. A firm needs to install a new software system. Activities and data (in weeks) are given in the following table:

Activity	Predecessors	Min	Most Likely	Max
A Requirements analysis	None	2	3	3
B Programming	A	6	7	7
C Hardware acquisition	A	3	3	3
D Train users	B	12	12	12
E Implement	B,C	3	5	5
F Test	E	1	1	1

Distributions are triangularly distributed with the given parameters. Due to the need to coordinate across departments, all simulated times are to be rounded up. Compute expected project completion time, mean project completion time, minimum and maximum project completion time, percentiles of project completion time, and the frequency distribution of project completion time

6. In the *Overbooking.xls* model, suppose that hotel management is interested in the number of customers who are sent to alternative hotels under various overbooking policies. Use the percentile charts to identify the number of overbookings by policy for 1,000 trials.

7. In *Call Volume.xls,* change the distribution of arriving call rates from triangular to exponential (with the same mean of 1). Based on 1,000 simulation runs with a controlled random number seed, compare Monday, Tuesday, and Wednesday overlay charts for the two systems, as well as overlay charts for Thursday, Friday, Saturday, and Sunday.

8. Using *Internal Rate of Return.xls,* experiment to see how the results might change when using different distributions. Using a common random number seed and 1,000 trials, compare the existing triangularly distributed data with the following distributions:

	Normal Mean	Std.	Lognormal Mean	Std.
Starting revenue	18,367	267	18,367	267
Revenue growth	10%	0.167%	10%	0.167%
Starting op cost	15,067	100	15,067	100
Op cost growth	10%	0.167%	105	0.167%
Total project cost	26,578	1,062	26,578	1,062

(The standard deviations are designed to match the triangular distribution, at least to the extent that the range for the triangular distribution is six standard deviations.) Compare the three mean rates of return, as well as the probabilities of obtaining less than 15 percent return.

9. The Hal Chase Investment Agency offers analysis for investors on various financial opportunities. The owner is a strong believer in the *random walk theory*, which can be used as the basis for a simulation model. This theory assumes that up-and-down changes in the Dow Jones average occur randomly from one year to the next according to some distribution. Assume that this change is normally distributed with a mean of 1.15 and a standard deviation of 0.1. Six investment opportunities are available, with the following characteristics. R(DJ) is the correlation of each opportunity with the Dow Jones average.

	Opp #1	*Opp #2*	*Opp #3*	*Opp #4*	*Opp #5*	*Opp #6*
Expected annual return	1.06	1.06	1.11	1.17	1.16	1.20
Standard deviation	0.02	0.03	0.04	0.06	0.05	0.07
R(DJ)	+0.2	−0.1	+0.1	+0.4	0	+0.1

The correlation with the Dow Jones measures the propensity of the option to rise and fall with the average price of stocks. Option 2 has negative correlation, meaning that as the Dow Jones rises, this opportunity tends to fall, and vice versa. The total return consists of the change due to change in the Dow Jones as well as an independent change.

a. Based on the expected value and standard deviation only, show that option 2 is dominated by option 1.

b. If the expected value and standard deviation were the only two considerations, portfolio theory suggests that no rational investor would place money on option 2. However, this analysis does not consider the correlation with the Dow Jones. Identify the proportion of times where option 2 is better than option 1, based on 100 trials.

c. Identify the mean return for each opportunity, as well as each minimum, maximum, and the probabilities of each opportunity exceeding a return of 1.10 and 1.15.

10. For the *Retirement Planning.xls* model, assume that the average raise after age 40 is triangularly distributed with minimum 0, most likely 2 percent, and maximum 4 percent; and that the average raise after age 50 is also triangularly distributed with minimum 0, most likely 0, and maximum 3 percent. Use common random number seeds and compare results against the original distribution of minimum 1 percent, most likely 4 percent, and maximum 5 percent for all ages. Find the mean and 90 percent certainty ranges for final pay and fund value at age 65, and hypothesize the distribution of final fund value based on visual inspection.

11. A 24-year-old dot-com entrepreneur wants to predict the value of his company after 5 years, when he plans to retire. He is going to bring his dot-com site out for an IPO, with which he hopes to make a killing. His investment lawyer tells him that the distribution for initial offers for sites of this nature is

Price	*Probability*
0 (nobody buys)	0.3
$1	0.25
$2	0.15
$3	0.1
$4	0.1
$5	0.1

The entrepreneur has 100,000 shares, of which his lawyer will take 20,000. He will sell 40,000 shares (with which he will run the company) and keep the other 40,000 as a long-term investment (5 years).

Each year there is a 0.3 probability that the company will fold, and its stock value go to zero. Should it survive a particular year, the change in value of the stock is expected to be lognormally distributed with a mean increase of 1.5 and standard deviation of 0.5.

    a. Compute the mean and maximum value of the stock, based on 1,000 simulation trials.

    b. Find the probability that the company will fold.

12. Using the *Cash Budget.xls* model, create a new cell to find the worst case of borrowing. (This will be the maximum of cells E28:J28.) Make this an assumption cell, and based on 1,000 trials, find the mean maximum borrowing amount, a 95 percent confidence interval for the mean, the worst maximum borrowing amount, and the probability of having to borrow.

13. The president of a new bank, the Fifth National Bank of Fargo, is quite concerned about cash flow. He expects depositors to contribute an average of $10,000 per day (lognormally distributed with mean $3,000), and he expects withdrawals to average $9,000 per day (normally distributed, standard deviation $2,000). The financial backers of this operation want to know how much initial capital will be required to provide a 99 percent chance that the new bank's balance will not fall below zero over its first 20 days of operation. Use the initial capitalization as a decision variable cell to estimate the required initial capitalization based on 1,000 trials. To do this, define a decision variable cell for initial investment (minimum $10,000, maximum $50,000, step $10,000). Define a forecast cell for the minimum balance—the minimum of the ending balance for all 20 days. When the simulation is complete, select *Extract Data* from the *Run* menu for minimum balance, which will list the 1,000 results for all five initial balances, sorted from the minimum. Identify the smallest initial balance with a positive minimum balance for the 11th smallest value, which will provide 99 percent assurance.

14. Use the file *Vision Research.xls* and use the 2D Simulation tool from the *CBTools* menu to test the sensitivity of gross profit if approved relative to growth rate of nearsightedness. You are presented a menu where you can assign each of the five assumption variables to either uncertainty or to variability. Assign growth rate of nearsightedness to the uncertainty list, and the other four variables to the variability list. Then examine the overlay and trend charts. (The overlay chart will be quite cluttered. Remove all forecasts, and then add forecasts for the 10th percentile, 50th percentile, and 100th percentiles.)

15. A toy company has a teddy bear stuffed with Styrofoam that they want to release in conjunction with a new movie. The proposed price of the toy is $8.00, and marketing expects to sell 900,000 units, following a normal distribution with a mean of 900,000 and the relatively high standard deviation of 300,000 (and a minimum of 0). Fixed production costs are estimated to be normally distributed with mean $700,000 and standard deviation $50,000. Per-unit variable costs are normally distributed with mean $3 and standard deviation $0.25. Selling expenses are lognormally distributed with mean $900,000 and standard deviation $50,000. General and administrative costs are fixed at $300,000. Identify the mean, standard deviation, and range for profit. Also find the 60 percent, 80 percent, and 90 percent probability intervals for expected profit. Find the probabilities of attaining profits greater than or equal to 0, $1 million, $2 million, $3 million, $4 million, and $5 million. Finally, examine the tornado chart to explain the relative impact of the assumption variables.

16. Using the *Garbage Recycling.xls* model, revise the mean price of oil to $30 per barrel (with the current standard deviation of $4). Based on 1,000 trials, create an overlay chart of the net present value of each of the three options. Use bootstrapping

to obtain 90 percent confidence intervals for the mean percentage of times that each of the three options is the best choice. Finally, construct a tornado chart for each option to explain the impact of the price of oil.

17. A company is considering two investments. Both cost $6 million to construct and have an 8-year life. Investment A is expected to generate an annual net cash flow of $1.25 million; investment B is expected to generate an annual net cash flow of $1.5 million. However, the distributions are expected to be triangular (A minimum $0.9 million, most likely $1.75 million, maximum $2.2 million; B minimum $0.6 million, most likely $1.5 million, maximum $3.4 million). Using a cost of capital of 18 percent per year, obtain the net present value discount of each year's net profit (assuming that the investment occurs in year 0). Find the mean expected net present value for each option, as well as the 70 percent, 80 percent, and 90 percent probability intervals for each mean. Also model the probability of each investment being the best choice, and obtain the mean and the 90 percent certainty range for each option's probability of being best (through bootstrapping). Finally, use an overlay chart for the net present value of A and B and compare relative risk.

18. Use the *Sales Projection.xls* model to compare the impact of the form of the distributional assumptions. Compare the mean, minimum, and maximum fourth quarter 2002 sales and trend charts for the original normally-distributed data against lognormally-distributed data with the same means and standard deviations. Use bootstrapping to compare the means and ranges of the estimated mean of fourth quarter sales 2002.

19. An MBA student is nearing graduation and considering starting her own computer consulting business. In assessing the market, she has categorized potential projects into five groups by size, with expected frequencies and durations:

Project Type	Frequency	Duration	
Minor	0.30	0.5 to 3 hours	Uniformly distributed
Small	0.35	3 to 10 hours	Uniformly distributed
Medium	0.20	10 to 25 hours	Uniformly distributed
Large	0.10	25 to 35 hours	Uniformly distributed
Major	0.05	35 to 50 hours	Uniformly distributed

She estimates the total number of projects per week, based on surveying similar firms in comparable cities, to have the following probability distribution:

Projects/Week	Probability
0	0.05
1	0.07
2	0.10
3	0.15
4	0.20
5	0.15
6	0.10
7	0.07
8	0.05
9	0.06

Her monthly expenses are estimated to be $2,300, and she would like to have an annual income that compares with her graduating classmates; that is, at least

$45,000. Find the probability of an annual income in excess of $45,000, $50,000, and $100,000. Determine the potential profitability of this business if the hourly consulting rates charged are $30, $40, or $50. Simplify the problem by assuming that over 52 weeks the distribution of project duration would be normal, with standard deviation equal to mean divided by the square root of 52 (as a result of the Central Limit Theorem).

20. Use the *Pharmaceutal Marketing Model.xls* and generate a tornado chart for the position in market, first production year, and market growth-rate assumptions. Which of these seem to impact net profit the most?

21. A manager has been told by his boss to install an enterprise resource planning (ERP) system. This system will cost $2 million for software immediately and will take approximately 2 years to install (lognormally distributed with mean of 2 years, standard deviation 0.5 years). Development costs per year are uncertain as well, but are expected to be lognormally distributed with a mean of $1 million per year and a standard deviation of $0.3 million per year. Development costs will be incurred by the firm until the installation is complete. The expense of operating this system will be $5 million the first year, with growth normally distributed at a rate of 10 percent per year (standard deviation 3 percent).

    Benefits during the first year of installation are expected to be at the rate of $6 million per year, normally distributed with a mean of $0.5 million. Management wants you to model the first year's benefit as being proportional to the time in operation during the first year. Subsequent benefits are to increase following the normal distribution with a mean increase of 30 percent per year, with a standard deviation of 30 percent.

    Identify the expected net present value of the system, a 95 percent confidence interval for net present value, and the probability that the project's net present value will be positive. In addition, find the mean time to project completion, and a 90 percent certainty range for this statistic.

22. Open the file *Distribution Strategies.run* and extract the percentile data for profit for the wholesale and the retail options and insert it into a spreadsheet. Plot these two inputs to compare relative profit.

23. Using the *Spring Reliability.xls* model, and based upon 10,000 trials, find the probability of attaining a reliability less than 1.0, 1.1, 1.2, or 1.3 for each of the three materials.

24. Open the file *Toxic Waste.xls*. Apply bootstrapping to identify a 90 percent certainty range for the number of people at risk out of a population of 1 million.

25. Oil field exploration involves high degrees of risk from a number of sources. Geological exploration often provides estimates of the thickness of the hydrocarbon deposit, estimates of the oil field area, saltwater saturation, and porosity of the oil-bearing material. A model for estimating the original oil in place (OOIP) is

$$OOIP = 7758hA(1 - Sw)\phi$$

    where h is the hydrocarbon thickness in feet; A, the productive area in acres; Sw, the saltwater saturation; and ϕ, the porosity. Build a simulation model and apply Crystal Ball to develop a 90 percent certainty range for the number of barrels of oil in a field. The field is estimated to be 50 feet thick and over 480 acres, both exponentially distributed. Saltwater saturation follows a triangular distribution with parameters 0.1, 0.6, and 1.0. Porosity also follows a triangular distribution with parameters 0.08, 0.2, and 0.47.

26. Use the *Electric Circuit.xls* model and try combinations for mean voltage of 10, 12, and 14 (all with standard deviation 2.5), along with mean resistance of 2, 3, and 4 (all with standard deviation of 0.8, minimum of 0.001). Find a 90 percent certainty range for each of these 9 combinations, and identify the combination that has the highest probability of falling within the specification of 4.0 ± 1 for current.

References

Camm, Jeffrey D. and James R. Evans. *Management Science: Modeling, Analysis, and Interpretation.* Cincinnati: South-Western, 1996.

Donnelly, James H. Jr. and Hohn M. Ivancevich. *Analysis for Marketing Decisions.* Homewood, IL: Richard D. Irwin, 1970.

Hertz, David B. and Howard Thomas. *Risk Analysis and Its Applications.* Chichester, U.K.: John Wiley & Sons, 1983.

Kotler, Philip. *Marketing Decision Making: A Model Building Approach.* New York: Holt, Rinehart and Winston, 1971.

Moore, Peter G. *Risk in Business Decision.* London: Longman Group Limited, 1972.

Zweig, Phillip L. "The High Art of Hedging at Merck," *Business Week,* October 31, 1994, p. 88.

Building System Simulation Models

Chapter Outline

In previous chapters, we have seen simple examples of system, or dynamic, simulation models. These include the Mantel Manufacturing problem, the casino game simulations of craps and blackjack, stock price simulation, and retirement planning. What makes these different from pure Monte-Carlo, or static, simulation models is that they involve the passage of time and an explicit representation of the sequence in which events occur. We used Monte-Carlo simulation as an approach to replicating these models in order to characterize the distribution of output variables in the same fashion as a static model.

More complex system simulation models—such as models of factory or service operations—involve the flow of some type of **entity** through the system. The entity might be a physical object, such as a job being processed in a factory or inventory being transferred from a warehouse to a customer. An entity might be a piece of information, such as a message sent through a communication system or a job awaiting processing at a computer center. Often, the model involves the simultaneous flow of many entities within the system. The goal of a system simulation model is to reproduce the activities that control the flow of entities and the logic by which events occur over time. System simulation models involving multiple entities are inherently more complex, making them more difficult, if not impossible, to implement on spreadsheets. In this chapter, we focus on simple types of system simulation models that can be implemented on spreadsheets in order to develop the basic concepts. In the next chapter, we introduce a commercial simulation package, ProcessModel, that facilitates the development and analysis of more complex models.

System Simulation Modeling Approaches

System simulation models can be implemented in several ways. First, we could describe the activities that occur during fixed intervals of time, such as a week, day, or hour. This is called **activity scanning.** For example, to model the operation of an inventory system, we could describe the sequence of events that occur during a specific time period: fulfilling customer demand, ordering new stock, and receiving stock that was ordered at an earlier time. Then we advance time to the next period and repeat. A second approach—**process-driven simulation**—describes the process through which entities in the system flow. For example, in a service system, customers arrive, wait in line if the server is busy, receive service, and then leave the system. A process-driven simulation models the logical sequence of events for each customer as he or she arrives to the system. A third approach is called **event-driven simulation.** With this approach, we describe the changes that occur in the system at the instant of time that each event occurs. Events are sequenced in chronological order and may not correspond to a natural flow of entities. In the service system example, for instance, the key events are the arrival of customers, the start of service, and the end of service. The arrival of the second customer might precede the starting time of service for the first customer. The simulation logic would describe what happens when customer one arrives first, the arrival of customer two second, the start of service for customer one third, and so on. Finally, for special situations in which variables change continuously over time, **continuous simulation** techniques are used.

Activity-scanning simulation models are generally easy to develop. We increment time by some fixed interval and describe what happens during the time interval. For example, the Mantel Manufacturing model (see Chapter 2) is an example of the activity-scanning approach. We incremented time by daily intervals and simulated the entire day's demand and production, looking only at the aggregate changes in inventory levels. By doing so, we essentially assumed that any production for that day could be used to fulfill demand during that day or, equivalently, that all demand is fulfilled at the end of the day. This might be appropriate for a mail-order operation in which daily orders are aggregated and packaged in the evening after all production has been completed. However, this approach might not be appropriate for a simulation model of a retail store that fulfills customers' orders immediately upon arrival at different times of the day. Thus, selecting the appropriate method depends on the purpose of the simulation model and the level of detail and accuracy needed in the analysis.

An important issue in activity-scanning models is selecting the size of the time interval. If it is too large, we may lose information due to the fact that many different activities occur during the time interval. (This is particularly important if we are interested in statistical information about *when* things happen.) If it is too small, then nothing may happen for a large number of intervals, causing the simulation model to be somewhat inefficient. In the next section, we illustrate a more complex example of activity-scanning simulation for inventory problems.

Simulating Inventory Systems Using Activity Scanning

Inventory is any resource that is set aside for future use. Inventory is necessary because the demand and supply of goods usually are not perfectly matched at any given time or place. Many different types of inventories exist. Examples include raw materials (such as coal, crude oil, cotton), semifinished products (aluminum ingots, plastic sheets, lumber), and finished products (cans of food, computer terminals, shirts). Inventories can also be human resources (standby crews and trainees), financial resources (cash on hand, accounts receivable), and other resources such as airplane seats.

Inventories represent a considerable investment for many organizations; thus, it is important that they be managed well. Excess inventories can cause a business to fail. On the other hand, not having inventory when it is needed can also result in business failure. Although many analytic models for managing inventories exist, the complexity of many practical situations often requires simulation. Before we study applications of simulation to inventory decisions, we will review some basic concepts and terminology.

BASIC CONCEPTS OF INVENTORY MODELING

The two basic inventory decisions that managers face are *how much* additional inventory to order or produce and *when* to order or produce it. Although it is possible to consider these two decisions separately, they are so closely related that a simultaneous solution is usually necessary. Typically, the objective is to minimize total inventory cost.

Total inventory cost can include four components: holding costs, ordering costs, shortage costs, and purchasing costs. **Holding costs,** or *carrying costs,* represent costs associated with maintaining inventory. These costs include interest incurred or the opportunity cost of having capital tied up in inventories; storage costs such as insurance, taxes, rental fees, utilities, and other maintenance costs of storage space; warehousing or storage operation costs, including handling, record keeping, information processing, and actual physical inventory expenses; and costs associated with deterioration, shrinkage, obsolescence, and damage. Total holding costs are dependent on how many items are stored and for how long they are stored. Therefore, holding costs are expressed in terms of *dollars associated with carrying one unit of inventory for one unit of time.*

Ordering costs represent costs associated with replenishing inventories. These costs are not dependent on how many items are ordered at a time, but on the number of orders that are prepared. Ordering costs include overhead, clerical work, data processing, and other expenses that are incurred in searching for supply sources, as well as costs associated with purchasing, expediting, transporting, receiving, and inspecting. In manufacturing operations, **setup cost** is the equivalent to ordering cost. Setup costs are incurred when a factory production line has to be shut down in order to reorganize machinery and tools for a new production run. Setup costs include the cost of labor and other time-related costs required to prepare for the new-product run. We usually assume that the ordering or setup cost is constant and is expressed in terms of *dollars per order.*

Shortage costs, or *stock-out costs,* are those costs that occur when demand exceeds available inventory in stock. A shortage may be handled as a *back order,* in which a customer waits until the item is available, or as a *lost sale*. In either case, a shortage represents lost profit and possible loss of future sales. Shortage costs depend on how much shortage has occurred and sometimes for how long. Shortage costs are expressed in terms of *dollar cost per unit of short item*.

Purchasing costs are what firms pay for the material or goods. In most inventory models, the price of materials is the same regardless of the quantity purchased; in this case, purchasing costs can be ignored. However, when price varies by quantity purchased, called the **quantity discount** case, inventory analysis must be adjusted to account for this difference.

The economic order quantity (EOQ) model is the simplest and most elementary inventory model. Although its assumptions often do not hold in practice, it is a good starting point for developing more realistic inventory models. The objective of the EOQ model is to determine the optimal order quantity that will minimize total inventory cost. The EOQ model assumes:

1. The demand for inventory is known with certainty. For example, inventory demand might be three units per day, and is assumed to be three units per day, every day.
2. Inventory replenishment is instantaneous (orders are received all at once) and occurs only when the inventory level reaches zero.
3. **Lead time**—the time between placement and receipt of an order—is constant.
4. A fixed order quantity, Q, is always ordered.
5. No shortages are incurred.
6. The holding cost per unit and ordering cost per order are constant.

The EOQ model applies to a *continuous review* inventory system. In a continuous review system, the inventory position is continuously monitored. **Inventory position** is defined as the amount of inventory on hand plus any amount on order but not received, less any back orders (which, under assumption 5, must be zero). Whenever the inventory position falls at or below a level r, called the **reorder point,** an order is placed for Q units. (Note that the reorder decision is based on the inventory position and *not* the inventory level. If we used the inventory level, orders would be placed continuously as the inventory level falls below r until the order is received.) When the order is received after the lead time, the inventory level jumps from zero to Q, and the cycle repeats. This is illustrated in Figure 6-1.

Development and solution of an analytical model for the EOQ decision can be found in any standard operations management or management science text.[1] We can easily show that the optimal order quantity (Q^*) minimizing total inventory cost is

$$Q^* = \sqrt{\frac{2 \times D \times OC}{HC}}$$

where Q = order quantity
$\quad OC$ = ordering cost per order
$\quad D$ = annual demand for items in inventory
$\quad HC$ = holding cost per unit per year

The total annual cost is given by

$$TC = Q(HC)/2 + D(OC)/Q$$

---

[1]See, for example, Jeffrey D. Camm and James R. Evans, *Management Science and Decision Technology,* Cincinnati, OH: South-Western Publishing Co., 2000.

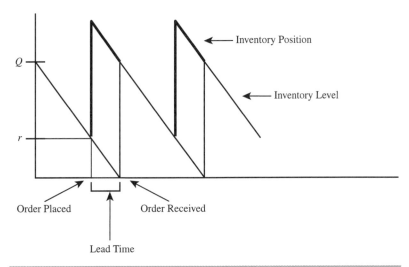

FIGURE 6-1 EOQ Inventory Process

where $Q(HC)/2$ represents the annual holding cost, and $D(OC)/Q$ is the annual ordering cost.

The EOQ model, as well as other classic analytical models, makes rather unrealistic assumptions about the constancy of demand rates and lead times. In real life, demand is usually uncertain, and the lead time can be variable as well. To protect against incurring shortages because of these uncertainties, safety stock is often maintained. **Safety stock** is an additional quantity kept in inventory above planned usage rates. Setting the safety stock level requires knowing the distribution of expected demand and the desired probability of not running out of stock. For instance, if demand during lead time is normal and has a mean of 200 units and a variance of 100 units, and management wants to have a 0.90 assurance of not incurring a shortage, then we must order a quantity x for which the area under the normal curve to the right is 0.10. This corresponds to $z = 1.282$. Thus $1.282 = (x - 200)/10$, or $x = 212.82$. Of this quantity, 200 units are expected to be used to cover demand. The extra 12.82 units (13) would be required to cover demand greater than average 90 percent of the time. The safety stock would thus be 13. The reorder point is defined as the expected demand during lead time plus safety stock. Thus, in this example, when inventory falls to 213 units, an order for Q units would be placed.

CONTINUOUS REVIEW MODEL WITH LOST SALES

In probabilistic situations, it is not clear what order quantities and reorder points should be used to minimize expected total inventory cost. In this section, we present a simulation model for addressing this question. We will assume that demand is Poisson distributed with a mean of 100 units per week; thus, the expected annual demand is 5,200 units.[2] It costs $0.20 to hold one unit for one week (HC = $10.40), and each order costs $50. Every unfilled demand is lost and is charged a penalty of $100 in estimated lost profit. The lead time between placing an order and the time the order is received is assumed to be constant and equal to 2 weeks. Therefore, the expected demand during lead time is 200 units. Orders are placed at the end of the week and received at the beginning of the week.

---

[2]For large values of λ, the Poisson distribution is approximately normal. Thus, this assumption is tantamount to saying that the demand is normally distributed with a mean of 100 and standard deviation of $\sqrt{100} = 10$. The Poisson is discrete, thus eliminating the need to round off normally-distributed random variates.

The EOQ model suggests an order quantity:

$$Q^* = \sqrt{\frac{2 \times 5{,}200 \times 50}{10.4}} = 224$$

For the EOQ policy, the reorder point should be equal to the lead time demand; that is, we place an order when the inventory position falls to 200 units. If the lead time demand is *exactly* 200 units, the order will arrive when the inventory level reaches zero, as illustrated in Figure 6-1. However, if demand fluctuates about a mean of 200 units, we would expect shortages to occur approximately half the time. Because of the high shortage costs, we would probably want to use either a larger reorder point or a larger order quantity. In either case, we will carry more inventory on average, which should result in a lower total shortage cost but a higher total holding cost. With a higher order quantity, we would order less frequently, thus incurring lower total ordering costs. However, the appropriate choice is not clear. We can use simulation to test various reorder point/order quantity policies.

Inventory models lend themselves well to activity-scanning approaches. To develop a simulation model, we need to step through the logic of how this inventory system operates. This is shown in Figure 6-2. Let us assume that no orders are outstanding initially and that the initial inventory level (*INV*) is equal to the order quantity (*Q*). Therefore, the beginning inventory position (*POS*) will be the same as the inventory level. We will use a 1-week time interval for the activity-scanning approach. As we noted, this approach does not capture detailed information about the timing of individual demands, for example, and forces us to make an assumption that orders are received at the beginning of the week. We could make the time interval 1 day, but this would require a larger spreadsheet and more computation.

At the beginning of the week, we check to see if any outstanding orders have arrived. If so, we add the order quantity *Q* to the current inventory level. Next, we determine the weekly demand, *D*, by generating an appropriate random variate, and check if sufficient inventory is on hand to meet this demand (is $D \leq INV$?). If not, then the number of lost sales is $D - INV$. We subtract the current inventory level from the inventory position, set *INV* to zero, and compute the lost-sales cost. If sufficient inventory is available, we satisfy all demand from stock and reduce both the inventory level and inventory position by *D*.

The next step is to check if the inventory position is at or below the reorder point. If so, we place an order for *Q* units and compute the order cost. The inventory position is increased by *Q*, but the inventory level remains the same. We then schedule a receipt of *Q* units to arrive after the lead time.

Finally, we compute the holding cost based on the inventory level at the end of the week (after demand is satisfied) and the total cost. If we wish to simulate another week, we return to the start of the week; otherwise, stop.

A portion of the spreadsheet model for this simulation is shown in Figure 6-3. The model follows the logic of the flowchart in Figure 6-2; however, it is necessary to use some logical functions in Excel in order to implement it correctly. This requires some explanation. The basic problem data are shown in the upper-left corner. The spreadsheet is designed so that lead times of 1 to 5 weeks can be specified. The beginning inventory position and inventory level for each week are equal to the ending levels for the previous week, except for the first week, which is specified in the problem data. The demands in column F were obtained using the Poisson distribution with a mean of 100 and implemented using the *CB.Poisson( )* function; this is easier than defining an assumption cell for each week within the Crystal Ball environment. Because all shortages

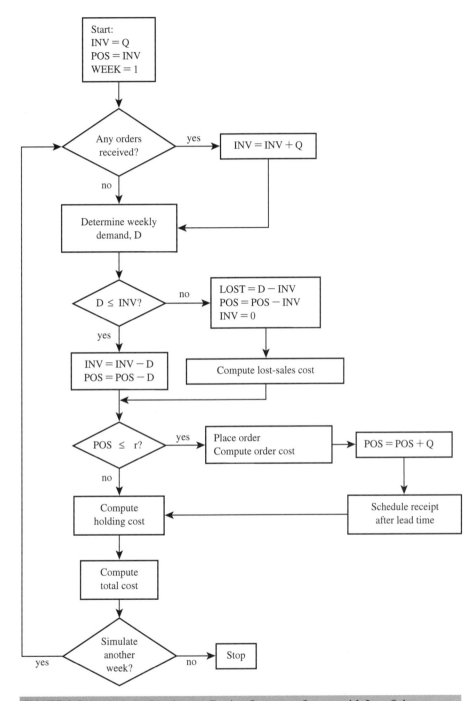

FIGURE 6-2 Logic for a Continuous Review Inventory System with Lost Sales

are lost sales, the inventory level cannot be negative. Thus, the ending inventory each week is computed as the maximum of zero and the beginning inventory level less demand plus any orders received. Lost sales are computed by checking if demand exceeds available stock and computing the difference.

In column I, we determine if an order should be placed by checking if the beginning inventory position less the weekly demand is at or below the reorder point. Ending inventory position is computed as beginning inventory position less weekly demand plus lost sales plus any order that may have been placed that week. This formula might not

FIGURE 6-3 Inventory Simulation Model—Lost-Sales Case

	A	B	C	D	E	F	G	H	I	J	K	L	M	N	O
1	**Inventory Simulation With Lost Sales**														
2				Order Quantity	224			Order Cost	$ 50						
3				Reorder Point	200			Holding Cost	$ 0.20						
4				Initial Inventory	224			Lost Sales Cost	$ 100						
5				Lead time(1-5)	2							Total Annual Costs			
6												$ 753	$1,050	$55,300	$ 57,103
7		Beg								Ending					
8		Inv	Beg	Order	Units		End	Lost	Order	Inv	Lead	Hold	Order	Short	Total
9	Week	Pos	Inv	Rec'd	Rec'd	Dmd	Inv	Sales	Placed?	Pos	time	Cost	Cost	Cost	Cost
10	1	224	224		0	98	126	0	YES	350	2	$25.20	$ 50	$ -	$ 75
11	2	350	126		0	113	13	0	NO	237		$ 2.60	$ -	$ -	$ 3
12	3	237	13		0	95	0	82	YES	448	2	$ -	$ 50	$ 8,200	$ 8,250
13	4	448	0	YES	224	105	119	0	NO	343		$23.80	$ -	$ -	$ 24
14	5	343	119		0	80	39	0	NO	263		$ 7.80	$ -	$ -	$ 8
15	6	263	39	YES	224	129	134	0	YES	358	2	$26.80	$ 50	$ -	$ 77
16	7	358	134		0	106	28	0	NO	252		$ 5.60	$ -	$ -	$ 6
17	8	252	28		0	109	0	81	YES	448	2	$ -	$ 50	$ 8,100	$ 8,150
18	9	448	0	YES	224	94	130	0	NO	354		$26.00	$ -	$ -	$ 26
19	10	354	130		0	95	35	0	NO	259		$ 7.00	$ -	$ -	$ 7
20	11	259	35	YES	224	119	140	0	YES	364	2	$28.00	$ 50	$ -	$ 78
21	12	364	140		0	97	43	0	NO	267		$ 8.60	$ -	$ -	$ 9
22	13	267	43		0	103	0	60	YES	448	2	$ -	$ 50	$ 6,000	$ 6,050
23	14	448	0	YES	224	104	120	0	NO	344		$24.00	$ -	$ -	$ 24

(a)

	B	C	D
10	=$E$4	=$E$4	
11	=J10	=G10	
12	=J11	=G11	=IF(AND(I10="YES",K10=1),"YES","")
13	=J12	=G12	=IF(OR(AND(I11="YES",K11=1),AND(I10="YES",K10=2)),"YES","")
14	=J13	=G13	=IF(OR(AND(I10="YES",K10=3),AND(I11="YES",K11=2),AND(I12="YES",K12=1)),"YES","")
15	=J14	=G14	=IF(OR(AND(I10="YES",K10=4),AND(I11="YES",K11=3),AND(I12="YES",K12=2),AND(I13="YES",K13=1)),"YES","")
16	=J15	=G15	=IF(OR(AND(I10="YES",K10=5),AND(I11="YES",K11=4),AND(I12="YES",K12=3),AND(I13="YES",K13=2),AND(I14="YES",K14=1)),"YES","")
17	=J16	=G16	=IF(OR(AND(I11="YES",K11=5),AND(I12="YES",K12=4),AND(I13="YES",K13=3),AND(I14="YES",K14=2),AND(I15="YES",K15=1)),"YES","")
18	=J17	=G17	=IF(OR(AND(I12="YES",K12=5),AND(I13="YES",K13=4),AND(I14="YES",K14=3),AND(I15="YES",K15=2),AND(I16="YES",K16=1)),"YES","")

(b)

	E
10	=IF(D10="YES",$E$2,0)
11	=IF(D11="YES",$E$2,0)
12	=IF(D12="YES",$E$2,0)
13	=IF(AND(I11="YES",K11=1),$E$2,0)+IF(AND(I10="YES",K10=2),$E$2,0)
14	=IF(AND(I10="YES",K10=3),$E$2,)+IF(AND(I11="YES",K11=2),$E$2,0)+IF(AND(I12="YES",K12=1),$E$2,0)
15	=IF(AND(I10="YES",K10=4),$E$2,0)+IF(AND(I11="YES",K11=3),$E$2,0)+IF(AND(I12="YES",K12=2),$E$2,0)+IF(AND(I13="YES",K13=1),$E$2,0)
16	=IF(AND(I10="YES",K10=5),$E$2,0)+ IF(AND(I11="YES",K11=4),$E$2,0)+IF(AND(I12="YES",K12=3),$E$2,0)+IF(AND(I13="YES",K13=2),$E$2,0)+IF(AND(I14="YES",K14=1),$E$2,0)
17	=IF(AND(I11="YES",K11=5),$E$2,0)+ IF(AND(I12="YES",K12=4),$E$2,0)+IF(AND(I13="YES",K13=3),$E$2,0)+IF(AND(I14="YES",K14=2),$E$2,0)+IF(AND(I15="YES",K15=1),$E$2,0)
18	=IF(AND(I12="YES",K12=5),$E$2,0)+ IF(AND(I13="YES",K13=4),$E$2,0)+IF(AND(I14="YES",K14=3),$E$2,0)+IF(AND(I15="YES",K15=2),$E$2,0)+IF(AND(I16="YES",K16=1),$E$2,0)

(c)

	F	G	H	I
10	=CB.Poisson(100)	=MAX(0,C10-F10+E10)	=IF(F10>C10+E10,IF(C10+E10>=0,F10-C10-E10,F10),0)	=IF(G10<=$E$3,"YES","NO")
11	=CB.Poisson(100)	=MAX(0,C11-F11+E11)	=IF(F11>C11+E11,IF(C11+E11>=0,F11-C11-E11,F11),0)	=IF(B11-F11<=$E$3,"YES","NO")
12	=CB.Poisson(100)	=MAX(0,C12-F12+E12)	=IF(F12>C12+E12,IF(C12+E12>=0,F12-C12-E12,F12),0)	=IF(B12-F12<=$E$3,"YES","NO")
13	=CB.Poisson(100)	=MAX(0,C13-F13+E13)	=IF(F13>C13+E13,IF(C13+E13>=0,F13-C13-E13,F13),0)	=IF(B13-F13<=$E$3,"YES","NO")
14	=CB.Poisson(100)	=MAX(0,C14-F14+E14)	=IF(F14>C14+E14,IF(C14+E14>=0,F14-C14-E14,F14),0)	=IF(B14-F14<=$E$3,"YES","NO")
15	=CB.Poisson(100)	=MAX(0,C15-F15+E15)	=IF(F15>C15+E15,IF(C15+E15>=0,F15-C15-E15,F15),0)	=IF(B15-F15<=$E$3,"YES","NO")
16	=CB.Poisson(100)	=MAX(0,C16-F16+E16)	=IF(F16>C16+E16,IF(C16+E16>=0,F16-C16-E16,F16),0)	=IF(B16-F16<=$E$3,"YES","NO")
17	=CB.Poisson(100)	=MAX(0,C17-F17+E17)	=IF(F17>C17+E17,IF(C17+E17>=0,F17-C17-E17,F17),0)	=IF(B17-F17<=$E$3,"YES","NO")
18	=CB.Poisson(100)	=MAX(0,C18-F18+E18)	=IF(F18>C18+E18,IF(C18+E18>=0,F18-C18-E18,F18),0)	=IF(B18-F18<=$E$3,"YES","NO")

(d)

	J	K	L	M	N	O
10	=B10-F10+H11+IF(I10="YES",$E$2,0)	=IF(I10="YES",$E$5,"")	=MAX(0,G10*$I$3)	=IF(I10="YES",$I$2,0)	=H10*$I$4	=SUM(L10:N10)
11	=B11-F11+H11+IF(I11="YES",$E$2,0)	=IF(I11="YES",$E$5,"")	=MAX(0,G11*$I$3)	=IF(I11="YES",$I$2,0)	=H11*$I$4	=SUM(L11:N11)
12	=B12-F12+H12+IF(I12="YES",$E$2,0)	=IF(I12="YES",$E$5,"")	=MAX(0,G12*$I$3)	=IF(I12="YES",$I$2,0)	=H12*$I$4	=SUM(L12:N12)
13	=B13-F13+H13+IF(I13="YES",$E$2,0)	=IF(I13="YES",$E$5,"")	=MAX(0,G13*$I$3)	=IF(I13="YES",$I$2,0)	=H13*$I$4	=SUM(L13:N13)
14	=B14-F14+H14+IF(I14="YES",$E$2,0)	=IF(I14="YES",$E$5,"")	=MAX(0,G14*$I$3)	=IF(I14="YES",$I$2,0)	=H14*$I$4	=SUM(L14:N14)
15	=B15-F15+H15+IF(I15="YES",$E$2,0)	=IF(I15="YES",$E$5,"")	=MAX(0,G15*$I$3)	=IF(I15="YES",$I$2,0)	=H15*$I$4	=SUM(L15:N15)
16	=B16-F16+H16+IF(I16="YES",$E$2,0)	=IF(I16="YES",$E$5,"")	=MAX(0,G16*$I$3)	=IF(I16="YES",$I$2,0)	=H16*$I$4	=SUM(L16:N16)
17	=B17-F17+H17+IF(I17="YES",$E$2,0)	=IF(I17="YES",$E$5,"")	=MAX(0,G17*$I$3)	=IF(I17="YES",$I$2,0)	=H17*$I$4	=SUM(L17:N17)
18	=B18-F18+H18+IF(I18="YES",$E$2,0)	=IF(I18="YES",$E$5,"")	=MAX(0,G18*$I$3)	=IF(I18="YES",$I$2,0)	=H18*$I$4	=SUM(L18:N18)

(e)

	K	L	M	N	O
7					
8	Lead	Hold	Order	Short	Total
9	time	Cost	Cost	Cost	Cost
10	=IF(I10="YES",$E$5,"")	=MAX(0,G10*$I$3)	=IF(I10="YES",$I$2,0)	=H10*$I$4	=SUM(L10:N10)
11	=IF(I11="YES",$E$5,"")	=MAX(0,G11*$I$3)	=IF(I11="YES",$I$2,0)	=H11*$I$4	=SUM(L11:N11)
12	=IF(I12="YES",$E$5,"")	=MAX(0,G12*$I$3)	=IF(I12="YES",$I$2,0)	=H12*$I$4	=SUM(L12:N12)
13	=IF(I13="YES",$E$5,"")	=MAX(0,G13*$I$3)	=IF(I13="YES",$I$2,0)	=H13*$I$4	=SUM(L13:N13)
14	=IF(I14="YES",$E$5,"")	=MAX(0,G14*$I$3)	=IF(I14="YES",$I$2,0)	=H14*$I$4	=SUM(L14:N14)
15	=IF(I15="YES",$E$5,"")	=MAX(0,G15*$I$3)	=IF(I15="YES",$I$2,0)	=H15*$I$4	=SUM(L15:N15)
16	=IF(I16="YES",$E$5,"")	=MAX(0,G16*$I$3)	=IF(I16="YES",$I$2,0)	=H16*$I$4	=SUM(L16:N16)
17	=IF(I17="YES",$E$5,"")	=MAX(0,G17*$I$3)	=IF(I17="YES",$I$2,0)	=H17*$I$4	=SUM(L17:N17)
18	=IF(I18="YES",$E$5,"")	=MAX(0,G18*$I$3)	=IF(I18="YES",$I$2,0)	=H18*$I$4	=SUM(L18:N18)

(f)

FIGURE 6-3 (*continued*)

appear to be obvious. However, it essentially incorporates both cases in Figure 6-2. Clearly, if there are no lost sales, the ending inventory position is simply the beginning position less the demand plus any order that may have been placed. If lost sales have occurred, computing the ending inventory position in this way would also have reduced it by the unfulfilled demand, which would be incorrect. Thus, we add back the number of lost sales to account for this. To see this better, examine row 12 in the spreadsheet. We have $POS = 237$, $INV = 13$, $D = 95$, and $LS = 82$. The ending inventory position should be $237 - 13 + 224 = 448$. The formula in cell J12 computes this as $237 - 95 + 82 + 224 = 448$. Observe that the negative demand plus lost sales ($-95 + 82$) is equal to the negative of the available inventory, 13.

The most complicated part of this spreadsheet is determining if an order is received and the amount received in columns D and E. First note that orders are placed at the end of the week and received at the beginning of the week, and that lead times of 1 to 5 weeks are allowed. Thus, in Figure 6-3, the order placed at the *end* of the first week with a lead time of 2 will arrive at the *beginning* of the fourth week. The spreadsheet is designed to allow lead times to vary; thus, the formula in cell D13, for example, must check if any order was placed either at the end of week 1 or week 2. Similarly, the formula for cells D16 and lower must step back to determine if an order was placed anywhere from 1 to 5 weeks earlier.

Consider the formula in cell D13, corresponding to the beginning of week 4. This formula consists of nested *IF, OR,* and *AND* logic functions. The *AND* function is true if all its arguments are true. Thus, if the content of cell I11 is "YES" and K11 is 1, an order was placed in week 2 with a lead time of 1, and will arrive at the beginning of week 4. However, an order can also arrive if it were ordered in week 1 and had a lead time of 2. This is captured by the function AND(I10 = "YES", K10 = 2). If either of these are true, then the contents of cell D13 should be "YES." This logic is captured by the *OR* function, which is true if any of its arguments are true. The *IF* function simply checks if any of these conditions hold and puts "YES" or a blank in the cell. In cell D16, for example, we use the same logic, only going back 5 periods. In column E, the logic is similar. If lead times vary, it may occur that more than one order might arrive at the same time. We check if orders were placed in prior weeks and add the order quantities.

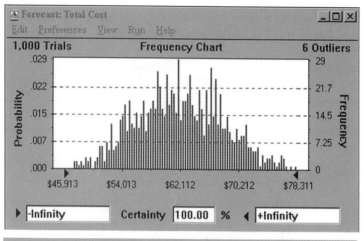

FIGURE 6-4 Total Cost Forecast Chart for Inventory Simulation

The total annual cost, cell O6, is defined as a forecast cell for Crystal Ball simulation. Figure 6-4 shows the forecast chart for 1,000 trials. You can see a large amount of variability in the total annual cost for one simulation. By viewing the forecast statistics, we find the sample mean to be $62,144 with a standard deviation of $6,183.

To find the best values of Q and r, we might experiment with different combinations using the spreadsheet. A better alternative is to use the Crystal Ball Decision Table tool. We define the order quantity and reorder point in cells E2 and E3 as Crystal Ball decision variables. In setting up the decision table, we incremented the order quantity from 200 to 400 in steps of 25, and the reorder point from 200 to 400 in steps of 50. Figure 6-5 shows the simulation results of the mean total annual cost for each of these combinations. We see that the total annual cost has a minimum around $Q = 325$ and $r = 350$. Figure 6-6 shows the forecast chart for this scenario. We might apply the Crystal Ball Decision Table tool again to refine the optimum within a smaller range of Q and r.

SKILLBUILDER EXERCISE

Open the file *Lost Sales Inventory Model.xls*. Validate this model intuitively by defining holding, order, and shortage costs as forecast cells, and examining the pattern of costs over various combinations of Q and r. For example, what happens as Q increases for a fixed r, or as r changes for a fixed Q? Is this what you expect to happen?

CONTINUOUS REVIEW MODEL WITH BACK ORDERS

Another common type of stock-out situation occurs when customers are willing to wait for an item that is temporarily out of stock; this is called a **back-order situation.** The principal difference between this and the lost-sales case is that the inventory level—from an accounting standpoint—can be negative. A negative inventory level represents accumulated back orders. When an order arrives, it goes first to satisfy any back orders. For

	A	B	C	D	E	F	G	H	I	J	K
	Trend Chart / Overlay Chart / Forecast Charts	Order Quantity (200)	Order Quantity (225)	Order Quantity (250)	Order Quantity (275)	Order Quantity (300)	Order Quantity (325)	Order Quantity (350)	Order Quantity (375)	Order Quantity (400)	
1											
2	Reorder Point (200)	$84,879	$60,714	$62,757	$45,489	$70,599	$97,515	$67,056	$38,513	$59,330	1
3	Reorder Point (250)	$23,768	$35,620	$25,689	$26,009	$11,730	$20,043	$28,084	$33,846	$11,033	2
4	Reorder Point (300)	$15,204	$11,127	$ 8,905	$ 6,053	$ 4,552	$ 3,493	$ 3,634	$ 3,553	$ 4,042	3
5	Reorder Point (350)	$12,827	$10,379	$ 7,878	$ 5,374	$ 3,597	$ 3,056	$ 3,061	$ 3,159	$ 3,259	4
6	Reorder Point (400)	$13,242	$10,788	$ 8,349	$ 5,867	$ 4,116	$ 3,547	$ 3,564	$ 3,645	$ 3,755	5
7		1	2	3	4	5	6	7	8	9	

FIGURE 6-5 Crystal Ball Decision Table Results

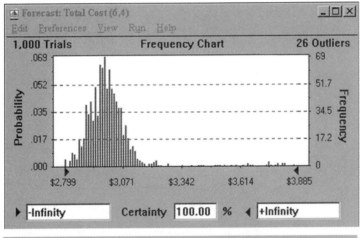

FIGURE 6-6 Crystal Ball Forecast Chart for $Q = 325, r = 350$

example, suppose the inventory level is -14 when an order of 20 units arrives. This means that 14 items are on back-order status. When the firm receives the order for 20 units, 14 of them will go to satisfy the current back orders, resulting in a net inventory level of 6.

Figure 6-7 shows the simulation logic for the back-order case. The only differences from Figure 6-2 occur when the demand is greater than the available inventory. If so, we compute the number of back orders as $D - INV$ and the back-order cost. Then, instead of a separate computation to adjust the inventory position and stock level, we branch to the same box as if there were sufficient inventory to satisfy demand and compute $INV = INV - D$ and $POS = POS - D$. Everything else is the same. Figure 6-8 shows a spreadsheet that implements this logic for a situation in which demand is Poisson distributed with a mean of 1.8 units.

In this model, we have allowed the lead time to vary probabilistically using a Crystal Ball custom distribution and the data in the range K4:L8. The other principal differences from the lost-sales case are that we allow the ending inventory level to be negative to reflect accumulated back orders, and the back-order computation is changed to account for a negative inventory level and ensure the correct number of back orders.

Open the file *Backorder Inventory Model.xls.* Apply the Crystal Ball Decision Table tool to find the best order quantity and reorder point combination.

FIGURE 6-7 Logic for a Continuous Review Inventory System with Back Orders

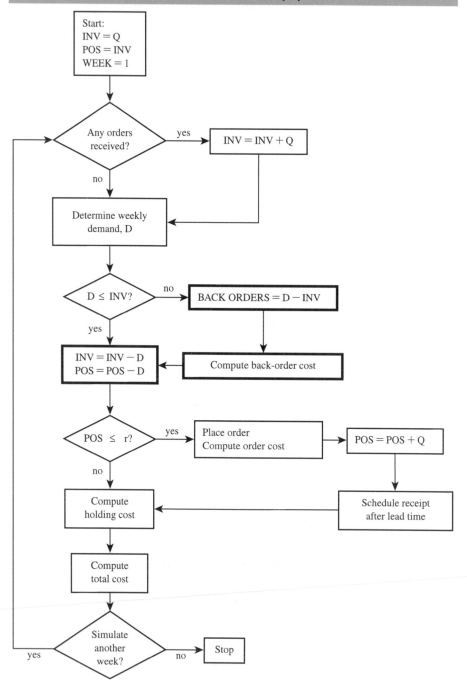

Simulating Waiting-Line Systems Using Process-Driven Models

Waiting lines occur in many important business operations. Most service systems, such as fast-food restaurants, banks, gasoline stations, and technical support telephone hot lines, involve customer waiting. Arrivals to the system and the service times vary probabilistically. The important issue in designing such systems involves the trade-off between customer waiting and system cost, usually determined by the number of servers. A design that balances the average demand with the average service rate will cause

FIGURE 6-8 Simulation Model for an Inventory System with Back Orders

	A	B	C	D	E	F	G	H	I	J	K	L	M	N	O
1	Inventory Simulation With Back Orders										Lead Time Distribution				
2				Order Quantity	5			Order Cost	$ 50		*Lead*				
3				Reorder Point	3			Holding Cost	$ 4.00		*Time*	*Prob.*			
4				Initial Inventory	5			Back Order Cost	$ 20		1	0.2			
5											2	0.3			
6											3	0.2			
7											4	0.2			
8											5	0.1			
9															
10													Total Annual Costs		
11												$ 312	$ 800	$ 580	$1,692
12		Beg								End					
13		Inv	Beg	Order	Units		End	Back	Order	Inv	Lead	Hold	Order	Short	Total
14	Week	Pos	Inv	Rec'd	Rec'd	Dmd	Inv	Orders	Placed?	Pos	time	Cost	Cost	Cost	Cost
15	1	5	5		0	0	5	0	NO	5		$ 20.00	$ -	$ -	$ 20
16	2	5	5		0	2	3	0	YES	8	1	$ 12.00	$ 50	$ -	$ 62
17	3	8	3		0	3	0	0	NO	5		$ -	$ -	$ -	$ -
18	4	5	0	YES	5	2	3	0	YES	8	5	$ 12.00	$ 50	$ -	$ 62
19	5	8	3		0	1	2	0	NO	7		$ 8.00	$ -	$ -	$ 8
20	6	7	2		0	3	-1	1	NO	4		$ -	$ -	$ 20	$ 20
21	7	4	-1		0	1	-2	1	YES	8	2	$ -	$ 50	$ 20	$ 70
22	8	8	-2		0	1	-3	1	NO	7		$ -	$ -	$ 20	$ 20
23	9	7	-3		0	0	-3	0	NO	7		$ -	$ -	$ -	$ -
24	10	7	-3	YES	10	1	6	0	NO	6		$ 24.00	$ -	$ -	$ 24
25	11	6	6		0	0	6	0	NO	6		$ 24.00	$ -	$ -	$ 24
26	12	6	6		0	1	5	0	NO	5		$ 20.00	$ -	$ -	$ 20

(a)

	A	B	C	D
11				
12		Beg		
13		Inv	Beg	Order
14	Week	Pos	Inv	Rec'd
15	1	=$E$4	=$E$4	
16	2	=J15	=G15	
17	3	=J16	=G16	=IF(AND(I15="YES",K15=1),"YES","")
18	4	=J17	=G17	=IF(OR(AND(I16="YES",K16=1),AND(I15="YES",K15=2)),"YES","")
19	5	=J18	=G18	=IF(OR(AND(I15="YES",K15=3),AND(I16="YES",K16=2),AND(I17="YES",K17=1)),"YES","")
20	6	=J19	=G19	=IF(OR(AND(I15="YES",K15=4),AND(I16="YES",K16=3),AND(I17="YES",K17=2),AND(I18="YES",K18=1)),"YES","")
21	7	=J20	=G20	=IF(OR(AND(I15="YES",K15=5),AND(I16="YES",K16=4),AND(I17="YES",K17=3),AND(I18="YES",K18=2),AND(I19="YES",K19=1)),"YES","")
22	8	=J21	=G21	=IF(OR(AND(I16="YES",K16=5),AND(I17="YES",K17=4),AND(I18="YES",K18=3),AND(I19="YES",K19=2),AND(I20="YES",K20=1)),"YES","")
23	9	=J22	=G22	=IF(OR(AND(I17="YES",K17=5),AND(I18="YES",K18=4),AND(I19="YES",K19=3),AND(I20="YES",K20=2),AND(I21="YES",K21=1)),"YES","")

(b)

	E
11	
12	
13	Units
14	Rec'd
15	=IF(D15="YES",$E$2,0)
16	=IF(D16="YES",$E$2,0)
17	=IF(D17="YES",$E$2,0)
18	=IF(AND(I16="YES",K16=1),$E$2,0)+IF(AND(I15="YES",K15=2),$E$2,0)
19	=IF(AND(I15="YES",K15=3),$E$2,)+IF(AND(I16="YES",K16=2),$E$2,0)+IF(AND(I17="YES",K17=1),$E$2,0)
20	=IF(AND(I15="YES",K15=4),$E$2,0)+IF(AND(I16="YES",K16=3),$E$2,0)+IF(AND(I17="YES",K17=2),$E$2,0)+IF(AND(I18="YES",K18=1),$E$2,0)
21	=IF(AND(I15="YES",K15=5),$E$2,0)+ IF(AND(I16="YES",K16=4),$E$2,0)+IF(AND(I17="YES",K17=3),$E$2,0)+IF(AND(I18="YES",K18=2),$E$2,0)+IF(AND(I19="YES",K19=1),$E$2,0)
22	=IF(AND(I16="YES",K16=5),$E$2,0)+ IF(AND(I17="YES",K17=4),$E$2,0)+IF(AND(I18="YES",K18=3),$E$2,0)+IF(AND(I19="YES",K19=2),$E$2,0)+IF(AND(I20="YES",K20=1),$E$2,0)

(c)

(continued)

	F	G	H	I	J
11					
12					End
13		End	Back	Order	Inv
14	Dmd	Inv	Orders	Placed?	Pos
15	=CB.Poisson(1.8)	=C15-F15+E15	=IF(F15>C15+E15,IF(C15+E15>=0,F15-C15-E15,F15),0)	=IF(G15<=$E$3,"YES","NO")	=B15-F15+IF(I15="YES",$E$2,0)
16	=CB.Poisson(1.8)	=C16-F16+E16	=IF(F16>C16+E16,IF(C16+E16>=0,F16-C16-E16,F16),0)	=IF(B16-F16<=$E$3,"YES","NO")	=B16-F16+IF(I16="YES",$E$2,0)
17	=CB.Poisson(1.8)	=C17-F17+E17	=IF(F17>C17+E17,IF(C17+E17>=0,F17-C17-E17,F17),0)	=IF(B17-F17<=$E$3,"YES","NO")	=B17-F17+IF(I17="YES",$E$2,0)
18	=CB.Poisson(1.8)	=C18-F18+E18	=IF(F18>C18+E18,IF(C18+E18>=0,F18-C18-E18,F18),0)	=IF(B18-F18<=$E$3,"YES","NO")	=B18-F18+IF(I18="YES",$E$2,0)
19	=CB.Poisson(1.8)	=C19-F19+E19	=IF(F19>C19+E19,IF(C19+E19>=0,F19-C19-E19,F19),0)	=IF(B19-F19<=$E$3,"YES","NO")	=B19-F19+IF(I19="YES",$E$2,0)
20	=CB.Poisson(1.8)	=C20-F20+E20	=IF(F20>C20+E20,IF(C20+E20>=0,F20-C20-E20,F20),0)	=IF(B20-F20<=$E$3,"YES","NO")	=B20-F20+IF(I20="YES",$E$2,0)
21	=CB.Poisson(1.8)	=C21-F21+E21	=IF(F21>C21+E21,IF(C21+E21>=0,F21-C21-E21,F21),0)	=IF(B21-F21<=$E$3,"YES","NO")	=B21-F21+IF(I21="YES",$E$2,0)

(d)

	K	L	M	N	O
11		=SUM(L15:L66)	=SUM(M15:M66)	=SUM(N15:N66)	=SUM(O15:O66)
12					
13	Lead	Hold	Order	Short	Total
14	time	Cost	Cost	Cost	Cost
15	=IF(I15="YES",CB.Custom($K$4:$L$8),"")	=MAX(0,G15*$I$3)	=IF(I15="YES",$I$2,0)	=H15*$I$4	=SUM(L15:N15)
16	=IF(I16="YES",CB.Custom($K$4:$L$8),"")	=MAX(0,G16*$I$3)	=IF(I16="YES",$I$2,0)	=H16*$I$4	=SUM(L16:N16)
17	=IF(I17="YES",CB.Custom($K$4:$L$8),"")	=MAX(0,G17*$I$3)	=IF(I17="YES",$I$2,0)	=H17*$I$4	=SUM(L17:N17)
18	=IF(I18="YES",CB.Custom($K$4:$L$8),"")	=MAX(0,G18*$I$3)	=IF(I18="YES",$I$2,0)	=H18*$I$4	=SUM(L18:N18)
19	=IF(I19="YES",CB.Custom($K$4:$L$8),"")	=MAX(0,G19*$I$3)	=IF(I19="YES",$I$2,0)	=H19*$I$4	=SUM(L19:N19)
20	=IF(I20="YES",CB.Custom($K$4:$L$8),"")	=MAX(0,G20*$I$3)	=IF(I20="YES",$I$2,0)	=H20*$I$4	=SUM(L20:N20)
21	=IF(I21="YES",CB.Custom($K$4:$L$8),"")	=MAX(0,G21*$I$3)	=IF(I21="YES",$I$2,0)	=H21*$I$4	=SUM(L21:N21)
22	=IF(I22="YES",CB.Custom($K$4:$L$8),"")	=MAX(0,G22*$I$3)	=IF(I22="YES",$I$2,0)	=H22*$I$4	=SUM(L22:N22)

(e)

FIGURE 6-8 (*continued*)

unacceptable delays. The decision is difficult because the marginal return for increasing service capacity declines. For example, a system that can handle 99 percent of expected demand will cost much more than a system designed to handle 90 percent of expected demand. In this section, we discuss the basic components of waiting-line models and illustrate a process-driven simulation model for a simple case.

BASIC CONCEPTS OF QUEUEING MODELS

A **waiting-line** (or **queueing**) **system** has three basic components: arrivals of entities to the system; waiting-lines (or queues); and the service facility.

Arrivals   Arrivals to a queueing system can occur in a number of different ways. Arrivals can be constant, as with an assembly line fed by a machine operating at a constant rate. Usually, however, arrivals occur randomly and are described by some probability distribution. The Poisson distribution is often used to describe the number of arrivals in a fixed time period. An important fact in waiting-line applications is *if the number of arrivals in an interval of time is Poisson, then the time between arrivals has an exponential distribution.*

Queues   If the service facility is busy when an entity arrives, the entity will wait in a line, or queue. Entities wait in the queue according to a decision rule that prescribes how they are to be served. This rule is referred to as the *queue discipline*. The most common queue discipline is *first-come, first-served (FCFS)*. Other decision rules found in real-world applications include last-come, first-served; random service; or some sort of priority decision rule.

Service Facility   The **service facility** consists of the servers that provide service. Many different configurations exist, such as a single server (an ATM), multiple servers (several bank tellers), or sequential servers (gasoline pumping followed by a car wash). Service rates typically vary according to some probability distribution. A common assumption in analytical models is that service times are exponential. Of course, assumptions about the arrival and service time distributions should be validated with empirical data. The distribution-fitting capability of Crystal Ball can be used to identify appropriate distributions.

The total system cost consists of *service costs* and *waiting costs*. A key objective of waiting-line studies is to minimize the total expected cost of the waiting-line system. As the level of service increases (for example, as the number of checkout counters in a grocery store increases), the cost of service increases. Simultaneously, customer waiting time will decrease and, consequently, expected waiting cost will decrease. Waiting costs are difficult to measure because they depend upon customers' perceptions to waiting. It is difficult to quantify how much revenue is lost because of long lines. However, managerial judgment, tempered by experience, can provide estimates of waiting costs. If waiting occurs in a work situation, such as workers waiting in line at a copy machine, the cost of waiting should reflect the cost of productive resources lost while waiting. Although we usually cannot eliminate customer waiting completely without prohibitively high costs, we can minimize the total expected system cost by balancing service and waiting costs.

ANALYTICAL MODELS

Many analytical models have been developed for predicting the characteristics of waiting-line systems. The ability to obtain analytical results depends on the assumptions made about the arrival distribution, service distribution, number of servers, queue discipline, system capacity, and population of customers. These models can be found in general textbooks on management science.

The outputs of these models, called **operating characteristics,** that are of interest to managers include:

1. The mean (expected) waiting time for each customer, W_q
2. The mean (expected) length of the waiting line, L_q
3. The mean time in the system for each customer (waiting plus being served), W
4. The mean number of customers in the system (including those waiting and those receiving service), L
5. The probability that the service facility will be idle (that is, zero units are in the system), P_0
6. The average utilization of the system (that is, the percent of time that servers are busy), ρ

Single-Server Model   The most basic queueing model assumes Poisson arrivals, exponential service times, a single server, and a FCFS queue discipline. If we define

λ = mean arrival rate, expressed in customer arrivals per time period ($1/\lambda$ = mean time between arrivals)

μ = mean service rate, expressed as customers served per time period ($1/\mu$ = mean service time)

then the operating characteristics are

$$\text{Average waiting time: } W_q = \frac{\lambda}{\mu(\mu - \lambda)}$$

$$\text{Average number in qeue: } L_q = \frac{\lambda^2}{\mu(\mu - \lambda)}$$

$$\text{Average time in system: } W = \frac{1}{\mu - \lambda}$$

$$\text{Average number in system: } L = \frac{\lambda}{\mu - \lambda}$$

$$\text{Probability server is idle: } P_0 = 1 - \frac{\lambda}{\mu}$$

$$\text{Percent busy: } \rho = \frac{\lambda}{\mu}$$

Note that the arrival rate, λ, must be less than the service rate, μ, or these formulas do not make sense. If λ is greater than or equal to μ, the queue will grow forever.

These analytical formulas provide long-term expected values for the operating characteristics; they do not describe short-term dynamic behavior of system performance. In a real waiting-line system, we typically see large fluctuations around the averages, and in systems in which the system begins empty, it may take a very long time to reach these expected performance levels. Simulation provides information about the dynamic behavior of waiting lines that analytical models cannot. In addition, simulation is not constrained by the restrictive assumptions necessary to obtain a mathematical solution. Thus, simulation has some important advantages over analytical approaches.

SINGLE-SERVER QUEUEING MODEL

Dirty Dan O'Callahan operates a car wash. Dan is in charge of finance, accounting, marketing, and analysis; his son is in charge of production. During the "lunch hour," which Dan defines as the period from 11 A.M. to 1 P.M., customers arrive randomly at an average of 15 cars per hour (or one car every 4 minutes). A car takes an average of 3 minutes to wash (or 20 cars per hour), but fluctuates quite a bit due to variations in hand-prepping. Dan does not understand how a line could possibly pile up when his son can work faster than the rate at which cars arrive. Although customers complain a bit, they do not leave if they have to wait. Dan is particularly interested to understand the waiting time, the number waiting, and how long his son is actually busy before considering improving his facility.

Because customers arrive randomly, a Poisson distribution is a good assumption to model the number of customers arriving each hour using a mean rate $\lambda = 15$. Recall from Chapter 3 that this implies that the time between arrivals is exponentially distributed. We will also assume that the service time has an exponential distribution with $\mu = 20$ cars per hour. Because there is only one server and customers are processed on a FCFS basis, this problem meets the assumptions for the analytical model presented in the previous section.

The operating characteristics in which Dan is interested and their expected values are

$$\text{Average waiting time: } W_q = \frac{\lambda}{\mu(\mu - \lambda)} = \frac{15}{20(20 - 15)} = 0.15$$

$$\text{Average number in queue: } L_q = \frac{\lambda^2}{\mu(\mu - \lambda)} = \frac{15^2}{20(20 - 15)} = 2.25$$

$$\text{Probability server is idle: } P_o = 1 - \frac{\lambda}{\mu} = 1 - \frac{15}{20} = .25$$

However, these expected values apply in the long run; they may not be representative of what to expect during the short lunchtime interval. We will compare these results with simulated results.

To develop a process-driven simulation model, consider the sequence of activities that each customer undergoes:

1. Customer arrives.
2. Customer waits for service if the server is busy.

3. Customer receives service.
4. Customer leaves the system.

In order to compute the waiting time, we need to know the time a customer arrived and the time service began; the waiting time is the difference. Similarly, to compute the server idle time, we need to know if the arrival time of the next customer is greater than the time at which the current customer completes service. If so, the idle time is the difference. To find the number in the queue, we note that when a customer arrives, then all prior customers who have not completed service by that time must still be waiting. We can make three other observations:

a. If a customer arrives and the server is idle, then service can begin immediately upon arrival.
b. If the server is busy when a customer arrives, then the customer cannot begin service until the previous customer has completed service.
c. The time that a customer completes service equals the time service begins plus the actual service time.

These observations provide all the information we need to run the simulation.
The logic of the simulation is shown in Figure 6-9. The definitions of the variables are

ARRIVAL.TIME(I) = arrival time of customer I
START.TIME(I) = time service for customer I is begun
COMPLETION.TIME(I) = time service for customer I is completed
WAIT.TIME(I) = waiting time for customer I
IDLETIME(I) = idle time incurred by server between customer I-1 and customer I
NUMBER.IN.QUEUE = number of customers waiting
TBA(I) = time between arrival from customer I-1 to customer I
ST(I) = service time for customer I

The simulation is initialized by setting the arrival and completion times of a fictitious "customer 0" to zero. As each new customer arrives, we generate the time since the last arrival, TBA(I), using the arrival time distribution. The arrival time for customer I is set to TBA(I) minutes after the previous customer arrival.

Next, we check if the completion time of customer I-1 is greater than the arrival time of customer I. If it is, then customer I begins service when customer I-1 completes service; if not, customer I can go immediately into service. In either case, we generate a service time, ST(I). Next, we compute the waiting time and service completion time for customer I, and compute any idle time the server may have incurred. Finally, we make a decision whether to repeat the process.

Finding the length of the queue when a customer arrives is a bit tricky. We use the Excel function *MATCH* to determine the last customer, say X, whose completion time is less than or equal to the arrival time of the current customer, say Y. All customers who arrived after customer X are still in the system (including, now, customer Y). Therefore, the system has $Y - X$ customers, one of whom is in service, leaving $Y - X - 1$ customers in the queue. This is the same as Y minus the value of the *MATCH* function.

Figure 6-10 shows a portion of the spreadsheet designed to simulate this problem (the entire simulation is for 500 customers). To generate the time between arrivals and service times, we use the Crystal Ball *CB. Exponential* function. Formulas for the arrival time, start time, completion time, wait time, and idle time follow the logic in Figure 6-9. Columns L through Q compute and maintain updated statistics of the average number

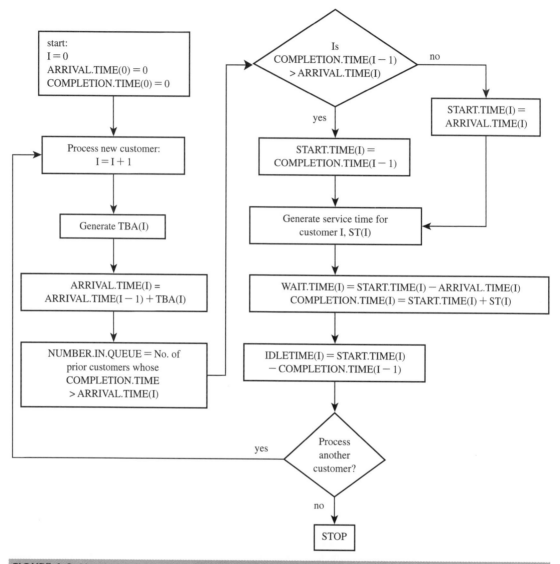

FIGURE 6-9 Simulation Logic for Single-Server Waiting-Line Model

in the queue, average wait time, and average percent idle time as each customer is processed. The final values in columns M, O, and Q, which represent the averages for the 500 customers, are shown in the *Simulation Results* section in cells B12:B14.

Using the Excel Chart tool, we may construct graphs of these performance measures to provide better insight into the dynamics of the queueing system. These are shown in Figures 6-11 through 6-14. Figure 6-11 shows that the number in the queue fluctuates significantly over time. During the simulation, Dan had as few as zero and as many as 14 customers waiting. However, if we look at the *average* number in the queue at any point in the simulation (see Figure 6-12), we see that the average eventually converges to the expected value calculated by the analytical model (the solid black line on the chart). This can be better understood if we realize that each simulation began with an *empty system.* Early arrivals to the system would not expect to wait very long, but when a short-term random surge of customers arrives, the queue begins to build up. (We call this the *warm-up* or *transient* period, which will be discussed further in the next

	A	B	C	D	E	F	G	H	I	J	K	L	M	N	O	P	Q
1	Dan's Car Wash				Arrival	No. In	Start	Service	Completion	Wait	Idle	Cum. No.	Avg. No.	Cum. Wait	Avg. Wait	Cum. Idle	Avg. %
2			Customer	TBA	Time	Queue	Time	Time	Time	Time	Time	in Queue	in Queue	Time	Time	Time	Idle Time
3	Mean arrival rate	15							0.00								
4	Mean service rate	20	1	0.0034	0.0034	0	0.0034	0.0167	0.0201	0.0000	0.0034	0.0000	0.0000	0.0000	0.0000	0.0034	17%
5			2	0.1174	0.1208	0	0.1208	0.0679	0.1886	0.0000	0.1007	0.0000	0.0000	0.0000	0.0000	0.1040	55%
6	Expected No. in Queue	2.250	3	0.0060	0.1268	1	0.1886	0.1097	0.2983	0.0619	0.0000	0.0060	0.0474	0.0619	0.0206	0.1040	35%
7	Expected Waiting Time	0.150	4	0.0939	0.2207	1	0.2983	0.0015	0.2999	0.0777	0.0000	0.0999	0.4528	0.1395	0.0349	0.1040	35%
8	Expected % idle time	25%	5	0.0240	0.2447	2	0.2999	0.0045	0.3044	0.0552	0.0000	0.1480	0.6047	0.1947	0.0389	0.1040	34%
9			6	0.1382	0.3829	0	0.3829	0.0056	0.3885	0.0000	0.0785	0.1480	0.3864	0.1947	0.0324	0.1826	47%
10	*Simulation Results*		7	0.0910	0.4739	0	0.4739	0.0807	0.5547	0.0000	0.0854	0.1480	0.3122	0.1947	0.0278	0.2680	48%
11			8	0.1323	0.6063	0	0.6063	0.0613	0.6676	0.0000	0.0516	0.1480	0.2441	0.1947	0.0243	0.3196	48%
12	No. in Queue	2.1469	9	0.0605	0.6668	1	0.6676	0.2491	0.9167	0.0008	0.0000	0.2085	0.3127	0.1955	0.0217	0.3196	35%
13	Waiting Time	0.1382	10	0.1276	0.7944	1	0.9167	0.0524	0.9691	0.1223	0.0000	0.3361	0.4231	0.3177	0.0318	0.3196	33%
14	% idle time	29%	11	0.0587	0.8531	2	0.9691	0.2325	1.2016	0.1160	0.0000	0.4534	0.5315	0.4337	0.0394	0.3196	27%
15			12	0.0888	0.9419	2	1.2016	0.1191	1.3206	0.2597	0.0000	0.6310	0.6700	0.6934	0.0578	0.3196	24%
16			13	0.0119	0.9538	3	1.3206	0.0663	1.3869	0.3668	0.0000	0.6668	0.6991	1.0602	0.0816	0.3196	23%
17			14	0.1007	1.0545	3	1.3869	0.0213	1.4082	0.3324	0.0000	0.9689	0.9188	1.3926	0.0995	0.3196	23%
18			15	0.0024	1.0569	4	1.4082	0.0113	1.4194	0.3513	0.0000	0.9784	0.9257	1.7439	0.1163	0.3196	23%
19			16	0.0709	1.1277	5	1.4194	0.0077	1.4271	0.2917	0.0000	1.3327	1.1817	2.0356	0.1272	0.3196	22%
20			17	0.0210	1.1488	6	1.4271	0.0378	1.4649	0.2784	0.0000	1.4590	1.2700	2.3139	0.1361	0.3196	22%
21			18	0.0556	1.2044	6	1.4649	0.0887	1.5536	0.2605	0.0000	1.7927	1.4885	2.5744	0.1430	0.3196	21%
22			19	0.0639	1.2683	7	1.5536	0.0416	1.5953	0.2853	0.0000	2.2399	1.7661	2.8598	0.1505	0.3196	20%
23			20	0.0099	1.2782	8	1.5953	0.1647	1.7600	0.3170	0.0000	2.3194	1.8145	3.1768	0.1588	0.3196	18%

(a)

	D	E	F	G	H	I	J	K
1		Arrival	No. In	Start	Service	Completion	Wait	Idle
2	TBA	Time	Queue	Time	Time	Time	Time	Time
3						0		
4	=CB.Exponential($B$3)	=D4	0	=MAXA(E4,I3)	=CB.Exponential($B$4)	=G4+H4	=G4-E4	=G4-I3
5	=CB.Exponential($B$3)	=E4+D5	=C5-MATCH(E5,$I$3:I4,1)	=MAXA(E5,I4)	=CB.Exponential($B$4)	=G5+H5	=G5-E5	=G5-I4
6	=CB.Exponential($B$3)	=E5+D6	=C6-MATCH(E6,$I$3:I5,1)	=MAXA(E6,I5)	=CB.Exponential($B$4)	=G6+H6	=G6-E6	=G6-I5
7	=CB.Exponential($B$3)	=E6+D7	=C7-MATCH(E7,$I$3:I6,1)	=MAXA(E7,I6)	=CB.Exponential($B$4)	=G7+H7	=G7-E7	=G7-I6
8	=CB.Exponential($B$3)	=E7+D8	=C8-MATCH(E8,$I$3:I7,1)	=MAXA(E8,I7)	=CB.Exponential($B$4)	=G8+H8	=G8-E8	=G8-I7
9	=CB.Exponential($B$3)	=E8+D9	=C9-MATCH(E9,$I$3:I8,1)	=MAXA(E9,I8)	=CB.Exponential($B$4)	=G9+H9	=G9-E9	=G9-I8
10	=CB.Exponential($B$3)	=E9+D10	=C10-MATCH(E10,$I$3:I9,1)	=MAXA(E10,I9)	=CB.Exponential($B$4)	=G10+H10	=G10-E10	=G10-I9

(b)

	L	M	N	O	P	Q
1	Cum. No.	Avg. No.	Cum. Wait	Avg. Wait	Cum. Idle	Avg. %
2	in Queue	in Queue	Time	Time	Time	Idle Time
3						
4	=F4*E4	=L4/E4	=J4	=N4/C4	=K4	=P4/I4
5	=F5*(E5-E4)+L4	=L5/E5	=N4+J5	=N5/C5	=P4+K5	=P5/I5
6	=F6*(E6-E5)+L5	=L6/E6	=N5+J6	=N6/C6	=P5+K6	=P6/I6
7	=F7*(E7-E6)+L6	=L7/E7	=N6+J7	=N7/C7	=P6+K7	=P7/I7
8	=F8*(E8-E7)+L7	=L8/E8	=N7+J8	=N8/C8	=P7+K8	=P8/I8
9	=F9*(E9-E8)+L8	=L9/E9	=N8+J9	=N9/C9	=P8+K9	=P9/I9
10	=F10*(E10-E9)+L9	=L10/E10	=N9+J10	=N10/C10	=P9+K10	=P10/I10

(c)

FIGURE 6-10 Simulation Model for Single-Server Queue

chapter.) As the number of customers increases, the number in the queue averaged over all customers begins to level off, reaching what is termed *steady state*. This is what the analytical results provide. However, during the lunch period, the car wash would probably not run long enough to reach steady state; therefore, the analytical results will never present an accurate picture of the system behavior. We see similar phenomena for the average waiting time and percent idle time in Figures 6-13 and 6-14.

Figure 6-11, which shows the number in the queue when each customer arrives, deserves further explanation. It is important to realize that this measure is an approximation because we compute it only when a new customer arrives and do not update it when a customer leaves. (This is a characteristic of the process-driven simulation approach; we will rectify this when discussing event-driven models later in this chapter.) We can see this more clearly by plotting the actual number in the queue for the first several minutes of operation based on the data in the spreadsheet. This is shown by the shaded area in Figure 6-15 on page 220. Customer 1 arrives at time 0.0034 and goes immediately into service and completes service at time 0.0201. Customer 2 arrives at 0.1208 and completes service at 0.1886. However, customer 3 arrives at .1268 and must wait until customer 2 is finished, and so on. The chart in Figure 6-11 displays only the points shown

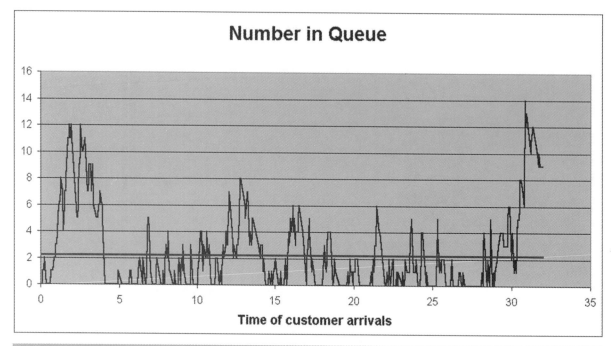

FIGURE 6-11 Chart of Number in Queue (approximation)

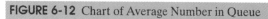

FIGURE 6-12 Chart of Average Number in Queue

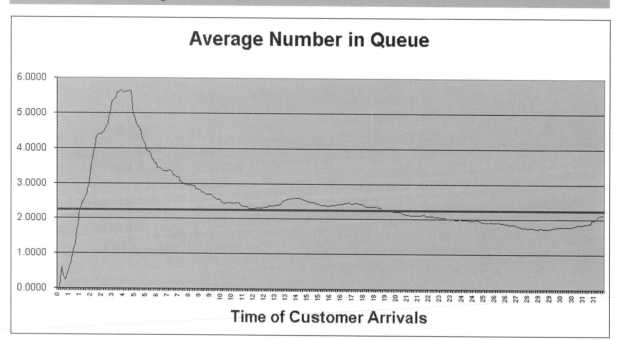

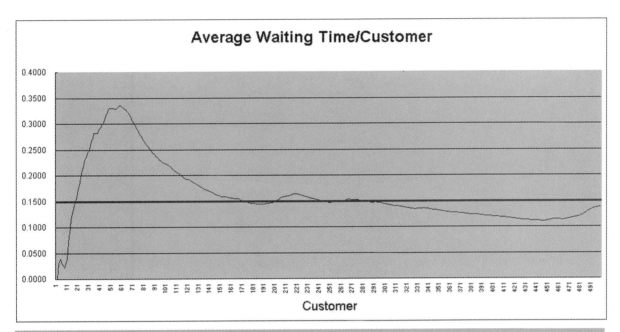

FIGURE 6-13 Chart of Average Waiting Time/Customer

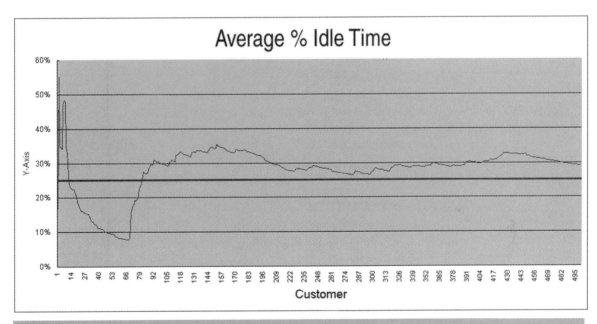

FIGURE 6-14 Chart of Average Percent Idle Time

by the solid dots. This results in only an approximation of the average number in the queue in Figure 6-12. The true average number in the queue at the time that customer 6 arrives (0.3829) is calculated by taking the shaded area and dividing by the length of time, or:

$[(0.1886 - 0.1268)*1 + (0.2447 - 0.2207)*1 + (0.2983 - 0.2447)*2 + (0.2999 - 0.2983)*1]/0.3829 = .5108$

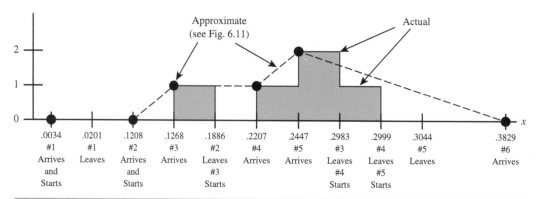

FIGURE 6-15 Graph of Number in Queue (actual versus approximate)

The spreadsheet, however, approximates this as 0.3864 (cell M9). Nevertheless, this does provide a means of understanding the dynamic behavior of the queue.

The results shown in Figure 6-10 represent a single trial. Figure 6-16 shows the results for 1,000 trials using Crystal Ball. The first thing we notice is that the distributions for the forecasts have considerable variability; that is, single trials of the simulation show dramatically different results. This is not unusual for system simulation models and shows the necessity of replicating a model many times to obtain good estimates of the unknown population parameters we wish to estimate.

SKILLBUILDER EXERCISE

Open the file *Queueing Simulation.xls.* Modify the spreadsheet to include *balking* behavior—when an arriving customer refuses to join the queue because it is too long. Select a value for the maximum queue length (say, 8) and count the number of balks per week as a forecast cell.

MULTIPLE-SERVER MODELS

Many queueing systems have more than one server. You have undoubtedly experienced these in banks or perhaps when calling computer technical support functions. Customers form a single line (usually behind a "Wait Here for the Next Available Server" sign) and are processed in order by the next server who is free. It is not difficult to extend the spreadsheet model in the last section to the multiple-server case.

To illustrate this, suppose that Dan's car wash is considering adding a second server to reduce customer waiting. When a customer arrives, a check must be made to see if any server is idle. If so, then the free server (or server 1 when both are idle) can process the customer immediately. If both servers are busy, the customer waits in the queue. The necessary changes in the logic of Figure 6-9 are rather straightforward, and a question at the end of this chapter addresses this.

Figure 6-17 shows an implementation of a two-server model on Excel. The spreadsheet is quite similar to the single-server model. An extra column, F, identifies which of the two servers is assigned to the customer. This is selected using an *IF* statement, selecting for each server the smallest of the finish times of the prior customer. The only other difference is that we need to identify the completion times for each server. This is

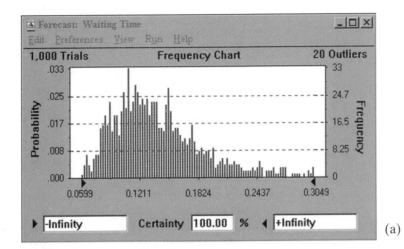

(a)

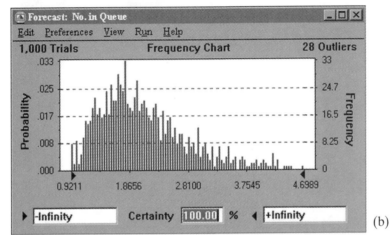

(b)

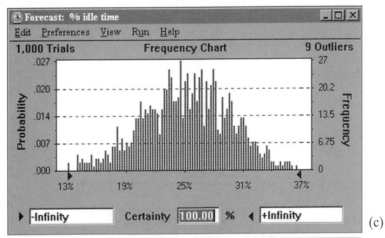

(c)

FIGURE 6-16 Crystal Ball Forecast Charts for Queueing Simulation Results

		Customer	TBA	Arrival Time	Server	Start Time	Service Time	Completion Time Customer	Completion Time Server 1	Completion Time Server 2	Wait Time	Cum. Wait Time	Avg. Wait Time
Dan's Car Wash - Two Server Model													
Mean arrival rate	15												
Mean service rate	20												
Simulation Results		Customer	TBA	Time	Server	Time	Time	Customer	Server 1	Server 2	Time	Time	Time
Waiting Time	0.008	1	0.024	0.024	1	0.024	0.052	0.076	0.076	0.000	0.000	0.000	0.000
		2	0.088	0.112	2	0.112	0.023	0.135	0.076	0.135	0.000	0.000	0.000
		3	0.007	0.119	1	0.119	0.150	0.269	0.269	0.135	0.000	0.000	0.000
		4	0.029	0.148	2	0.148	0.016	0.163	0.269	0.163	0.000	0.000	0.000
		5	0.009	0.157	2	0.163	0.033	0.196	0.269	0.196	0.006	0.006	0.001
		6	0.017	0.175	2	0.196	0.014	0.210	0.269	0.210	0.022	0.028	0.005
		7	0.005	0.179	2	0.210	0.051	0.261	0.269	0.261	0.031	0.059	0.008
		8	0.158	0.337	2	0.337	0.015	0.352	0.269	0.352	0.000	0.059	0.007
		9	0.086	0.424	1	0.424	0.036	0.460	0.460	0.352	0.000	0.059	0.007
		10	0.065	0.488	2	0.488	0.077	0.565	0.460	0.565	0.000	0.059	0.006

(a)

TBA	Arrival Time	Server	Start Time	Service Time	Customer	Server 1
=CB.Exponential($B$3)	=D7	=1	=E7	=CB.Exponential($B$4)	=G7+H7	=IF(F7=1,I7,0)
=CB.Exponential($B$3)	=E7+D8	=2	=E8	=CB.Exponential($B$4)	=G8+H8	=J7
=CB.Exponential($B$3)	=E8+D9	=IF(J8<K8,1,2)	=MAX(E9,MIN(J8,K8))	=CB.Exponential($B$4)	=G9+H9	=IF(F9=1,I9,J8)
=CB.Exponential($B$3)	=E9+D10	=IF(J9<K9,1,2)	=MAX(E10,MIN(J9,K9))	=CB.Exponential($B$4)	=G10+H10	=IF(F10=1,I10,J9)
=CB.Exponential($B$3)	=E10+D11	=IF(J10<K10,1,2)	=MAX(E11,MIN(J10,K10))	=CB.Exponential($B$4)	=G11+H11	=IF(F11=1,I11,J10)
=CB.Exponential($B$3)	=E11+D12	=IF(J11<K11,1,2)	=MAX(E12,MIN(J11,K11))	=CB.Exponential($B$4)	=G12+H12	=IF(F12=1,I12,J11)

(b)

Server 2	Wait Time	Cum. Wait Time	Avg. Wait Time
=IF(F7=2,I7,0)	=G7-E7	=L7	=M7/C7
=I8	=G8-E8	=M7+L8	=M8/C8
=IF(F9=2,I9,K8)	=G9-E9	=M8+L9	=M9/C9
=IF(F10=2,I10,K9)	=G10-E10	=M9+L10	=M10/C10
=IF(F11=2,I11,K10)	=G11-E11	=M10+L11	=M11/C11
=IF(F12=2,I12,K11)	=G12-E12	=M11+L12	=M12/C12

(c)

FIGURE 6-17 Two-Server Queueing Simulation Model

done in columns H and I in a similar manner as was done in the single-server case. For this example, we have not included the percent idle time for each server; this is left as an exercise.

Figure 6-18 shows the Crystal Ball forecast chart for the waiting time. You can see that the average waiting time has dropped significantly by adding a second server. As a practical matter, however, Dan must consider the trade-off between this improvement in customer service and the additional cost of the extra server. In this way, simulation can help answer the "What if?" questions that form the basis for the best decision.

SKILLBUILDER EXERCISE

Open the file *Queueing Simulation—Two Server Model.xls.* Modify the spreadsheet to calculate the idle time for each server.

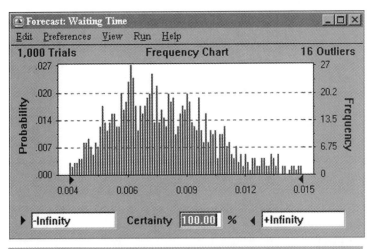

FIGURE 6-18 Crystal Ball Forecast Chart for Waiting Time (2-server model)

Event-Driven Simulation Models

Although activity-scanning and process-driven approaches to simulation work well for certain types of problems, they do not provide the flexibility required to model many practical problems. For example, in a manufacturing system, many different types of events occur. Jobs arrive for processing, materials must be moved from one work center to another, jobs begin and end at individual work centers, and machines break down and must be repaired. With an activity-scanning approach, so many things can occur within a time interval that crucial information may be lost because the exact times of their occurrence are not known. For instance, in the inventory example, we assumed that all demand occurred at the end of the week and did not record the times at which individual demands occurred. In addition, every possible activity must be scanned in each time period, even if it does not occur (as in the case of reordering). For complex problems, this can waste a significant amount of computing time. Similarly, a process orientation may not be applicable; for instance, the breakdown of a machine is a random occurrence that is not associated with the flow of a job through a factory.

In this section, we introduce event-driven simulation. An **event** is an occurrence in a system at which changes to a system occur. Events take place at an instance of time. For example, in an inventory model, an event might be the time at which a demand occurs, or the time when an order is received. In a queueing model, the system changes when a customer arrives and when a customer departs the system. With event-driven simulation, a system is modeled by defining the events that occur in the system, and describing the logic that takes place at these times. Events are processed in chronological order, and simulated time is advanced from one event to the next. This is a very efficient approach from a computational point of view. Most popular commercial simulation languages are based on event-driven simulation. We will also describe some of these languages in this chapter.

AN EVENT-DRIVEN SIMULATION FOR A SINGLE-SERVER QUEUE

In a single-server queueing situation, the only occurrences that change the system are the arrival of a customer and the end of a service. When a customer arrives, the system changes because the number of customers in the system has increased. Similarly, when service is completed, the customer departs, and the number of customers in the system decreases. Thus, arrivals and departures represent the two events that drive the system.

In an event-driven simulation, we must describe the logical sequence of activities that happen when each event occurs. Figure 6-19 shows the logic for an arrival event. When an arrival occurs, we first schedule the next arrival by generating a random variate from the interarrival time distribution and adding it to the current time. This ensures that we have an event scheduled to occur in the future to keep the simulation going. (Shortly, we will describe how we handle these "future" events.) When a customer arrives, we check to see if the server is currently busy. If not, the customer can immediately go into service, and we schedule a departure event by generating a random variate from the service time distribution and adding it to the current time. We then mark the server as busy so that we can collect statistics on server utilization. If the server is busy when the customer arrives, the customer is placed in the queue. This logic follows exactly what you would encounter if you arrived at a drive-up window, for instance.

Figure 6-20 shows the logic for a departure event. In this case, put yourself in the place of the server. After a service is completed, we check to see if any customers are waiting. If there are none, the server is idle and awaits the arrival of the next customer. If the queue is not empty, the first customer waiting in the queue goes into service.

Earlier we stated that events are processed in chronological order. To keep track of future events and control the simulation, we use a control routine, shown in Figure 6-21. Future events and their times of occurrence are placed in a calendar file, which stores them in chronological order. As each event is completed, the control routine removes the next event from the calendar file, advances the simulated time to the event time, and calls the appropriate event routine. To stop the simulation, we can place an end-of-simulation event in the control routine at the start of the simulation. This stores the ending time of the simulation; when it occurs, the simulation is terminated.

To illustrate this approach, we will use the data from the spreadsheet simulation for Dan's car wash, shown in Figure 6-10. (Slight differences are due to rounding.) We actually set the foundation for an event-driven simulation when we discussed the calculation of the number in the queue in Figure 6-15. Table 6-1 shows the time between arrivals and service time for each customer. We will assume that these are the random variates generated whenever the simulation logic requires us to do so.

FIGURE 6-19 Logic for Arrival Event

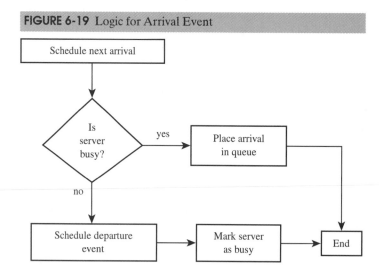

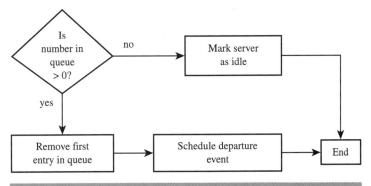

FIGURE 6-20 Logic for Departure Event

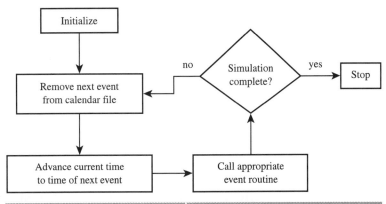

FIGURE 6-21 Control Routine Logic

TABLE 6-1   Arrival and Service Time Data for Dan's Car Wash Example

Customer	Time Between Arrivals	Service Time
1	0.0034	0.0167
2	0.1174	0.0679
3	0.0060	0.1097
4	0.0939	0.0015
5	0.0240	0.0045
6	0.1382	0.0056
7	0.0910	0.0807
8	0.1323	0.0613
9	0.0605	0.2491
10	0.1276	0.0524

We will arbitrarily initialize the control routine with an end-of-simulation event at time 30 (or whatever length of simulated time we wish) and the first arrival at time 0.0034. Thus, at time 0, the calendar file would be

Event	Time	Current Time = 0
Arrival	0.0034	
Endsim	30.000	

The control routine removes the first event, an arrival at time 0.0034; advances the simulation clock to 0.0034; and calls the arrival event routine. We schedule the next arrival at the current time plus the time between arrivals from Table 6-1: $0.0034 + 0.1174 = 0.1208$. We place this event in the calendar file. Because the server is idle, the customer can go immediately into service. We schedule a departure event at the current time plus the service time: $0.0034 + 0.0167 = 0.0201$. Because we maintain the calendar file sorted in chronological order, it looks like:

Event	*Time*	*Current Time = 0.0034*
Departure	0.0201	
Arrival	0.1208	
Endsim	30.000	

The server is marked as busy, and we return to the control routine.

The next event is the departure of the first customer at time 0.0201. Because no other customers are waiting, the server becomes idle. The calendar file is

Event	*Time*	*Current Time = 0.0201*
Arrival	0.1208	
Endsim	30.000	

The next event is an arrival at time 0.1208. We generate the next arrival at time $0.1208 + 0.0060 = 0.1268$. Because the server is idle, we schedule a departure at time $0.1208 + 0.0679 = 0.1886$ and mark the server as busy. The calendar file is

Event	*Time*	*Current Time = 0.1208*
Arrival	0.1268	
Departure	0.1886	
Endsim	30.000	

The next event removed from the calendar file is the arrival at time 0.1268. We schedule the next arrival at time 0.2207, and note that the server is busy. Therefore, this customer is placed in the queue file. The calendar file is

Event	*Time*	*Current Time = 0.1268*
Departure	0.1886	
Arrival	0.2207	
Endsim	30.000	

We would continue this process until we reach the end-of-simulation event.

Table 6-2 summarizes the sequence of events for the portion of the simulation we have described, as well as the number in the system and queue, server status, and idle time. Note that whenever an arrival event occurs, the number in the system is increased by 1; when there is a departure, the number decreases by 1. Server status refers to the status immediately after the event occurs. Thus, if the server is idle and an arrival occurs, the status changes to busy. If the number in the system drops to zero, the server becomes idle. Idle time is computed whenever the server status changes from idle to busy. You

TABLE 6-2 Summary of Events for Dan's Car Wash Simulation Example

Event	Event Type	Clock Time	Number in System	Number in Queue	Server Status	Idle Time
1	Arrival	0.0034	1	0	busy	0.0034
2	Departure	0.0201	0	0	idle	
3	Arrival	0.1208	1	0	busy	0.1007
4	Arrival	0.1268	2	1	busy	
5	Departure	0.1886	1	0	busy	

should work through the remainder of this table using the data in Table 6-1 to verify your understanding of the event-driven approach. From this table, you can easily see how a graph of the number in the queue or system, such as the one in Figure 6-15, can be derived.

Using the principles we have described, it is relatively easy to construct an event-driven spreadsheet simulation model for the single-server waiting-line simulation, shown in Figure 6-22. Columns C through I are identical to Table 6-2. To initialize the simulation, we schedule an arrival as the first event. As the simulation progresses, we use the following logic. In column D, we check whether the time of the next scheduled departure is less than the time of the next scheduled arrival (from columns J and K in the previous row) and list the appropriate event. In column E, the clock time is updated to the appropriate time. This logic follows the control routine in Figure 6-21. In column F, if the event is a departure, we decrease the number in the system by 1; otherwise, we increase it by 1. The number in the queue (column G) is one less than the number in the system or zero, whichever is greater. If the number in the system is zero, the server is marked as idle in column F; otherwise, busy. Whenever the server status changes from idle to busy, we compute the idle time from the last departure until the current arrival in column I. Columns J and K essentially mimic the calendar file. In column J, we schedule a new arrival if the current event is an arrival; otherwise, we simply copy the next arrival time from the previous row. When the server is idle, no departure is scheduled, and a large number (99999) is used in column K to designate this. Columns L and M are used to compute data needed for calculating the average number in the system and in the queue along the same lines as our discussion of Figure 6-15. Figure 6-23 shows a chart of the number in the system (a chart for number in queue is also available in the Excel workbook) that is correct, as opposed to the approximation we used in Figure 6-11.

The only complicated logic in this spreadsheet is in column K for determining the next departure time. The formula for cell K9 follows, broken down into its three nested *IF* statements:

```
(part 1): =IF(AND(D9="depart",F9>0),E9+CB.Exponential($B$4),
(part 2): IF(AND(D9="depart",F9=0),99999,
(part 3): IF(AND(D9="arrival",H8="idle"),E9+CB.Exponential($B$4),K8)))
```

In part 1, if the current event is a departure and cell F9 is greater than 0, indicating that a customer is still waiting, we schedule the next departure at the current time plus an exponential random variate. In part 2, if the current event is a departure and the system is empty, the next departure time is set to an arbitrarily high number. (This corresponds to a calendar file with no scheduled departures.) Finally, in part 3, if the current event is an arrival and the server is idle, we schedule the next departure. If both conditions of the *AND* function are not true, which can only be the case if the current event is an arrival and the server is busy, we copy the already-scheduled next departure time from the previous row. Studying this logic is a good way to better understand the procedures associated with event-driven simulation models.

(a)

	A	B	C	D	E	F	G	H	I	J	K	L	M
1	Dan's Car Wash Event-Driven Simulation												
2													
3	Mean arrival rate	15											
4	Mean service rate	20											
5												Temporary data	
6	*Simulation Results*		Event	Type	Clock	No. in	No. in	Server	Idle	Next	Next	used to compute	
7	Avg. No. in System	1.659			Time	System	Queue	Status	Time	Arrival	Departure	simulation results	
8	Avg. No. in Queue	0.996	1	arrival	0.14	1	0	busy	0.00	0.17	0.17		
9	% Idle Time	0.329	2	depart	0.17	0	0	idle	0.00	0.17	99999.00	0.029	0.000
10			3	arrival	0.17	1	0	busy	0.01	0.20	0.24	0.000	0.000
11			4	arrival	0.20	2	1	busy	0.00	0.22	0.24	0.022	0.000
12			5	arrival	0.22	3	2	busy	0.00	0.28	0.24	0.053	0.026
13			6	depart	0.24	2	1	busy	0.00	0.28	0.26	0.048	0.032
14			7	depart	0.26	1	0	busy	0.00	0.28	0.28	0.036	0.018
15			8	arrival	0.28	2	1	busy	0.00	0.34	0.28	0.026	0.000
16			9	depart	0.28	1	0	busy	0.00	0.34	0.29	0.004	0.002
17			10	depart	0.29	0	0	idle	0.00	0.34	99999.00	0.004	0.000

(b)

	D	E	F	G	H	I
6	Type	Clock	No. in	No. in	Server	Idle
7		Time	System	Queue	Status	Time
8	arrival	=CB.Exponential($B$3)	1	0	busy	=IF(H8="idle",E8,0)
9	=IF(K8<J8,"depart","arrival")	=MIN(J8,K8)	=IF(D9="depart",F8-1,F8+1)	=MAX(F9-1,0)	=IF(F9=0,"idle","busy")	=IF(H8="idle",E9-E8,0)
10	=IF(K9<J9,"depart","arrival")	=MIN(J9,K9)	=IF(D10="depart",F9-1,F9+1)	=MAX(F10-1,0)	=IF(F10=0,"idle","busy")	=IF(H9="idle",E10-E9,0)
11	=IF(K10<J10,"depart","arrival")	=MIN(J10,K10)	=IF(D11="depart",F10-1,F10+1)	=MAX(F11-1,0)	=IF(F11=0,"idle","busy")	=IF(H10="idle",E11-E10,0)
12	=IF(K11<J11,"depart","arrival")	=MIN(J11,K11)	=IF(D12="depart",F11-1,F11+1)	=MAX(F12-1,0)	=IF(F12=0,"idle","busy")	=IF(H11="idle",E12-E11,0)
13	=IF(K12<J12,"depart","arrival")	=MIN(J12,K12)	=IF(D13="depart",F12-1,F12+1)	=MAX(F13-1,0)	=IF(F13=0,"idle","busy")	=IF(H12="idle",E13-E12,0)

(c)

	J
6	Next
7	Arrival
8	=IF(D8="arrival",E8+CB.Exponential($B$3),"")
9	=IF(D9="arrival",E9+CB.Exponential($B$3),J8)
10	=IF(D10="arrival",E10+CB.Exponential($B$3),J9)
11	=IF(D11="arrival",E11+CB.Exponential($B$3),J10)
12	=IF(D12="arrival",E12+CB.Exponential($B$3),J11)
13	=IF(D13="arrival",E13+CB.Exponential($B$3),J12)

(d)

	K
6	Next
7	Departure
8	=E8+CB.Exponential($B$4)
9	=IF(AND(D9="depart",F9>0),E9+CB.Exponential($B$4),IF(AND(D9="depart",F9=0),99999,IF(AND(D9="arrival",H8="idle"),E9+CB.Exponential($B$4),K8)))
10	=IF(AND(D10="depart",F10>0),E10+CB.Exponential($B$4),IF(AND(D10="depart",F10=0),99999,IF(AND(D10="arrival",H9="idle"),E10+CB.Exponential($B$4),K9)))
11	=IF(AND(D11="depart",F11>0),E11+CB.Exponential($B$4),IF(AND(D11="depart",F11=0),99999,IF(AND(D11="arrival",H10="idle"),E11+CB.Exponential($B$4),K10)))
12	=IF(AND(D12="depart",F12>0),E12+CB.Exponential($B$4),IF(AND(D12="depart",F12=0),99999,IF(AND(D12="arrival",H11="idle"),E12+CB.Exponential($B$4),K11)))
13	=IF(AND(D13="depart",F13>0),E13+CB.Exponential($B$4),IF(AND(D13="depart",F13=0),99999,IF(AND(D13="arrival",H12="idle"),E13+CB.Exponential($B$4),K12)))
14	=IF(AND(D14="depart",F14>0),E14+CB.Exponential($B$4),IF(AND(D14="depart",F14=0),99999,IF(AND(D14="arrival",H13="idle"),E14+CB.Exponential($B$4),K13)))

(e)

	L	M
4	Next order	Used for computing
5	receipt	average stock level
6	99999	=B6*F6
7	=IF(J7<=$B$5,F7+CB.	=I6*(F7-F6)
8	=IF(J8<=$B$5,F8+CB.	=I7*(F8-F7)
9	=IF(J9<=$B$5,F9+CB.	=I8*(F9-F8)
10	=IF(J10<=$B$5,F10+C	=I9*(F10-F9)
11	=IF(J11<=$B$5,F11+C	=I10*(F11-F10)
12	=IF(J12<=$B$5,F12+C	=I11*(F12-F11)
13	=IF(J13<=$B$5,F13+C	=I12*(F13-F12)

FIGURE 6-22 Event-Driven Queueing Simulation Model

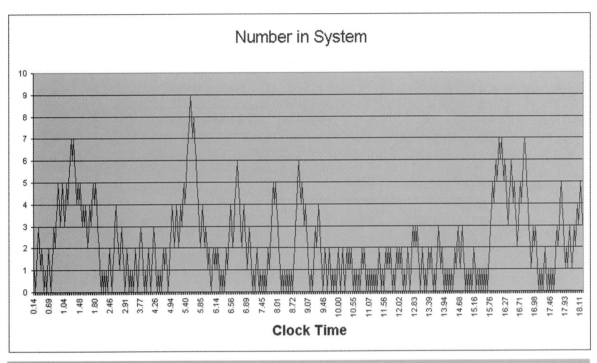

FIGURE 6-23 Chart of Number in System

AN EVENT-DRIVEN INVENTORY SIMULATION MODEL

Earlier in this chapter, we presented a simulation model for simulating a continuous review inventory system with lost sales using an activity-scanning approach. In this section, we show how to simulate this problem using an event-driven approach. The major difference is that, instead of scanning each week to determine the value of the weekly demand and if any orders are received, we let events correspond to:

1. The demand for an individual item
2. The receipt of an order

This requires that the demand be characterized by a probability distribution representing the time between individual demands, not the number of items demanded during a time period. This approach is particularly useful for high-value, low-demand items and infrequent ordering situations, particularly when it is important to capture the exact amount of time that items remain in inventory in order to compute accurate holding costs.

In the logic that we will discuss, we will also include the calculation of average inventory and the average level of safety stock (the stock level just before an order arrives). Figures 6-24 and 6-25 show the logic for the demand and receipt-of-order events. When a demand occurs, we schedule the next demand at the current time plus a random variate drawn from the time-between-demand distribution. Next, we check if any stock is available. If no stock is available, we increase lost sales by 1 and quit. Otherwise, we increase the number sold by 1, and decrease both the stock level and inventory position by 1. We also record the current time and number in inventory. This allows us to compute the average number of units in inventory in a manner similar to the way we computed the average number in the queue for the Dan's car wash simulation. Next, we

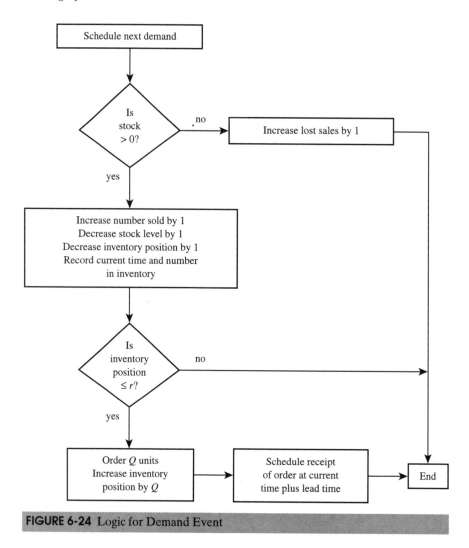

FIGURE 6-24 Logic for Demand Event

check if the inventory position has reached the reorder point. If so, we order Q units and increase the inventory position by the order quantity. Then we schedule a receipt-of-order event at the current time plus the lead time (which may be a random variate).

The receipt-of-order event is simple. Prior to "receiving" the order, we record the current time and number of units in inventory and the amount of safety stock. Then we increase the stock level by Q. The simulation is driven by the same control routine shown in Figure 6-21.

Figure 6-26 shows a spreadsheet implementation for this approach. In this model, we assume that the demand is Poisson distributed with a mean of 10 units per week; equivalently, the time between individual demands is exponential with a mean of 1/10 weeks. We also assume that the lead time is uniformly distributed between 0.2 and 0.5 weeks, the order quantity is $Q = 8$, and the reorder point is $r = 4$. This implementation also assumes that at most one order will be outstanding at any one time. Therefore, if the lead time is too long, the results may be in error.

Columns K and L list the times of the next demand and next order receipt as would be recorded in the calendar file; the smaller of these two determines the next event in

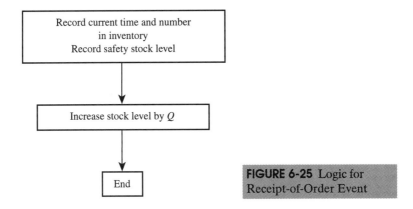

FIGURE 6-25 Logic for Receipt-of-Order Event

column E and the current clock time in column F. In column G, if the event is a demand and the stock level is positive, the total sold is incremented by 1. If the stock level is 0, and a demand occurs, the total lost sales is incremented by 1 in column H.

Column I records the current stock level, either decreasing it by 1 when a demand occurs, or increasing it by Q when an order receipt occurs. Figure 6-27 shows the stock level as a function of time. In column J, the inventory position is increased by Q whenever an order was placed at the previous event; decreased by 1 if a simple demand

FIGURE 6-26 Event-Driven Inventory Simulation Model

	A	B	C	D	E	F	G	H	I	J	K	L	M
1	**Event-Driven Inventory Simulation**												
2													
3	Mean Demand Rate	10	units/week										
4	Order Quantity	8				Clock	Total	Total	Stock	Inventory	Next	Next order	Used for computing
5	Reorder Point	4		Event	Type	Time	sold	lost	level	position	demand	receipt	average stock level
6	Initial Inventory	8		1	demand	0.003	1	0	7	7	0.130	99999.000	0.0217
7				2	demand	0.130	2	0	6	6	0.231	99999.000	0.8931
8	*Simulation Results*			3	demand	0.231	3	0	5	5	0.339	99999.000	0.6049
9	Lost Sales/Week	0.933		4	demand	0.339	4	0	4	4	0.373	0.793	0.5386
10	Average Stock Level	5.106		5	demand	0.373	5	0	3	11	0.402	0.793	0.1370
11				6	demand	0.402	6	0	2	10	0.414	0.793	0.0859
12				7	demand	0.414	7	0	1	9	0.477	0.793	0.0255
13				8	demand	0.477	8	0	0	8	0.505	0.793	0.0628
14				9	demand	0.505	8	1	0	8	0.573	0.793	0.0000
15				10	demand	0.573	8	2	0	8	0.590	0.793	0.0000
16				11	demand	0.590	8	3	0	8	0.898	0.793	0.0000
17				12	receipt	0.793	8	3	8	8	0.898	99999.000	0.0000
18				13	demand	0.898	9	3	7	7	0.957	99999.000	0.8388

(a)

	D	E	F	G	H
4			Clock	Total	Total
5	Event	Type	Time	sold	lost
6	1	demand	=CB.Exponential($B$3)	1	0
7	2	=IF(K6<L6,"demand","receipt")	=MIN(K6,L6)	=IF(AND(E7="demand",I6>0),G6+1,G6)	=IF(AND(E7="demand",I6=0),H6+1,H6)
8	3	=IF(K7<L7,"demand","receipt")	=MIN(K7,L7)	=IF(AND(E8="demand",I7>0),G7+1,G7)	=IF(AND(E8="demand",I7=0),H7+1,H7)
9	4	=IF(K8<L8,"demand","receipt")	=MIN(K8,L8)	=IF(AND(E9="demand",I8>0),G8+1,G8)	=IF(AND(E9="demand",I8=0),H8+1,H8)
10	5	=IF(K9<L9,"demand","receipt")	=MIN(K9,L9)	=IF(AND(E10="demand",I9>0),G9+1,G9)	=IF(AND(E10="demand",I9=0),H9+1,H9)

(b)

	I	J
4	Stock	Inventory
5	level	position
6	=B6-1	=B6-1
7	=IF(E7="demand",MAX(0,I6-1),IF(E7="receipt",I6+$B$4,I6))	=IF(AND(E7="demand",J6=$B$5),J6-1+$B$4,IF(AND(E7="demand",I6=0),J6,IF(E7="receipt",I7,J6-1)))
8	=IF(E8="demand",MAX(0,I7-1),IF(E8="receipt",I7+$B$4,I7))	=IF(AND(E8="demand",J7=$B$5),J7-1+$B$4,IF(AND(E8="demand",I7=0),J7,IF(E8="receipt",I8,J7-1)))
9	=IF(E9="demand",MAX(0,I8-1),IF(E9="receipt",I8+$B$4,I8))	=IF(AND(E9="demand",J8=$B$5),J8-1+$B$4,IF(AND(E9="demand",I8=0),J8,IF(E9="receipt",I9,J8-1)))
10	=IF(E10="demand",MAX(0,I9-1),IF(E10="receipt",I9+$B$4,I9))	=IF(AND(E10="demand",J9=$B$5),J9-1+$B$4,IF(AND(E10="demand",I9=0),J9,IF(E10="receipt",I10,J9-1)))

(c)

(continued)

	K	L	M
4	Next	Next order	Used for computing
5	demand	receipt	average stock level
6	=F6+CB.Exponential($B$3)	99999	=B6*F6
7	=IF(E7="demand",F7+CB.Exponential($B$3),K6)	=IF(J7<=$B$5,F7+CB.Uniform(0.2,0.5),IF(E7="receipt",99999,L6))	=I6*(F7-F6)
8	=IF(E8="demand",F8+CB.Exponential($B$3),K7)	=IF(J8<=$B$5,F8+CB.Uniform(0.2,0.5),IF(E8="receipt",99999,L7))	=I7*(F8-F7)
9	=IF(E9="demand",F9+CB.Exponential($B$3),K8)	=IF(J9<=$B$5,F9+CB.Uniform(0.2,0.5),IF(E9="receipt",99999,L8))	=I8*(F9-F8)
10	=IF(E10="demand",F10+CB.Exponential($B$3),K9)	=IF(J10<=$B$5,F10+CB.Uniform(0.2,0.5),IF(E10="receipt",99999,L9))	=I9*(F10-F9)

(d)

FIGURE 6-26 (*continued*)

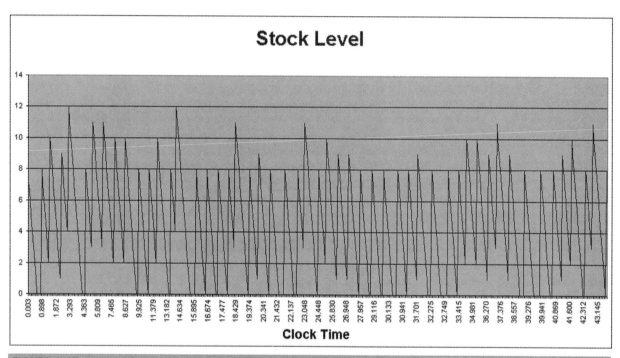

FIGURE 6-27 Chart of Stock Level

occurred; or left the same if the current event is a receipt or if stock is zero. In column K, the next demand is determined by adding an exponential random variate to the current clock time, in a manner similar to that used in the Dan's car wash example. Finally, in column L, we record the time of the next order receipt. This is set to 99999 if no order is currently scheduled in the calendar file, or the current time plus the lead time if an order is pending. Three nested *IF* statements are used. In the first part, we check if the inventory position has been reached. If so, we schedule the next order receipt by adding a uniform random variate to the current clock time. In the second part, we check if the current clock time is less than the time of the next order receipt from the previous event; if so, we know that the time of the next order receipt will not change. If neither of these conditions is true and the current event is a receipt, we copy the time from the previous row. We could use Crystal Ball to find the distribution of the average number of lost sales or stock level, as shown in the forecast charts in Figure 6-28.

You can see that the logic necessary to handle an event-driven simulation on a spreadsheet is more complex and requires more ingenuity than that of activity-scanning or process-driven approaches. Next, we discuss some of the features of more complicated event-driven simulations and commercial languages that facilitate their modeling and analysis.

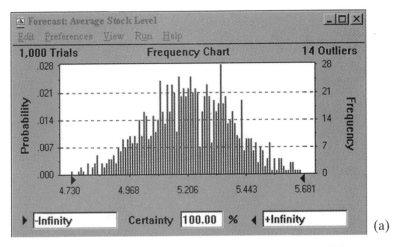

(a)

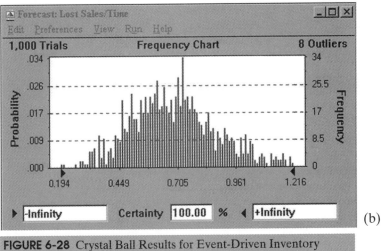

(b)

FIGURE 6-28 Crystal Ball Results for Event-Driven Inventory
Simulation

Entities and Attributes

Often, the information we need to compute an important output statistic is a characteristic of the entities that move through the system. We saw this in the process-driven
simulation for Dan's car wash. Each customer has an associated waiting time. We may
compute the waiting time of a customer whenever the customer begins service, as long
as we know the time that the customer arrived. In the event-driven simulation described
in the previous section, it is not clear how to compute waiting times because the simulation does not keep track of this information. To see how to do this, let us return to the
logic for the arrival and departure events in Figures 6-19 and 6-20.

In Figure 6-19, we see that if the server is not busy, the waiting time of the arriving
customer is zero. However, if the server is busy, we need to record the arrival time of
the customer in order to compute the waiting time when the customer eventually begins service. The arrival time is unique to each customer; that is, it is an attribute of the
customer. In general, an **attribute** is a characteristic associated with an entity. (Other examples of attributes are the quantity ordered of an order entity in an inventory system,
the priority of a job entity arriving at a manufacturing department, or the number of
items purchased by a shopper entity in a supermarket.) Attributes are necessary when

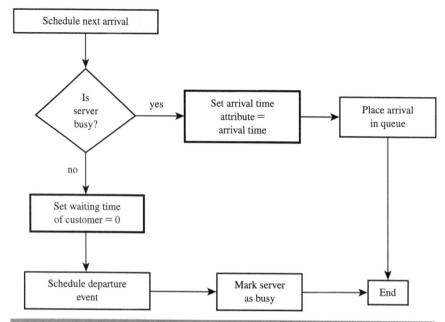

FIGURE 6-29 Logic for Arrival Event with Waiting-Time Computation

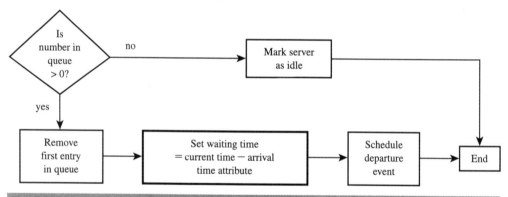

FIGURE 6-30 Logic for Departure Event with Waiting-Time Computation

we must access information associated with different entities at different points of the simulation process. Figures 6-29 and 6-30 show the modifications to Figures 6-19 and 6-20 required to incorporate waiting-time computations into the simulation logic. The boxes with heavy lines represent the changes in the logic.

Because attributes are associated with entities, they can easily be incorporated into process-driven spreadsheet simulations as we saw in the previous chapter, because each row corresponds to an entity. For an event-driven simulation, this is more difficult to do using a spreadsheet; however, commercial simulation software generally captures both process- and event-oriented approaches together.

Continuous Simulation Modeling

Many models contain variables that change continuously. One example would be a model of an oil refinery. The amount of oil moving between various stages of production is clearly a continuous variable. In other models, changes in variables occur gradually (though discretely) over an extended time period; however, for all intents and purposes, they may be treated as continuous. An example would be the amount of inventory at a warehouse in a production-distribution system over several years. As customer demand is fulfilled, inventory is depleted, leading to factory orders to replenish the stock. As orders are received from suppliers, the inventory increases. Over time, particularly if orders are relatively small and frequent, as we see in just-in-time environments, the inventory level can be represented by a smooth, continuous function.

Continuous variables are often called *state variables*. A continuous simulation model defines equations for relationships among state variables so that the dynamic behavior of the system over time can be studied. To simulate continuous systems, we use an activity-scanning approach whereby time is decomposed into small increments. The defining equations are used to determine how the state variables change during an increment of time. A specific type of continuous simulation is called *system dynamics,* which dates back to the early 1960s and a classic work by Jay Forrester of Massachusetts Institute of Technology. System dynamics focuses on the structure and behavior of systems that are composed of interactions among variables and feedback loops. A system dynamics model usually takes the form of an influence diagram that shows the relationships and interactions among a set of variables.

To gain an understanding of system dynamics and how continuous simulation models work, let us develop a model for the cost of medical care. Doctors and hospitals charge more for services, citing the rising cost of research, equipment, and insurance rates. Insurance companies cite rising court awards in malpractice suits as the basis for increasing their rates. Lawyers stress the need to force professionals to provide their patients with the best care possible and use the courts as a means to enforce patient rights. The medical cost system has received focused attention from those paying for medical care and from government officials.

Let us suppose that we are interested in how medical rates (MEDRATE) are influenced by other factors, specifically:

1. The demand for medical service (DEMAND)
2. Insurance rates (INSRATE)
3. Population levels (POPLVL)
4. Medical-related lawsuits (MEDSUIT)
5. Avoidance of risk by doctors (RISK)

Figure 6-31 shows an influence diagram of how these factors might relate to one another. For example, rates rise as the demand for medical service increases and as insurance rates rise (as indicated by the arrows pointed into MEDRATE). The demand is influenced by the population level and its growth rate. Also, increasing rates have a negative influence on demand, meaning that as rates rise, the demand will decrease. Insurance rates increase as medical lawsuits increase and drop as doctors avoid taking risks. At the same time, lawsuits increase as medical rates increase but also decline with risk avoidance. Some of these influences do not occur immediately, as noted by the "delay" factors in the figure. It might take about 1 year before some variables actually influence others.

We may express these relationships quantitatively through a set of equations that describe how each variable changes from one year to the next (that is, year $t-1$ to year t). At time $t=0$, we index all variables to 1.0. We will assume that the population level grows each year by a value, *GROWTH(t),* that is normally distributed with a mean of 0.05 and a standard deviation of 0.03. This is expressed by the equation:

$$POPLVL(t) = POPLVL(t-1) + GROWTH(t)$$

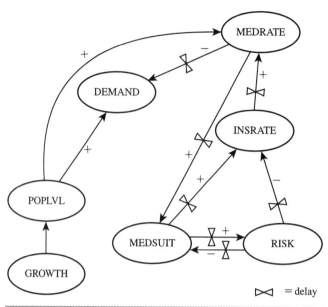

FIGURE 6-31 Influence Diagram for the Cost of Medical Services

The demand for medical service increases with the population and decreases with the rate of increase in the cost of medical service, lagged by 1 year. Thus, demand is computed by the formula:

$$DEMAND(t) = POPLVL(t) - [MEDRATE(t - 1) - MEDRATE(t - 2)]$$

The cost of medical service increases with the change in population level and a portion (80 percent) of the increase in insurance rates, lagged by 1 year:

$$MEDRATE(t) = MEDRATE(t - 1) + POPLVL(t) - POPLVL(t - 1) + .8* [INSRATE(t - 1) - INSRATE(t - 2)]$$

Insurance rates increase by a fraction (10 percent) of the previous year's level of lawsuits and decrease with any increases in doctors' adoption of safer practices to avoid risk:

$$INSRATE(t) = INSRATE(t - 1) + .10*MEDSUIT(t - 1) - [RISK(t - 1) - RISK(t - 2)]$$

Increase in lawsuits is proportional to the increased costs of medical service and inversely proportional to risk avoidance, both lagged by 1 year:

$$MEDSUIT(t) = MEDSUIT(t - 1) + [MEDRATE(t - 1) - 1]/RISK(t - 1)$$

Finally, the avoidance of risk increases as a proportion (10 percent) of the increase in the level of lawsuits, based on the previous year:

$$RISK(t) = RISK(t - 1) + .10*[MEDSUIT(t - 1) - 1]$$

Figure 6-32 presents a spreadsheet model for simulating this system. This simulation model is deterministic, as none of the variables are assumed to be uncertain. Figure 6-33 shows a graph of each of the variables over the 30-year period of the

	Time period	Population growth	Population level	Med. Service demand	Medical rate	Insurance rate	Medical lawsuits	Risk avoidance
3	0		1	1	1	1	1	1
4	1	0.050	1.050	1.050	1.050	1	1	1
5	2	0.050	1.100	1.050	1.100	1.1	1.050	1
6	3	0.050	1.150	1.100	1.230	1.205	1.150	1.005
7	4	0.050	1.200	1.070	1.364	1.315	1.379	1.020
8	5	0.050	1.250	1.116	1.502	1.438	1.736	1.058
9	6	0.050	1.300	1.162	1.650	1.574	2.210	1.131
10	7	0.050	1.350	1.202	1.809	1.721	2.785	1.252
11	8	0.050	1.400	1.241	1.977	1.879	3.431	1.431
12	9	0.050	1.450	1.282	2.153	2.043	4.113	1.674
13	10	0.050	1.500	1.324	2.334	2.211	4.802	1.985
14	11	0.050	1.550	1.368	2.519	2.380	5.474	2.366
15	12	0.050	1.600	1.415	2.704	2.547	6.116	2.813
16	13	0.050	1.650	1.465	2.888	2.712	6.722	3.325
17	14	0.050	1.700	1.516	3.069	2.872	7.290	3.897
18	15	0.050	1.750	1.569	3.248	3.029	7.821	4.526
19	16	0.050	1.800	1.622	3.423	3.182	8.318	5.208
20	17	0.050	1.850	1.675	3.596	3.332	8.783	5.940
21	18	0.050	1.900	1.728	3.765	3.478	9.220	6.718
22	19	0.050	1.950	1.780	3.933	3.622	9.632	7.540
23	20	0.050	2.000	1.833	4.098	3.763	10.021	8.403
24	21	0.050	2.050	1.885	4.261	3.902	10.389	9.305
25	22	0.050	2.100	1.937	4.422	4.039	10.740	10.244
26	23	0.050	2.150	1.989	4.581	4.174	11.074	11.218
27	24	0.050	2.200	2.041	4.739	4.307	11.393	12.225
28	25	0.050	2.250	2.092	4.896	4.439	11.699	13.265
29	26	0.050	2.300	2.143	5.051	4.570	11.992	14.335
30	27	0.050	2.350	2.194	5.206	4.699	12.275	15.434
31	28	0.050	2.400	2.246	5.359	4.827	12.547	16.561
32	29	0.050	2.450	2.297	5.512	4.955	12.811	17.716
33	30	0.050	2.500	2.347	5.664	5.081	13.065	18.897

(a)

	Time period	Population growth	Population level	Med. Service demand	Medical rate	Insurance rate	Medical lawsuits	Risk avoidance
3	0		1	1	1	1	1	1
4	1	=CB.Normal(0.05,0.03)	=C3+B4	=D3+B4	=E3+B4	1	1	1
5	2	=CB.Normal(0.05,0.03)	=C4+B5	=C5-(E4-E3)	=E4+C5-C4+0.8*(F4-F3)	=F4+0.1*G4-(H4-H3)	=G4+(E4-1)/H4	1
6	3	=CB.Normal(0.05,0.03)	=C5+B6	=C6-(E5-E4)	=E5+C6-C5+0.8*(F5-F4)	=F5+0.1*G5-(H5-H4)	=G5+(E5-1)/H5	=H5+0.1*(G5-1)
7	4	=CB.Normal(0.05,0.03)	=C6+B7	=C7-(E6-E5)	=E6+C7-C6+0.8*(F6-F5)	=F6+0.1*G6-(H6-H5)	=G6+(E6-1)/H6	=H6+0.1*(G6-1)
8	5	=CB.Normal(0.05,0.03)	=C7+B8	=C8-(E7-E6)	=E7+C8-C7+0.8*(F7-F6)	=F7+0.1*G7-(H7-H6)	=G7+(E7-1)/H7	=H7+0.1*(G7-1)
9	6	=CB.Normal(0.05,0.03)	=C8+B9	=C9-(E8-E7)	=E8+C9-C8+0.8*(F8-F7)	=F8+0.1*G8-(H8-H7)	=G8+(E8-1)/H8	=H8+0.1*(G8-1)
10	7	=CB.Normal(0.05,0.03)	=C9+B10	=C10-(E9-E8)	=E9+C10-C9+0.8*(F9-F8)	=F9+0.1*G9-(H9-H8)	=G9+(E9-1)/H9	=H9+0.1*(G9-1)

(b)

FIGURE 6-32 Spreadsheet Model for Continuous Simulation Example

simulation. Based on our assumptions, the population has increased by 150 percent. However, the demand for medical service has not quite reached that level, dampened by a fivefold increase in the cost of medical service. Insurance rates have increased five times, and lawsuits have increased 13 times (a compounded rate of 9 percent per year); risk avoidance practices have increased an average of over 10 percent per year.

System dynamics has been applied to the analysis of material and information flows in logistics systems, sales and marketing problems, social organizations, ecology, and many other fields. System dynamics was quite popular among researchers and practitioners until the early 1970s. The concept was brought back to the attention of business in the 1990s by Peter Senge through his book *The Fifth Discipline,* which explores the role and importance of systems thinking in modern organizations and has formed the basis for many contemporary concepts in supply chain management.

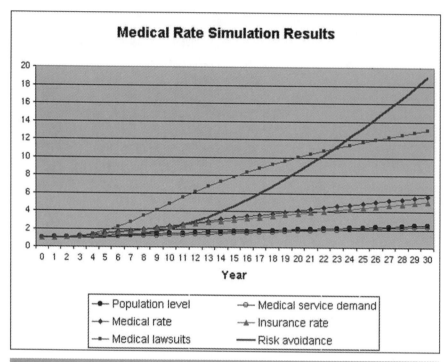

FIGURE 6-33 Dynamic Behavior of Variables in Medical Rate Simulation

Simulation in Practice

IMPROVING DISTRIBUTION COSTS AT CIBA-GEIGY USING SIMULATION[3]

Ciba-Geigy, headquartered in Switzerland, is a major manufacturer of chemicals, plastics, pharmaceuticals, and agricultural products. The logistics department of one of the company's divisions was interested in trying to reduce distribution costs by reallocating safety stocks for finished goods inventory and introducing an alternative distribution policy. The system that had been used involved meeting demands from decentralized regional centers or warehouses. Safety stock is critical in highly competitive markets where profit margins are low. The proposal was based on the idea that if all demands were met from a central location, much of the waste inherent in safety stocks would be reduced, while the same probability of satisfying demands (the service level, defined as orders delivered on time/total orders) could be maintained. The study also addressed distribution policies to wholly-owned subsidiaries, called *group companies*. Shipping finished goods directly to them immediately after production considerably shortens distribution time, but increases the risk that large amounts of stock might be allocated to the wrong company at the wrong time.

An event-driven simulation model of the problem was developed, which allowed relaxing typical assumptions such as identical regional centers and normally distributed lead time distributions that mathematical modeling approaches required. The choice of simulation also turned out to be fortunate, because the production planning process had to be incorporated into the model, something that would have been infeasible to do in

---

[3]U. Fincke and W. Vaessen, "Reducing Distribution Costs in a Two-Level Inventory System at Ciba-Geigy," *Interfaces* Vol. 18, no. 6, 1988, pp. 92–104.

a conventional mathematical analysis. The simulation model was used to determine the effects of different safety-stock levels in the parent company and group companies on service levels, average stock levels, overall distribution costs, margins, and tied-up capital; any advantages of a direct shipment distribution policy; and the influence of subsidiary order periods on average cycle and safety-stock levels. The model was tested on past demand data, which helped to validate the model.

A key feature of the model was its integration of modules for simulating the material and information flows and for calculating cost-based performance measures and service levels. An example of the simulation results is illustrated in Figure 6-34. This figure shows that more safety stocks in the parent company (*pc*) reduce safety stocks in the group companies (*gc*), but increase the total amount of safety stock. Each curve was formed from the output of 14 simulation experiments over 10 years of simulated time, fixing the customer service level at 99 percent. Simultaneously, overall distribution costs decreased as the parent company service level increased.

Simulation experiments also showed that direct shipment would make it possible to reduce capital tied up in the distribution chain and the annual distribution costs by more than 10 percent. However, Ciba-Geigy meanwhile had built a highly automated central warehouse and distribution center. This new center made direct shipment unnecessary, because handling times were cut in half. Taking advantage of lower interest rates in Switzerland, where the central warehouse was located, also impacted the relative performance of the alternatives. The study was cited as resulting in savings of over 50 million Swiss francs.

SECURITY CHECKPOINT IMPROVEMENT THROUGH SIMULATION[4]

Westinghouse-Hanford is a government contractor in the nuclear industry located in Hanford, Washington. Before anyone can enter the facility, they must pass through a security gate, with a 4-lane entry road and a 2-lane exit. This creates a bottleneck that can reduce productive work time, as well as cause a safety problem.

The existing system involved a single-file, first-come first-served pattern of two parallel lines to the guard facility. Traffic arrived at the rate of 7 buses and 285 cars and vans

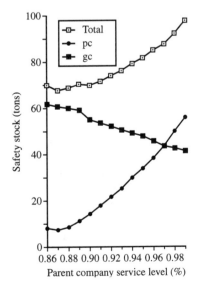

FIGURE 6-34 Example of Simulation Results

Source: Reprinted by permission, U. Fincke and W. Vaessen, "Reducing Distribution Costs in a Two-Level Inventory System at Ciba-Geigy," *Interfaces,* Vol. 18, no. 6, 1988, pp. 92–104. Copyright 1988, The Institute of Management Sciences and the Operations Research Society of America (currently INFORMS), 2 Charles Street, Suite 300, Providence, RI 02904 USA.

---

[4]E. G. Landauer and L. C. Becker, "Reducing Waiting Time at Security Checkpoints," *Interfaces,* Vol. 19, no. 5, 1989, pp. 57–65.

in a 1-hour period. Security guards entered the vehicle to check drivers and passengers. Typically, one to three guards were available. The line for cars and vans would sometimes extend beyond the waiting area, which could hold 40 vehicles. When waiting vehicles spilled over onto the highway, a serious traffic hazard arose. Buses had a separate line (the fourth lane) and received priority. Although three lanes were available for cars, only one was used, as there were problems merging traffic when more than one lane was used. The situation was deemed intolerable by both guards and drivers. Private vehicles had a second gate available, although additional time and distance (as well as another wait) were involved. When lines were long, cars would sometimes balk, or go to the second gate.

In the afternoon, private vehicles used the left lane, and buses used the right. Van pools picked up passengers in the right lane and then merged into the private vehicle lane. Buses had to wait for van pools to blend into private vehicle traffic.

A study was conducted to determine how the queue length could be minimized while also minimizing the number of security guards. A separate study of the afternoon situation was also conducted to get buses through the gate as soon as possible while again minimizing the number of guards.

Two alternatives were generated initially. First, the number of security guards could be increased, although this was a more expensive alternative. A second proposal was to use all of the available lanes for traffic approaching the gate. However, several difficulties with this approach existed. First, guards entered passenger cars from the driver's side and vans from the passenger side, and posting guards between two lanes of traffic would cause a safety hazard. Second, the company wanted the number of lines to be constant to avoid confusion. Third, guards would have to cross traffic to get to buses when they arrived.

Data were gathered on the interarrival rate, the nature of the queue, and the service rate. The service capacity matched the arrival rate during the hour of interest, so analytical queuing analysis was not practical. Additionally, the number of servers available varied depending on the arrival of buses and the number of security guards on duty that day. Consequently, a simulation model was developed.

The first step was to validate the model. Figure 6-35 shows the actual and simulated data for queue length with a single-service line, showing a very close fit. Next, scenarios were developed to reflect alternative options. The first scenario increased the number of security guards to three while maintaining the single lane of traffic. The simulation showed that this system would have a maximum line length of 28 vehicles, an improvement over the base case. In the second scenario, vehicles were channeled into two lines, each with their own security guard. When buses arrived, one lane of private vehicles would wait while the bus was processed. This alternative had a maximum queue length of 14 vehicles and a waiting time of 12 minutes.

Because the second scenario improved the system at no extra cost, it was tested on real traffic. Observed line length and time matched the simulated data quite closely, with a maximum line length of 21 and waiting time of 14 minutes. Further, traffic increased from 284 to 345 vehicles per day, as fewer private vehicles balked to the second gate.

The analysts observed that the afternoon system, where two security guards checked private vehicles sequentially, involved significant idle time for one of the guards. The study team recommended that van pools not merge with the private vehicle lane, but rather exit using the bus lane. One security guard was proposed for each lane. When this was proposed, an objection was that this new system would delay the buses. However, van pools usually arrived before the buses. When the proposal was implemented, buses were found to have to wait less than 30 seconds and usually left 5 to 7 minutes earlier than they had previously.

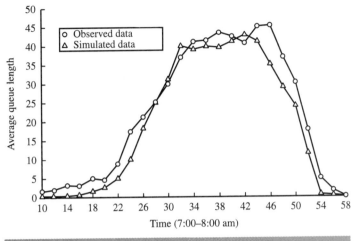

FIGURE 6-35 Queue Length with a Single-Service Line

Source: Reprinted by permission, E. G. Landauer and L. C. Becker, "Reducing Waiting Time at Security Checkpoints," *Interfaces,* Vol. 19, no. 5, 1989, pp. 57–65. Copyright 1989, The Institute of Management Sciences and the Operations Research Society of America (currently INFORMS), 2 Charles Street, Suite 300, Providence RI 02904 USA.

The simulation study demonstrated that increasing the number of servers in a queueing situation was not always the best approach. Increasing the number of service channels with the same number of servers drastically improved the results without incurring additional costs.

A SYSTEM DYNAMICS MODEL FOR INSURANCE EVALUATION[5]

One of the major problems facing society today is the high cost of insurance. This is especially true in the medical field, which has experienced high growth in both medical costs of services and the cost of professional liability. The New York State Insurance Department had the responsibility of recommending means to improve the medical malpractice environment in their state. The premium rates were becoming prohibitive, jeopardizing the willingness of physicians to work in New York. Furthermore, insurance companies were withdrawing coverage because even the high rates were insufficient to cover expected losses. Doctors were pressuring the agency to lower premium rates while insurance companies wanted to raise them. A task force was formed to generate improved policy options, such as reducing insurance company reserve requirements, shifting to no-fault insurance, or placing a cap on dollar awards to malpractice victims.

Stakeholder groups included health care consumers, hospitals, lawyers, malpractice victims, property and casualty insurers, and doctors. Three public decision conferences were held, where stakeholders and decision analysts built analytic models to support the decision-making process. A system dynamics simulation model was used to estimate implications of adopting various policy options.

The system dynamics simulation was used to estimate the long-term implications of decision alternatives. This approach was chosen because it would expose the nature of the medical malpractice insurance system, make controversial assumptions explicit, and provide a common framework that would help policy makers develop a shared understanding

---

[5]Adapted from P. Reagan-Cirincione, S. Schuman, G. P. Richardson, and S. A. Dorf, "Decision Modeling: Tools for Strategic Thinking," *Interfaces,* Vol. 21, no. 6, 1991, pp. 52–65.

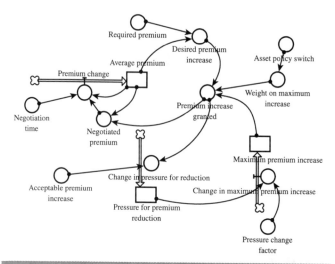

FIGURE 6-36 Portion of System Dynamics Simulation Model

Source: Reprinted by permission, P. Reagan-Cirincione, S. Schuman, G. P. Richardson, and S. A. Dorf, "Decision Modeling: Tools for Strategic Thinking," *Interfaces,* Vol. 21, no. 6, 1991, pp. 52–65. Copyright 1991, The Institute of Management Sciences and the Operations Research Society of America (currently INFORMS), 2 Charles Street, Suite 300, Providence RI 02904 USA.

of the problem. In addition, the model would predict the long-term impact of policy changes on key elements of the system. Figure 6-36 shows a typical portion of the simulation model for the insurance premium-setting process. Output variables of the model were solvency of insurance companies (measured by assets, liabilities, and surplus) and the average insurance premium paid by doctors and by others.

The model was used to replicate the actual behavior of the system over the previous 10 years, providing validation and confidence in the approach. Continuing the simulation into the future also yielded the expected outcomes: increasing premiums and insolvent insurers (see Figure 6-37). One option was a "pay-as-you-go" policy in which

FIGURE 6-37 Simulation Results for Current System

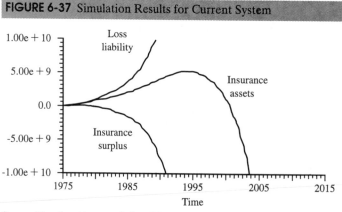

Source: Reprinted by permission, P. Reagan-Cirincione, S. Schuman, G. P. Richardson, and S. A. Dorf, "Decision Modeling: Tools for Strategic Thinking," *Interfaces,* Vol. 21, no. 6, 1991, pp. 52–65. Copyright 1991, The Institute of Management Sciences and the Operations Research Society of America (currently INFORMS), 2 Charles Street, Suite 300, Providence RI 02904 USA.

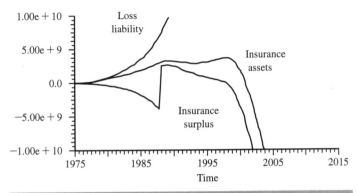

FIGURE 6-38 Simulation Results for Pay-As-You-Go Policy Option

Source: Reprinted by permission, P. Reagan-Cirincione, S. Schuman, G. P. Richardson, and S. A. Dorf, "Decision Modeling: Tools for Strategic Thinking," *Interfaces,* Vol. 21, no. 6, 1991, pp. 52–65. Copyright 1991, The Institute of Management Sciences and the Operations Research Society of America (currently INFORMS), 2 Charles Street, Suite 300, Providence RI 02904 USA.

the stock of assets would only be required to cover a flow of payments over some number of years. This would immediately lower the asset requirement for medical malpractice insurance, cure the insolvency problem, and allow doctors' premiums to drop significantly. However, the simulation model showed that this alternative would provide only short-term relief and would not prevent long-term collapse of the system (see Figure 6-38).

The primary strength of the simulation was better understanding of the complicated interrelationships and feedback loops involved. It also provided a means to explore options and better understand the nature of the system dynamics. The model suggested that no single policy option would provide a suitable solution. As a result, the task force developed three multifaceted policy options that were evaluated using multiattribute utility theory.

Questions and Problems

1. Explain how system simulation models differ from Monte-Carlo simulation models.
2. Describe the differences among activity-scanning, process-driven, and event-driven approaches to simulation.
3. Consider the following scenario. An airport having a single runway and a 3-gate terminal is being proposed for a small city. Actual arrival and departure times will vary from their scheduled times anywhere from 10 minutes early to 60 minutes late due to a variety of factors such as weather and air traffic control. On the runway, arriving aircraft have priority over departures. The time to taxi from the runway to the terminal after landing or from the terminal to the runway before a departure is 5 minutes. Landings and/or takeoffs must have a minimum separation of 2 minutes. Thus, a plane cannot depart if an arrival is scheduled within 2 minutes. Each arriving airplane waits for its assigned gate if it is currently occupied. Discuss how a simulation model might be developed using an activity-scanning, process-driven, or event-driven approach. What advantages and disadvantages would each approach have for this problem?
4. Suppose that in the Mantel Manufacturing example the ending inventory is charged at a rate of $0.50 per unit per day, and the cost of an additional shift is $500. Determine the minimum cost policy for scheduling a second shift.

5. For the inventory model with lost sales, determine the best order quantity–reorder point combination to minimize average annual cost given the lead time distribution.

6. Using the *Lost Sales Inventory Model.xls* model, make cells E2 (order quantity) and E3 (reorder point) decision cells. Set the order quantity to range between 200 to 400 in increments of 25, and the reorder point from 200 to 400 in increments of 50. Find the optimal combination of order quantity and reorder point.

7. Use the back-order simulation model (*Backorder Inventory Model.xls*) to determine the best order quantity–reorder point combination to minimize expected total annual costs.

8. Use the *Backorder Inventory Model.xls* model and set the annual holding cost, annual order cost, and annual shortage cost as forecast variables. Then create an overlay chart to examine the relative importance of these three costs to total annual inventory cost.

9. Modify the spreadsheet *Backorder Inventory Model.xls* to allow a fixed number, k, of back orders, at which point the remaining sales are lost.

10. Modify the spreadsheet *Backorder Inventory Model.xls* to reflect the situation in which 80 percent of the shortages are back orders and the rest are lost sales (where each is chosen randomly).

11. Describe the basic elements of waiting-line systems and how they affect simulation models.

12. An airport ticket counter has a service rate of 180 per hour, exponentially distributed, with Poisson arrivals at the rate of 120 per hour.
    a. Find the operating characteristics of this system by using the analytical formulas presented in this chapter.
    b. Determine the average number in queue, average waiting time, and percent idle time using the spreadsheet model. Use Crystal Ball to quantify the variability in the simulation results, and compute 90 percent confidence intervals for these statistics.
    c. Compare the results from a. and b. and explain any differences.

13. Suppose that the arrival rate to a waiting-line system is 10 customers per hour (exponentially distributed). Using simulation, analyze how the average waiting time changes as the service rate varies from 2 to 10 customers per hour (exponentially distributed) in increments of 2. (You can use *Queuing Simulation.xls*.)

14. A machine shop has a large number of machines that fail regularly. One repairperson is available to fix them. Each machine fails on average every 3 hours, with time between failures being exponential. Repair time has the distribution:

Time	Probability
15 min.	0.1
30 min.	0.2
45 min.	0.3
50 min.	0.4

Simulate the system to determine the average time that machines spend waiting for repair and the average percent of time the repairperson is busy. Obtain averages and 90 percent certainty ranges based on 1,000 trials.

15. A trash collection company uses 15-ton trucks. The trucks leave the landfill at 5:00 A.M. and travel to their collection routes, pick up garbage until the truck is full, return to the landfill to dump the load, and repeat this process until 2:00 P.M. At 2:00 P.M., they return to the landfill, empty their load, and wait until the next morning. The amount of trash collected each hour varies according to the distribution:

Tons/Quarter Hour	Probability
.25	0.05
.50	0.25
.75	0.30
1.00	0.25
1.25	0.15

Similarly, the time to travel to or from the landfill to the routes has the distribution:

Time (min.)	Probability
15	0.10
30	0.45
45	0.30
60	0.15

Develop a process-driven simulation model (both in flowchart and spreadsheet form). The model should simulate a full day's work. Replicate the simulation using Crystal Ball and determine the productivity of the trucks (that is, the amount of trash collected per day).

16. A small factory has two workstations: mold/trim and assemble/package. Jobs to the factory arrive at an exponentially-distributed rate of 1 every 10 hours. Time at the mold/trim station is exponential with a mean of 7 hours. Each job proceeds to the assemble/package station and requires an average of 5 hours, again exponentially distributed. Based on 20 jobs and 1,000 trials, estimate the average time in system, average waiting time at mold and trim, and average waiting time at assembly and packaging (along with 90 percent confidence intervals for each of the three statistics).

17. Modify the factory simulation in problem 16 to include two molding/trim servers (both with the original time distribution).

18. Modify the *Queueing Simulation—Two Server Model.xls* spreadsheet to include the calculation of the percent idle time for each server.

19. Donna Sweigert, an independent tax preparer, is getting ready for the oncoming season. She works in 5- or 6-hour blocks, during which an average of 30 customers arrive. Arrivals are distributed exponentially with an arrival rate of five per hour. Each tax season consists of 78 such sessions, beginning with 5 per week in mid-January and finishing with 7 per week in April. Revenue averages $50 per customer whose tax is prepared. Donna was swamped last year and feels that some good profit potential was lost. Customers who waited 15 minutes left and went to other tax preparers. As a result, she is considering other options:

Option 1: She works alone, completing an average of six tax returns per hour, exponentially distributed.
Option 2: She hires a partner, who would split the work as well as the revenue. This partner would have the same service rate of six returns per hour, again exponentially distributed.
Option 3: Donna attends a class to make her more efficient and expects to be able to service eight customers per hour. The cost of this course is $500.
Option 4: Buy an expert system tax preparation package that takes 10 minutes to prepare each return (constant). The cost of this software is $1,000.

The single-server system has the capacity of serving six customers per hour, exponentially distributed. Simulate 30 customers using all 4 systems. Replicate the simulation

100 times using Crystal Ball to get the distributions of results. Identify the option that is expected to yield the greatest profit for Donna.

20. Define an *event*. How are events used to drive simulation models?
21. Explain the function of a calendar file and a control routine in event-driven simulations.
22. Arrival and processing times in a waiting-line system are as follows:

Arrival Time	Service Time
116	20
202	25
214	51
235	3
329	24
336	4
442	58
553	80
580	10
696	29

Simulate the system manually using an event-driven approach. Show the status of the calendar file when the next event is to be selected, and summarize your results in a table.

23. A manufacturing department has three machines and one repairperson. Machines fail after running approximately 3 hours, exponentially distributed. Repair times have the following distribution:

Repair Time	Probability
15 minutes	0.1
30 minutes	0.2
45 minutes	0.3
60 minutes	0.3
75 minutes	0.1

Simulate this situation using an event-driven approach to determine the average waiting time of machines awaiting repair.

24. What is an *attribute?* How are attributes used in simulation?
25. Explain how continuous simulation and system dynamics models are constructed.
26. A fast-food restaurant has two parallel drive-through windows. Customers have a preference for window 1 if neither is occupied or if the waiting lines are equal. At all other times, the customer chooses the shortest line. After a customer has entered the system, he or she remains there until receiving service. However, the customer may change lines if he or she is the last customer in the line and the other line has two fewer cars. Because of the physical configuration of the system, only three cars may wait in each line. If the system is full, any arriving customers simply leave.
    a. Using only arrival, end-of-service, and end-of-simulation events, construct logical flowcharts for simulating this system using an event-driven approach.
    b. Modify this logic to have the model compute statistics on percent of time windows are occupied, average number of customers in the system, time between departures from each window, and average time the customer is in the system.

27. A small manufacturing facility produces custom-molded plastic parts. The process consists of two sequential operations: mold/trim and assemble/package. Jobs arrive at a mean rate of five jobs every 2 hours. For any job, the mold/trim process is exponential, with a mean of 15 minutes, and the assemble/package operation is also exponential, with a mean of 30 minutes. The mold/trim workstation has room to store only four jobs awaiting processing. The assemble/package station has room to store only two waiting jobs. Jobs are transported from the first to the second workstation in 10 minutes. If the mold/trim workstation is full when a job arrives, the job is subcontracted. If the assemble/package workstation is full when a job is completed at the mold/trim operation, the mold/trim operation cannot process any further jobs until space is freed up in the assemble/package area.
    a. Using events defined as arrivals to the facility, end-of-services at each workstation, and an end-of-simulation event, construct logical flowcharts for simulating this system using the event-driven approach.
    b. Generate arrival and service times on a spreadsheet, and manually conduct a simulation for 20 jobs. Assume that at the start of the simulation, both workstations are busy with end-of-services scheduled at time 1; three units are in the queue of the mold/trim area; and no units are waiting in the assemble/package area.
28. The time between demands for an item has a mean of 1/5th week, exponentially distributed. Ordering lead time is constant at 3 weeks. A periodic review policy has been suggested by which the inventory position is checked at 2-week intervals and an order is placed so that the inventory position at the time of ordering is increased to 36 if the inventory position is 18 or less. Any stock-outs are lost sales.
    a. Define events and construct logical flowcharts for an event-driven simulation.
    b. Generate demands and manually simulate to determine average number of lost sales per week and average safety stock.
29. Consider the continuous system dynamics model of medical rates in Figure 6-31. A proposal has been made to improve the system by limiting medical rate and/or insurance rate increases to a maximum of 5 percent per year. Modify the spreadsheet to simulate each of the following scenarios and discuss the results:
    a. Limit medical rate increases to 5 percent per year only.
    b. Limit insurance rate increases to 5 percent per year only.
    c. Limit both medical rate and insurance rate increases to 5 percent per year.
30. The "cobweb" model in economics assumes that the quantity demanded of a particular product in a specified time period depends on the price in that period. The quantity supplied depends on the price in the preceding period. Also, the market is assumed to be cleared at the end of each period. These assumptions can be expressed in the following equations:

$$S(t) = c + dP(t - 1) + v(t)$$
$$D(t) = a - bP(t - 1) + w(t)$$
$$P(t) = \frac{a - c - dP(t - 1) - v(t) + w(t)}{b} + u(t)$$

where $P(t)$ is the price in period t, $D(t)$ is the demand in period t, and $S(t)$ is the quantity supplied in period t. The variables $u(t)$, $v(t)$, and $w(t)$ are random variables with mean zero and some variance.
    a. Draw an influence diagram.
    b. Suppose that $a = 10,000$, $b = 2$, $c = 0$, $d = 0.1$, $u(t)$ is normal with variance 1, $v(t)$ is normal with variance 0.5, $w(t)$ is normal with variance 0.2, and $P(0) = 4,738$. Simulate this model for 50 time periods. (Note that prices are not allowed to be less than zero.)

    c. Examine the effect on *P(t)* of increasing the variance of *v(t)* from 0.5 to 10 and from 10 to 15.

    d. Compare the results from part b. with the assumptions that u, v, and w are fixed at zero.

References

Brennan, J. E., B. L. Golden, and H. K. Rappoport. "Go with the Flow: Improving Red Cross Bloodmobiles Using Simulation Analysis," *Interfaces,* Vol. 22, no. 5, 1992, pp. 1–13.

Johnson, L. C., and D. C. Montgomery. *Operations Research in Production Planning, Scheduling, and Inventory Control.* New York: John Wiley & Sons, 1974.

Orlicky, J. *Material Requirements Planning.* New York: McGraw-Hill, 1975.

Peterson, R., and E. A. Silver. *Decision Systems for Inventory Management and Production Planning,* 2d ed. New York: John Wiley & Sons, 1985.

CHAPTER

7

Output Analysis and Experimentation for Systems Simulation

Chapter Outline

- Dynamic Behavior in Systems Simulation
  Output Analysis for Terminating Systems
  Output Analysis for Nonterminating Systems
  The Method of Batch Means
- Comparing Policies and Systems
  Independent Sampling
  Correlated Sampling
  Comparing Several Systems
- Experimental Design
  Simple Factorial Experiments
- Simulation in Practice
  Simulation in Waste-Processing System Design
  Improving Florida Driver's Licensing Offices Through Simulation
- Questions and Problems

In Chapter 3, we discussed a variety of statistical issues associated with interpreting the results of Monte-Carlo simulations. Most of these issues, such as computing a confidence interval for a population mean, are straightforward applications of classical statistical techniques. We are able to use these techniques because repeated observations in Monte-Carlo simulations are independent. For systems simulation models, however, statistical analysis issues are not as straightforward because observations are not always independent.

For example, the waiting times of successive customers in a queue do not constitute a random sample, because if one customer waits a long time, it is highly likely that the following customer will also wait a long time. Similarly, in an inventory model, the ending stock level one week becomes the beginning stock level for the next week, causing the set of ending inventory values not to be independent. When the value of one random variable has some influence on the value of the next, the sequence of random variables is said to be **autocorrelated** (that is, correlated with itself). Dealing with autocorrelated data requires different approaches than classical statistics. In this chapter, we discuss several key issues of output analysis for systems simulation models that exhibit autocorrelation, and we will also address issues of comparing policy alternatives and designing simulation experiments.

Dynamic Behavior in Systems Simulation

In developing systems simulation models, we generally are interested in one of two things: *short-term behavior* or *steady-state behavior*. For instance, consider the Dan's car wash waiting-line model that we studied in the previous chapter. If the facility is open from 9 A.M. to 7 P.M., we would not only be interested in averages over the entire day, but also in short-term dynamic behavior, such as the time it takes until a certain queue length builds up, or perhaps the maximum length of the queue during the day. This is an example of a **terminating system,** one that runs for some duration of time until some natural event stops or closes the system. To simulate a terminating system, we must characterize the **initial conditions** that define the beginning of the simulation. For instance, when the car wash opens, the initial conditions would be that the system is initially empty. However, suppose we are simulating the daily operation of a factory that operates with one shift, but for which orders accumulate during the evening. At the beginning of each day, we would want to set initial conditions that include a representative number of orders received during the previous evening.

A **nonterminating system** is one that runs continuously or over a very long period of time; for example, simulation of an automated manufacturing facility that runs three shifts a day, 7 days a week, or an Internet service provider. For a nonterminating system simulation, we would be interested in the steady-state performance of the system, as measured by the expected queue lengths or waiting times over the long term, and which do not depend on initial conditions as in a terminating system. Because the system runs continuously, there is no natural termination event; nevertheless, we must stop the simulation artificially after some period of time. This brings up the question of how long to run the simulation in order to ensure that the results are representative of the true steady-state behavior. In addition, autocorrelation introduces biases in trying to estimate steady-state results. Thus, the length of the simulation run, measured by time or the number of entities processed, will affect the results.

OUTPUT ANALYSIS FOR TERMINATING SYSTEMS

For most simulation studies, it is more appropriate to define an ending clock time of the simulation rather than the number of entities that are processed, because this is the way we would naturally view a system. Figure 7-1 shows a modification of the queueing simulation spreadsheet for the Dan's car wash example that incorporates an ending simulation time as an input (cell B5). We have added a column to denote the system status as either "Open" or "Closed," depending on the arrival time of each customer. In the forecast cells, we use the Excel *MATCH* function to identify the row at which the system is first closed and the *INDEX* function to identify the current value of the forecast statistic at this time. Figure 7-2 shows the Crystal Ball results for running the system for only 2 hours. In comparing these results with Figure 6-16, we see that the average number in the queue (1.0948) and average waiting time (0.0752) are much smaller; the average percent idle time (32.14 percent) is larger. This makes sense if you realize that we started with an empty system, so the customers arriving at the beginning will generally not have to wait; with less chance of queueing, the server will have more idle time. As we saw in Figures 6-12 through 6-14, it takes quite a bit of time for the system to settle down at the

	Dan's Car Wash - Terminating Model				Arrival	System	No. In	Start	Service	Completion	Wait	Idle
			Customer	TBA	Time	Status	Queue	Time	Time	Time	Time	Time
2											0.00		
3	Mean arrival rate	15											
4	Mean service rate	20	1	0.1046	0.1046	Open	0	0.1046	0.0453	0.1499	0.0000	0.1046
5	Simulation Run Length	1.0	2	0.0294	0.1339	Open	1	0.1499	0.0136	0.1634	0.0159	0.0000
6	(hours)		3	0.1484	0.2823	Open	0	0.2823	0.1449	0.4272	0.0000	0.1189
7			4	0.0872	0.3695	Open	1	0.4272	0.0077	0.4350	0.0578	0.0000
8	Expected No. in Queue	2.250	5	0.0492	0.4187	Open	2	0.4350	0.0839	0.5188	0.0163	0.0000
9	Expected Waiting Time	0.150	6	0.0757	0.4943	Open	1	0.5188	0.0165	0.5353	0.0245	0.0000
10	Expected % idle time	25%	7	0.0049	0.4992	Open	2	0.5353	0.0395	0.5748	0.0361	0.0000
11			8	0.1350	0.6342	Open	0	0.6342	0.0332	0.6674	0.0000	0.0594
12	*Simulation Results*		9	0.0322	0.6664	Open	1	0.6674	0.1460	0.8134	0.0010	0.0000
13			10	0.0798	0.7463	Open	1	0.8134	0.0073	0.8207	0.0671	0.0000
14	No. in Queue	0.5526	11	0.2621	1.0084	Closed	0	1.0084	0.0674	1.0758	0.0000	0.1877
15	Waiting Time	0.0219	12	0.0804	1.0887	Closed	0	1.0887	0.0189	1.1076	0.0000	0.0129
16	% idle time	34%	13	0.0909	1.1797	Closed	0	1.1797	0.0161	1.1957	0.0000	0.0721
17			14	0.0219	1.2015	Closed	0	1.2015	0.0138	1.2153	0.0000	0.0058
18			15	0.0898	1.2913	Closed	0	1.2913	0.0242	1.3155	0.0000	0.0760

(a)

	B	C	D	E	F	G
4	20	1	=CB.Exponential($B$3)	=D4	Open	0
5	1	2	=CB.Exponential($B$3)	=E4+D5	=IF(E5<$B$5,"Open","Closed")	=C5-MATCH(E5,$J$3:J4,1)
6		3	=CB.Exponential($B$3)	=E5+D6	=IF(E6<$B$5,"Open","Closed")	=C6-MATCH(E6,$J$3:J5,1)
7		4	=CB.Exponential($B$3)	=E6+D7	=IF(E7<$B$5,"Open","Closed")	=C7-MATCH(E7,$J$3:J6,1)
8	=$B$3^2/($B$4*($B$4-$B$3))	5	=CB.Exponential($B$3)	=E7+D8	=IF(E8<$B$5,"Open","Closed")	=C8-MATCH(E8,$J$3:J7,1)
9	=B8/B3	6	=CB.Exponential($B$3)	=E8+D9	=IF(E9<$B$5,"Open","Closed")	=C9-MATCH(E9,$J$3:J8,1)
10	=1-B3/B4	7	=CB.Exponential($B$3)	=E9+D10	=IF(E10<$B$5,"Open","Closed")	=C10-MATCH(E10,$J$3:J9,1)
11		8	=CB.Exponential($B$3)	=E10+D11	=IF(E11<$B$5,"Open","Closed")	=C11-MATCH(E11,$J$3:J10,1)
12		9	=CB.Exponential($B$3)	=E11+D12	=IF(E12<$B$5,"Open","Closed")	=C12-MATCH(E12,$J$3:J11,1)
13		10	=CB.Exponential($B$3)	=E12+D13	=IF(E13<$B$5,"Open","Closed")	=C13-MATCH(E13,$J$3:J12,1)
14	=INDEX($N$3:$N$503,MATCH("Closed",$F$4:$F$503,0))	11	=CB.Exponential($B$3)	=E13+D14	=IF(E14<$B$5,"Open","Closed")	=C14-MATCH(E14,$J$3:J13,1)
15	=INDEX($P$3:$P$503,MATCH("Closed",$F$4:$F$503,0))	12	=CB.Exponential($B$3)	=E14+D15	=IF(E15<$B$5,"Open","Closed")	=C15-MATCH(E15,$J$3:J14,1)
16	=INDEX($R$3:$R$503,MATCH("Closed",$F$4:$F$503,0))	13	=CB.Exponential($B$3)	=E15+D16	=IF(E16<$B$5,"Open","Closed")	=C16-MATCH(E16,$J$3:J15,1)

(b)

	H	I	J	K	L	M	N	O	P	Q	R
4	=MAXA(E4,J3)	=CB.Exponential($B$4)	=H4+I4	=H4-E4	=H4-E4	=G4*E4	=M4/E4	=K4	=O4/C4	=L4	=Q4/J4
5	=MAXA(E5,J4)	=CB.Exponential($B$4)	=H5+I5	=H5-E5	=H5-E5	=G5*(E5-E4)+M4	=M5/E5	=O4+K5	=O5/C5	=Q4+L5	=Q5/J5
6	=MAXA(E6,J5)	=CB.Exponential($B$4)	=H6+I6	=H6-E6	=H6-E6	=G6*(E6-E5)+M5	=M6/E6	=O5+K6	=O6/C6	=Q5+L6	=Q6/J6
7	=MAXA(E7,J6)	=CB.Exponential($B$4)	=H7+I7	=H7-E7	=H7-E7	=G7*(E7-E6)+M6	=M7/E7	=O6+K7	=O7/C7	=Q6+L7	=Q7/J7
8	=MAXA(E8,J7)	=CB.Exponential($B$4)	=H8+I8	=H8-E8	=H8-E8	=G8*(E8-E7)+M7	=M8/E8	=O7+K8	=O8/C8	=Q7+L8	=Q8/J8
9	=MAXA(E9,J8)	=CB.Exponential($B$4)	=H9+I9	=H9-E9	=H9-E9	=G9*(E9-E8)+M8	=M9/E9	=O8+K9	=O9/C9	=Q8+L9	=Q9/J9

(c)

FIGURE 7-1 Queueing Simulation—Terminating Model

average value. This is why we cannot rely on the analytical results to describe the behavior of terminating systems and why simulation is valuable in such situations.

The analysis of a terminating system is the same as for any Monte-Carlo simulation model; we need to run the model for multiple trials to obtain a distribution of key statistics. Then we may use classical statistical methods to develop point estimates and confidence intervals because each trial is an independent sample. For the 2-hour simulation summarized in Figure 7-2, we found the following (based on 1,000 trials):

	Number in Queue	*Waiting Time*	*Percent Idle Time*
Mean	1.0948	0.0752	32.14%
Standard deviation	1.1745	0.0740	18.29%
Standard error	0.0371	0.0023	0.58%

Therefore, a 95 percent confidence interval for the mean number in the queue would be

$$1.0948 \pm 1.96(0.0371) \text{ or } [1.022, 1.168]$$

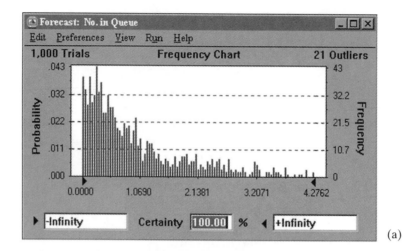

(a)

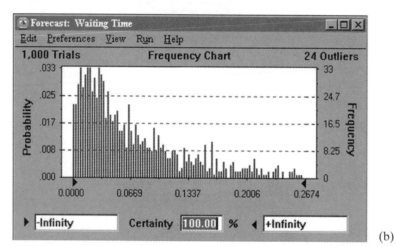

(b)

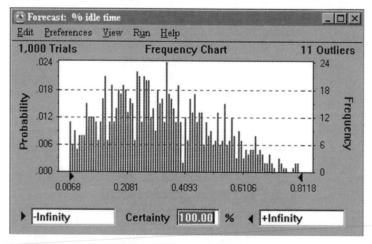

(c)

FIGURE 7-2 2-Hour Terminating Simulation Results

OUTPUT ANALYSIS FOR NONTERMINATING SYSTEMS

To understand the key issues involved with nonterminating systems simulation, let us revisit the Dan's car wash example. If you look back at Figures 6-12 through 6-14, you will observe that the statistics eventually approach the steady-state expected values, but they fluctuate significantly from the expected values during the early portion of the simu- lation run. It appears that steady-state is reached after about 200 customers have moved through the system or, equivalently, after about 10 hours of simulated time. The period during which these performance measures fluctuate before stabilizing is called the **tran- sient state.** The transient bias is a result of the initial conditions—recall that we ran this simulation starting with an empty system. Note that the output statistics computed over the course of the entire simulation include data from the transient period. If we stop a systems simulation model prematurely, the output statistics may not reflect the true steady-state values, as we observed in our discussion of terminating systems.

There are three ways of dealing with transient behavior in estimating steady-state statistics. First, we can run the simulation for a long time to reduce the relative impact of the transient bias. Without some experimental analysis, it may be difficult to know how long is "long enough." Another approach is to begin the simulation with initial con- ditions (for example, the number in the queue) that are representative of steady-state. It is difficult to include initial conditions in a process-driven model because we need to specify departure times for each of the initial customers; however, this is quite easy to do in an event-driven model. Figure 7-3 shows a modification to the event-driven model in Figure 7-1. Cell B7 allows an input of the initial number in the system. The first event is an initialization event. If the initial number in the system is zero, this event simply schedules the first arrival and sets the server to idle status. If the initial number in the system is greater than zero, the server is set to busy status, and a departure event is also scheduled for the entity in service with the remainder waiting in the queue.

To illustrate the effect that initial conditions can have on a simulation, we ran a 5-hour simulation starting with both an empty system and with four initial customers in the sys- tem. The results are as follows:

INITIAL CONDITION: EMPTY SYSTEM

	Number in Queue	*Percent Idle Time*
Mean	1.803	28%
Standard deviation	1.461	10%
Standard error	0.046	0.33%

INITIAL CONDITION: FOUR CUSTOMERS IN SYSTEM

	Number in Queue	*Percent Idle Time*
Mean	2.22	24%
Standard deviation	1.976	11%
Standard error	0.062	0.34%

(a)

	A	B	C	D	E	F	G	H	I	J	K	L
1	Dan's Car Wash Event-Driven Simulation - Terminating Model With Initial Conditions											
2												
3	Mean arrival rate	15										
4	Mean service rate	20										
5	Simulation Run Length	2.0										
6	(hours)		Event	Type	Clock	System	No. in	No. in	Server	Idle	Next	Next
7	Initial Number in System	2			Time	Status	System	Queue	Status	Time	Arrival	Departure
8			1	Initialize	0.000	Open	2	1	busy	0.000	0.048	0.018
9	Simulation Results		2	depart	0.018	Open	1	0	busy	0.000	0.048	0.026
10			3	depart	0.026	Open	0	0	idle	0.000	0.048	99999.000
11	Avg. No. in System	2.397	4	arrival	0.048	Open	1	0	busy	0.022	0.071	0.087
12	Avg. No. in Queue	1.599	5	arrival	0.071	Open	2	1	busy	0.000	0.112	0.087
13	% Idle Time	20%	6	depart	0.087	Open	1	0	busy	0.000	0.112	0.124
14			7	arrival	0.112	Open	2	1	busy	0.000	0.131	0.124
15			8	depart	0.124	Open	1	0	busy	0.000	0.131	0.223
16			9	arrival	0.131	Open	2	1	busy	0.000	0.166	0.223
17			10	arrival	0.166	Open	3	2	busy	0.000	0.179	0.223

(b)

	B	C	D	E	F
8		1	Initialize	0	Open
9		2	=IF(L8<K8,"depart","arrival")	=MIN(K8,L8)	=IF(E9<$B$5,"Open","Closed")
10		3	=IF(L9<K9,"depart","arrival")	=MIN(K9,L9)	=IF(E10<$B$5,"Open","Closed")
11	=INDEX($N$9:$N$507,MATCH("Closed",F8:F507,0)-1)	4	=IF(L10<K10,"depart","arrival")	=MIN(K10,L10)	=IF(E11<$B$5,"Open","Closed")
12	=INDEX($P$9:$P$507,MATCH("Closed",F8:F507,0)-1)	5	=IF(L11<K11,"depart","arrival")	=MIN(K11,L11)	=IF(E12<$B$5,"Open","Closed")
13	=INDEX($Q$9:$Q$507,MATCH("Closed",F8:F507,0)-1)	6	=IF(L12<K12,"depart","arrival")	=MIN(K12,L12)	=IF(E13<$B$5,"Open","Closed")
14		7	=IF(L13<K13,"depart","arrival")	=MIN(K13,L13)	=IF(E14<$B$5,"Open","Closed")

(c)

	G	H	I	J	K
8	=B7	=MAX(G8-1,0)	=IF(G8>0,"busy","idle")	=IF(I8="idle",E8,0)	=E8+CB.Exponential($B$3)
9	=IF(D9="depart",G8-1,G8+1)	=MAX(G9-1,0)	=IF(G9=0,"idle","busy")	=IF(I8="idle",E9-E8,0)	=IF(D9="arrival",E9+CB.Exponential($B$3),K8)
10	=IF(D10="depart",G9-1,G9+1)	=MAX(G10-1,0)	=IF(G10=0,"idle","busy")	=IF(I9="idle",E10-E9,0)	=IF(D10="arrival",E10+CB.Exponential($B$3),K9)
11	=IF(D11="depart",G10-1,G10+1)	=MAX(G11-1,0)	=IF(G11=0,"idle","busy")	=IF(I10="idle",E11-E10,0)	=IF(D11="arrival",E11+CB.Exponential($B$3),K10)
12	=IF(D12="depart",G11-1,G11+1)	=MAX(G12-1,0)	=IF(G12=0,"idle","busy")	=IF(I11="idle",E12-E11,0)	=IF(D12="arrival",E12+CB.Exponential($B$3),K11)
13	=IF(D13="depart",G12-1,G12+1)	=MAX(G13-1,0)	=IF(G13=0,"idle","busy")	=IF(I12="idle",E13-E12,0)	=IF(D13="arrival",E13+CB.Exponential($B$3),K12)
14	=IF(D14="depart",G13-1,G13+1)	=MAX(G14-1,0)	=IF(G14=0,"idle","busy")	=IF(I13="idle",E14-E13,0)	=IF(D14="arrival",E14+CB.Exponential($B$3),K13)

(d)

	L
8	=IF(G8>0,E8+CB.Exponential($B$4),99999)
9	=IF(AND(D9="depart",G9>0),E9+CB.Exponential($B$4),IF(AND(D9="depart",G9=0),99999,IF(AND(D9="arrival",I8="idle"),E9+CB.Exponential($B$4),L8)))
10	=IF(AND(D10="depart",G10>0),E10+CB.Exponential($B$4),IF(AND(D10="depart",G10=0),99999,IF(AND(D10="arrival",I9="idle"),E10+CB.Exponential($B$4),L9)))
11	=IF(AND(D11="depart",G11>0),E11+CB.Exponential($B$4),IF(AND(D11="depart",G11=0),99999,IF(AND(D11="arrival",I10="idle"),E11+CB.Exponential($B$4),L10)))
12	=IF(AND(D12="depart",G12>0),E12+CB.Exponential($B$4),IF(AND(D12="depart",G12=0),99999,IF(AND(D12="arrival",I11="idle"),E12+CB.Exponential($B$4),L11)))
13	=IF(AND(D13="depart",G13>0),E13+CB.Exponential($B$4),IF(AND(D13="depart",G13=0),99999,IF(AND(D13="arrival",I12="idle"),E13+CB.Exponential($B$4),L12)))
14	=IF(AND(D14="depart",G14>0),E14+CB.Exponential($B$4),IF(AND(D14="depart",G14=0),99999,IF(AND(D14="arrival",I13="idle"),E14+CB.Exponential($B$4),L13)))

FIGURE 7-3 Queueing Simulation—Terminating Model with Initial Conditions

Recall from Chapter 6 that the expected values are $L_q = 2.25$ and percent idle time $= 25$ percent. With an empty system, a 5-hour run results in a much smaller number in queue and larger idle time, as we had discussed previously. However, by setting the initial conditions representative of the steady-state results, we obtained averages much closer to the analytical values without running the simulation for a long period of time. Unfortunately, it is not always possible to identify such conditions in advance.

A third approach is to run the simulation for some period of time—a "warm-up" period—and then eliminate all the data from the transient state, using only data captured after this time. With whatever method we choose, we may then use independent trials to develop point and interval estimates. Most practitioners choose the second approach of eliminating data from the transient period; commercial software has this capability built in.

THE METHOD OF BATCH MEANS

Developing confidence intervals for performance measures requires repeated trials. Many complex system simulation models may take hours to run, even with the speed of modern computers. Thus, making a sufficient number of trials to develop confidence intervals with an acceptable precision may not be feasible. In addition, we waste a considerable amount of data, because we must remove the transient period from every replication.

One approach to obtain replications of data from one long simulation run as opposed to repeating the simulation run itself is called the *method of batch means*. The method of batch means works as follows:

1. Run the simulation long enough to remove any transient effects and provide a sufficient amount of data representative of steady-state.
2. Divide the remaining length of the simulation run into subintervals of time corresponding to "batches" of data.
3. Compute average performance measures for each batch, and use classical techniques to develop estimates, treating the batch means as independent replications.

To illustrate the batch means approach, we ran one 1,000-customer run of the Dan's car wash example and computed the average waiting time for the last 500 customers in batches of 100, with the following results:

Batch	Average Waiting Time
1	0.0512
2	0.2536
3	0.1797
4	0.2080
5	0.0899
Mean	0.1565
Standard deviation	0.0839
Standard error	0.0375

A 95 percent confidence interval would be (using t-values because the sample size is small):

$$\bar{x} \pm t_{.025,4} s/\sqrt{n} \quad \text{or} \quad 0.1565 \pm (2.776)0.0375$$

Thus, the confidence interval is (0.1461, 0.1670), which includes the true expected value of 0.15. However, a Monte-Carlo simulation of the batch means (see the Excel file *Queueing Simulation—Batch Means.xls*) shows considerable variability (see Figure 7-4). This might be due to the fact that the sample size is small and there is significant sampling error, or that the last 500 customers may not always be representative of steady-state and that the simulation needs to be run for a longer time.

In reality, successive batches will exhibit some autocorrelation. For example, if the system is congested at the end of one batch, it will be congested for some time at the beginning of the next batch. One way of reducing the autocorrelation effect is to eliminate every other batch (for example, use only the odd-numbered batches), although this reduces the number of independent observations and results in weaker confidence intervals. Doing this, along with repeating the entire simulation run a few times, will provide

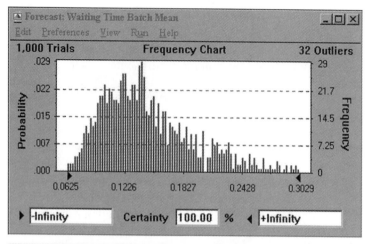

FIGURE 7-4 Distribution of Batch Means Obtained Through Monte-Carlo Sampling

better results and still avoid the necessity of replicating the entire simulation a large number of times.

Comparing Policies and Systems

A common application of simulation analysis is to compare two or more systems. For example, we might wish to compare the effect of speeding up service by adding some automated technology or increasing the number of servers in a queueing system. Or, we might want to compare different order quantity–reorder point inventory decision rules. As before, confidence intervals are the principal means of making comparisons. To illustrate this, let us return to the Dave's Candies example that we studied in Chapter 2. We will compare the policy of ordering 70 boxes with the policy of ordering 80 boxes.

INDEPENDENT SAMPLING

The first approach we describe, called **independent sampling,** consists of repeating the simulation for each order quantity by using independent demands; that is, a different stream of demands for each policy. Using the *Daves Candies Monte Carlo Simulation.xls* spreadsheet (see Figure 2-21), we would generate 100 trials for an order quantity of 70 and compute the mean and standard deviation of profit. Then, we would generate a *new* set of demands and rerun the simulation for an order quantity of 80. The results we obtained are

Order Quantity	Mean	Standard Deviation
70	247.70	68.572
80	259.20	96.061

If the sample size is large enough, a $100(1-\alpha)$ percent confidence interval for the difference in means is:

$$\bar{x}_1 + \bar{x}_2 \pm z_{\alpha/2} \sqrt{\frac{s_1^2}{n_1} + \frac{s_2^2}{n_2}}$$

The term under the square root represents the variance of the difference in means. Thus, for the results shown here, a 90 percent confidence interval would be

$$259.20 - 247.70 \pm 1.645\sqrt{\frac{(96.061)^2}{100} + \frac{(68.572)^2}{100}} = (-7.915, 30.915)$$

Because zero is contained in the confidence interval, we cannot reject the hypothesis that the means are equal. Therefore, we would conclude that there is no significant difference between ordering 70 boxes versus 80 boxes.

As we discussed in Chapter 3, we may reduce the size of the confidence interval by taking a larger sample. Another way to accomplish this *without* increasing the sample size is to use correlated sampling for the simulations.

CORRELATED SAMPLING

Correlated sampling uses the same stream of random numbers to generate random variates for both systems under consideration. This technique is also referred to as *common random numbers*. (Recall in Crystal Ball that you have the option of specifying a random number seed to control the stream of random numbers.) Because the same random numbers are used for each alternative, the results are no longer independent, but are positively correlated. Theoretically, this reduces the variance of difference in means. Specifically, if the sample sizes are identical, variance of the difference in means is reduced by

$$\frac{2\rho_{12}\sigma_1\sigma_2}{n}$$

where ρ_{12} is the correlation coefficient between the pairs of observations.

To compute a $100(1 - \alpha)$ percent confidence interval with correlated data, we first compute the pairwise differences between the two systems under consideration. Then we compute the sample mean, $\bar{D}$, and sample standard deviation, s_D, of the differences. The confidence interval is $\bar{D} \pm z_{\alpha/2}s_D/\sqrt{n}$. If this interval contains zero, then we would conclude that no significant difference exists between the two systems.

Figure 7-5 shows a modified spreadsheet for the Dave's Candies example for comparing the 70- and 80-unit ordering policies with correlated sampling. Column H computes the pairwise differences. For 100 replications, we found an average difference of -6.00 and standard deviation of 28.76. Therefore, a 90 percent confidence interval is

$$-6.00 \pm 1.645(28.76/\sqrt{100}) = [-10.73, -1.27]$$

Note that the confidence interval is much tighter than for independent sampling and suggests that a significant difference does exist between the alternatives because zero is not in the interval.

Correlated sampling is a means of **variance reduction.** One advantage of correlated sampling is that it provides an efficient way of comparing any number of alternatives at once. Other techniques for reducing the variance in simulation output exist, but they are beyond the scope of this book.

COMPARING SEVERAL SYSTEMS

In many simulation studies, it is desired to compare more than two alternatives. For example, we might wish to compare several alternatives to the mean performance of an existing system, or we might want to compare all possible combinations against one another. We can do these by constructing confidence intervals for pairwise comparisons using either independent or correlated sampling as described earlier. Another objective

	A	B	C	D	E	F	G	H
1	**Dave's Candies Simulation**			**Simulation Results**		**Order Quantity**		
2				Trial	Demand	70	80	Difference
3	Selling price	$ 12.00		1	90	$ 315.00	$ 360.00	$ (45.00)
4	Cost	$ 7.50		2	60	$ 255.00	$ 240.00	$ 15.00
5	Discount price	$ 6.00		3	60	$ 255.00	$ 240.00	$ 15.00
6				4	40	$ 135.00	$ 120.00	$ 15.00
7	Demand	Probability		5	40	$ 135.00	$ 120.00	$ 15.00
8	40	1/6		6	70	$ 315.00	$ 300.00	$ 15.00
9	50	1/6		7	70	$ 315.00	$ 300.00	$ 15.00
10	60	1/6		8	70	$ 315.00	$ 300.00	$ 15.00
11	70	1/6		9	80	$ 315.00	$ 360.00	$ (45.00)
12	80	1/6		10	40	$ 135.00	$ 120.00	$ 15.00
13	90	1/6		11	40	$ 135.00	$ 120.00	$ 15.00
14				12	70	$ 315.00	$ 300.00	$ 15.00
15				13	60	$ 255.00	$ 240.00	$ 15.00
16				14	90	$ 315.00	$ 360.00	$ (45.00)
17				15	80	$ 315.00	$ 360.00	$ (45.00)
18	**Random Number Range**		**Demand**	16	70	$ 315.00	$ 300.00	$ 15.00
19	0	1/6	40	17	40	$ 135.00	$ 120.00	$ 15.00
20	1/6	1/3	50	18	50	$ 195.00	$ 180.00	$ 15.00
21	1/3	1/2	60	19	90	$ 315.00	$ 360.00	$ (45.00)
22	1/2	2/3	70	20	90	$ 315.00	$ 360.00	$ (45.00)
23	2/3	5/6	80	21	40	$ 135.00	$ 120.00	$ 15.00
24	5/6	1	90	22	90	$ 315.00	$ 360.00	$ (45.00)

FIGURE 7-5 Modification of Dave's Candies Simulation for Correlated Sampling

might be to select the best alternative. This requires advanced statistical methods that are beyond the scope of this book.

To illustrate this, we will use the inventory simulation model with back orders that we developed in Chapter 6 (see Figure 6-8). In this model, daily demand is Poisson distributed with a mean of 1.8. Let us assume that the existing inventory system has a policy of ordering $Q = 7$ units whenever the inventory position falls to $r = 3$ or less. We wish to compare this system with three alternatives: $Q = 7, r = 5; Q = 10, r = 3;$ and $Q = 10, r = 5$. To do this, we will construct confidence intervals for $\mu_1 - \mu_2, \mu_1 - \mu_3,$ and $\mu_1 - \mu_4$.

We will replicate the inventory system 10 times for each policy and use common random numbers to generate the demand. Therefore, we will use the methodology described for correlated sampling and develop our confidence intervals from the differences of paired observations. Instead of using the *CB.Poisson* function in the spreadsheet model to generate demands, we use the *Tools/Data Analysis/Random Number Generation* option in Excel. We then recalculate the spreadsheet for each option, because recalculation will not change the demand stream set by the *Random Number Generation* procedure. Then, we would change the demand stream and evaluate each option.

The results are shown in Table 7-1. We will construct 99 percent confidence intervals. When comparing multiple systems, this guarantees a high level of confidence that all confidence intervals contain the true mean being estimated.[1] Because we have 10 tri-

---

[1] In statistical theory, this is a result of the Bonferroni inequality, which states that if, for example, three 95 percent confidence intervals are constructed, the overall confidence that all three contain the true mean can be as low as $1 - 3(.05) = .85$. Thus, if each confidence level is 99 percent, the overall confidence level is $1 - 3(.01) = .97$.

TABLE 7-1   Simulation Results for Comparing Inventory System Policies

	Average Daily Cost				*Observed Differences*		
	Q = 7, r = 3	*Q = 7, r = 5*	*Q = 10, r = 3*	*Q = 10, r = 5*			
Replication	*1*	*2*	*3*	*4*	*(1) − (2)*	*(1) − (3)*	*(1) − (4)*
1	1856	1852	**1494**	1788	4	362	68
2	1674	1536	**1384**	1644	138	290	30
3	1756	1790	1568	**1496**	−34	188	260
4	1424	1712	**1288**	1724	−288	136	−300
5	1714	1694	1620	**1568**	20	94	146
6	1712	1688	**1478**	1660	24	234	52
7	1842	1884	1674	**1666**	−42	168	176
8	1662	1810	**1490**	1714	−148	172	−52
9	1574	1506	**1466**	1478	68	108	96
10	2076	1794	1836	**1716**	282	240	360
				Mean	2.40	199.20	83.60
				Std. Dev.	153.44	83.56	179.43
				Std. Error	48.52	26.42	56.74

als, we use a t-value with an upper tail of 0.005 and 9 degrees of freedom, which is 3.250. The confidence intervals are constructed as the mean difference plus or minus 3.250 times the standard error. Therefore, we have

$$-155.29 \leq \mu_1 - \mu_2 \leq 160.09$$
$$113.335 \leq \mu_1 - \mu_3 \leq 285.065$$
$$-127.665 \leq \mu_1 - \mu_4 \leq 241.145$$

Note that the first and third confidence intervals contain zero; thus, we would conclude that no significant difference exists between the first and second policy, and between the first and fourth policy. However, the confidence interval for $\mu_1 - \mu_3$ lies completely above zero, which provides strong evidence that policy 3 is significantly better than policy 1, because its average daily cost is smaller.

A simple, though not rigorous, approach to comparing several systems when using common random numbers is simply to count the number of cases in which a particular system is best, and divide by the total number of replications. This provides an estimate of the probability that a particular system shows the best performance. For example, the costs shown in bold in Table 7-1 are the lowest for each replication. We see that the policy of $Q = 10$ and $r = 3$ is best in six of the 10 replications. Similarly, the policy $Q = 10$ and $r = 5$ is best in four of the 10. The first two policies are never the best. This provides strong evidence that an order quantity of $Q = 10$ is better than one with $Q = 7$, at least for reorder points of 3 and 5. Of course, we should probably use a larger number of replications in order to allow for extreme occurrences.

This approach can provide better insight for decision making than a formal statistical analysis. For example, statistical comparison showed that no significant difference exists between policies 1 and 4. Nevertheless, policy 4 is better than policy 1 in eight of the 10 cases. This is because the statistical analysis incorporates only the mean and standard deviation into the comparison, rather than the relative system performance under controlled circumstances (i.e., using common random numbers).

Experimental Design

An important part of the simulation process is designing effective and efficient experiments. This requires understanding the purpose of the simulation study, which might be to compare the means and variances of output variables under different system conditions, to determine the importance or effect of different variables, or to identify an optimal set of variables. An experiment consists of a set of *factors*—the input parameters and assumptions in a model, and *responses*—output performance measures. In a simple waiting-line model, for instance, factors might be the number of servers, type of queue discipline, arrival rate, and service rate. Responses might be the average time in the system, average system utilization, average length of the waiting line, probability that the server is idle, and so on.

Experimental design, often referred to as *design of experiments (DOE),* was developed by R. A. Fisher in England in the 1920s. A designed experiment is a test or series of tests that enables the experimenter to compare two or more methods to determine which is better or to determine levels of controllable factors to optimize the yield of a process or minimize the variability of a response variable. We actually performed a simple experimental design in the last section when we studied the differences between two systems by changing the ordering policy, with controllable factors being the order quantity and reorder point. Many simulation studies involve the analysis of several factors. For instance, in a queuing situation, you may wish to vary the speed of service and the number of servers simultaneously. Experimental design is concerned with identifying the factors that have significant effects on the responses.

SIMPLE FACTORIAL EXPERIMENTS[2]

Let us consider a queueing system for which we wish to determine the effects of two key factors: service rate and service distribution. The arrival rate is 9 per hour, exponentially distributed. The service rate is either 12 jobs per hour or 15 jobs per hour, and the service distribution is either exponential or constant. Therefore, we have two factors, each having two levels, giving a total of $2^2 = 4$ experimental combinations. The response of interest is the average time a job spends in the system.

We simulate each combination for a fixed number of replications, say 50. The average times that customers spend in the system and standard deviations for each combination are

Factor 1	Factor 2	Average Time in System	Standard Deviation
12 jobs/hour	Exponential	$W = 0.323$ hours	0.171
12 jobs/hour	Constant	$W = 0.206$ hours	0.075
15 jobs/hour	Exponential	$W = 0.118$ hours	0.014
15 jobs/hour	Constant	$W = 0.118$ hours	0.022

Based on the average time, the best combination appears to be one with the faster service rate. The worst combination appears to be the one with the slower service rate and exponential distribution. The difference of the average time is $0.323 - 0.118 = 0.205$

---

[2]This discussion assumes that readers are familiar with basic concepts of regression and analysis of variance and may be skipped as appropriate.

hours. We would like to determine how much of this gain is due to the service rate alone and how much is due to eliminating the variance in service time.

One approach to quantify the average impact of factors on the response variable is least squares regression analysis. A simple two-factor regression model relating the time in system to the service rate and service distribution is

$$W = \beta_0 + \beta_1 X + \beta_2 Y + \beta_{12} XY$$

where X is an indicator (dummy) variable corresponding to the service rate (a rate of 12/hour corresponds to $X=0$ and 15/hour to $X=1$), and Y corresponds to the service distribution (the exponential corresponds to $Y=0$ and the constant distribution to $Y=1$). The model includes an interaction term for the service rate and distribution.

Using the regression option from the *Data Analysis Toolpak* in Excel, we obtain the following results:

	Variable	Parameter Estimate	T for H₀ Parameter = 0	Prob > |T|
	Intercept	0.32253	24.20	0.0001
X	Service rate	−0.20411	−10.83	0.0001
Y	Distribution	−0.11634	−6.17	0.0001
XY	Interaction	0.11620	4.36	0.0001

The intercept represents the average time in the system when X and Y are both zero. The negative regression coefficients for X and Y suggest that the time in the system will be reduced by increasing the service rate and moving to a constant distribution. Both service rate and distribution were significant at reasonable probability levels, although changing the service rate was more significant. The interaction effect was also significant, indicating that the gains for increasing the service rate provided most if not all of the benefit gained from moving to a constant distribution.

We might extend this experiment by adding a third factor representing the number of servers. This adds an additional dummy variable Z, where $Z=0$ corresponds to a one-server system, and $Z=1$ corresponds to a two-server system. This increases the number of experimental combinations to $2^3 = 8$. The three-factor regression model is

$$W = \beta_0 + \beta_1 X + \beta_2 Y + \beta_3 Z + \beta_{12} XY + \beta_{13} XZ + \beta_{23} YZ + \beta_{123} XYZ$$

This model includes all two-way interactions as well as the single three-way interaction.

Fifty replications of each combination gave the following results

Factor 1	Factor 2	Factor 3	Average Time in System	Standard Deviation
12 jobs /hour	Exponential	1 server	$W = 0.323$	0.171
12 jobs/hour	Constant	1 server	$W = 0.206$	0.075
12 jobs/hour	Exponential	2 servers	$W = 0.099$	0.015
12 jobs/hour	Constant	2 servers	$W = 0.099$	0.015
15 jobs/hour	Exponential	1 server	$W = 0.118$	0.014
15 jobs/hour	Constant	1 server	$W = 0.118$	0.022
15 jobs/hour	Exponential	2 servers	$W = 0.068$	0.006
15 jobs/hour	Constant	2 servers	$W = 0.068$	0.006

The results from the regression model were

	Variable	Parameter Estimate	T for H₀ Parameter = 0	Prob > |T|
	Intercept	0.32253	33.95	0.0001
X	Service rate	−0.20411	−15.19	0.0001
Y	Distribution	−0.11634	−8.66	0.0001
Z	Servers	−0.22381	−6.51	0.0001
XY	Interaction	0.11620	6.12	0.0001
XZ	Interaction	0.17339	9.12	0.0001
YZ	Interaction	0.11634	6.12	0.0001
XYZ	Interaction	−0.11620	−4.32	0.0001

As before, the intercept represents the average time in the system for the case when all dummy variables are zero; that is, a service rate of 12/hour, exponentially-distributed service, and one server. The regression coefficients for individual variables indicate the change in the response variable from this base case if the factor were changed from the low to the high level. For instance, increasing the service rate from 12/hour to 15/hour is expected to lower the average time in the system by 0.20411 hours. Likewise, changing from exponential service to constant service times lowers the time in the system by 0.11634 hours. Finally, increasing the number of servers from one to two lowers average time in the system by 0.22381 hours. All three factors individually are significant, as are the interaction effects for all combinations of factors. This implies that there are overlapping effects. That is, the gains from a faster system include at least some of the gains from adding an extra server and changing the distribution of service time.

Simulation in Practice

SIMULATION IN WASTE-PROCESSING SYSTEM DESIGN[3]

A new waste-processing system was to be installed in Zagreb, Croatia. Large vehicles would deliver bulk waste, which was unloaded into bays. Small vehicles would move the waste into a storage facility, from which a crane transferred it to an incinerator. After incineration, ash and dross were transported to a depositing area, and trucks removed them from the system. Important considerations in the design included vehicle capacities, vehicle arrival rates, weighing machines, storage facility capacity, the number and capacity of cranes, and the capacity of incinerators. Because the system did not yet exist, a simulation model was developed to analyze design alternatives.

The key performance measure was system efficiency, measured by the proportion of waste incinerated and transported from the system. Seven initial factors were identified. Researchers conducted a dozen preliminary experiments and, as a result, determined that only two factors would be most significant: the number of incinerators and the number of cranes. Further study revealed that no more than two cranes would be required, and a reasonable number of incinerators to have was two to four.

A full factorial experiment with two factors (cranes and incinerators) at two levels (one or two cranes, and two or four incinerators) would involve four experiments: one crane with either two or four incinerators, and two cranes with either two or four incinerators. The researchers in this project modified this experimental design somewhat in order to gain better insight about the influencing factors. The following combinations of factors were studied:

[3]V. Ceric and V. Hlupic, "Modeling a Solid Waste-Processing System by Discrete Event Simulation," *Journal of the Operational Research Society,* Vol. 44, no. 2, 1993, pp. 107–114.

A. 1 crane and 2 incinerators
B. 1 crane and 3 incinerators
C. 2 cranes and 3 incinerators
D. 2 cranes and 4 incinerators

Five independent replications were simulated for each of these combinations. Each simulation began with an empty system and ran for 6 days of simulated time. Common random numbers were used to reduce variance. The output variable was the proportion of incinerated waste to the total quantity brought to the system.

Statistically, three comparisons were analyzed:

Comparison 1: A Versus C
Comparison 2: B Versus D
Comparison 3: A Versus D

The main effects and interaction effects obtained from the experiments are shown here:

Effect	*Comparison 1*	*Comparison 2*	*Comparison 3*
Incinerators	0.262	0.093	0.356
Cranes	0.025	0.144	0.118
Interaction	0.025	0.093	0.118

The main effects (i.e., incinerators and cranes) show the average change in system response caused by changing the factors from the first combination in the comparison to the second. Thus, for the first comparison, moving from one crane and two incinerators (combination A) to two cranes and three incinerators (combination C) results in an average increase in the output measure of 0.262.

These results indicated that when there is a smaller number of incinerators, an additional one causes a larger average increase in the system response than when there is a larger number (as exhibited by the differences between comparisons 1 and 2). That is, moving from two incinerators to three had a greater impact than moving from three incinerators to four. This suggested that two incinerators were insufficient to handle required work and that further analyses should examine either three or four incinerators. The interaction effects suggested that adding a crane had a bigger impact when there was a larger number of incinerators, indicating that adding a crane would not be useful unless at least one incinerator was added. The researchers concluded that two cranes and either three or four incinerators should be used.

More detailed estimates of system performance were obtained through additional simulation runs for these configurations, which also measured waiting times, crane utilization, and incinerator utilization. From these experiments, they recommended that the system should have two cranes and four incinerators.

IMPROVING FLORIDA DRIVER'S LICENSING OFFICES THROUGH SIMULATION[4]

At the state of Florida's driver's licensing offices, increases in population growth over a period of years resulted in long waiting lines and extended service times, reaching hours in some cases. In response to this undesirable situation, the Florida legislature and the Department of Highway Safety and Motor Vehicle Management instituted a number of

---

[4]T. D. Clark, Jr., D. H. Hammond, and K. L. Cossick, "Management Policies to Improve the Effectiveness of Multistation Service Organizations," *Decision Sciences,* Vol. 23, no. 4, 1992, pp. 1009–1022.

plans, which included extending renewal time to 6 years, and increasing examination staff by 25 percent. These policy changes, however, did not completely alleviate the problem. Consequently, a group of researchers developed a simulation model as a basis for analyzing motor vehicle licensing offices. An office in Miami provided a scenario to check the validity of the model.

Extensive data collection efforts were conducted to determine service times, transaction patterns, and arrival distributions. The system consisted of five workstations, shown in Figure 7-6. The office operated from 7:00 A.M. until 6:00 P.M. with servers taking rest and lunch breaks during the day. Customers usually line up before the office opens. After opening, the mean arrival rate varies throughout the day. At 5:00 P.M., the outer door is closed, but service continues until all customers inside the office are served. Based on typical customer requirements, 13 customer patterns, each with a distinct routing pattern through five workstations, were identified. Because of the complexity of the system, a SLAM II model[5] was developed to analyze the existing situation, as well as any proposed improvements to the system.

A number of options were considered to improve office system performance. One approach was to smooth demand through better job scheduling, such as use of reserva-

FIGURE 7-6 Driver's Licensing Office

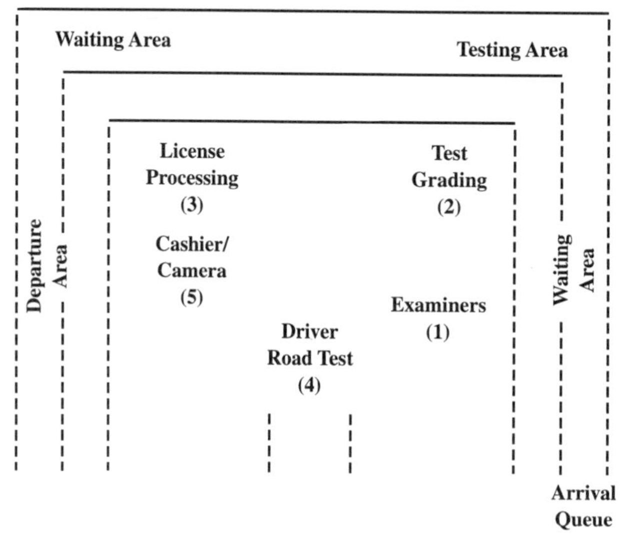

Station 1: Examiners greet applicants, determine eligibility, complete license renewal forms, check vision, distribute written test forms, and handle financial responsibility cases.

Station 2: Examiners grade written tests and assign road and oral tests.

Station 3: Examiners process license reinstatements and type original license forms.

Station 4: Examiners administer road test.

Station 5: Examiners operate the camera and cash register.

---

[5]SLAM II is a commercial systems simulation language that was developed by Pritsker & Associates of West Lafayette, Indiana.

tions or appointments. Labor scheduling was also considered, seeking to match server capacity to customer demand across the workday. This could be accomplished through flextime, compressed work schedules, or the use of part-time labor. A third option was to have job flexibility, allowing workers to be shifted from idle service facilities to bottleneck service facilities.

The first use of the simulation model was to identify the factors that most influenced the system. Four factors were considered:

1. Demand management options—existing, smoothed mean, and restricted arrivals during lunch hour
2. Labor scheduling—existing, and two others using flextime or compressed-week schedules
3. Labor assignment—existing, and a reconfiguration where an examiner was reassigned from initial examining to the last station
4 Job flexibility—three levels of cross-training

This yielded a $3 \times 3 \times 2 \times 3$ research design, or 54 combinations. Twelve runs were made for each combination of factors, with each run a simulation of the full 11-hour day. The dependent variable of interest was average customer time in the system. Analysis of variance (ANOVA) was used to interpret the results.

The ANOVA analysis revealed that five policy combinations were best at reducing average customer time in the system. Four of these five policy combinations had the smoothed-demand element. There was significant interaction between the labor-assignment factor and the labor-scheduling factor, as well as between labor assignment and job flexibility. This led to further study of policy options that included demand management and job flexibility, as shown in Figure 7-7. This experiment confirmed that both smoothed-demand rates and job-flexibility options were effective in lowering average time in the system. Implementation of the job-flexibility option would require the purchase of additional office equipment, but equipment costs would be more than offset by the savings realized by not hiring additional labor required to produce the same results.

FIGURE 7-7 Experimental Design for Demand/Job Flexibility Policy

Level	Factor 1. Demand Management
1	The current arrival pattern: nonstationary Poisson arrival.
2	A uniform arrival pattern: block scheduling.

Level	Factor 2. Management Policy
1	Current staffing (10 people) with the current assistance pattern.
2	Seven workers pooled at station 1. Driver testers assisted until needed at station 4. Moving servers when required.
3	Same as level 2, except that one station 1 worker moved to the driver tester pool after lunch break. Moving servers when required.
4	All 10 workers were pooled as general-purpose examiners at station 1. Moving servers when required.

Questions and Problems

1. Explain why the waiting times of successive customers in a queue are autocorrelated. Would the inventory level on successive days in an order quantity–reorder point simulation model also be autocorrelated? Why or why not?
2. Explain the difference between terminating and nonterminating simulations. Provide some examples different from the text.
3. Use the worksheet *Queueing Simulation—Terminating Model.xls* and find the mean number and the mean time in the system (and their 90 percent probability intervals) for a system operating 8 hours and having exponentially-distributed arrivals with a mean interarrival time of 6 minutes, normally-distributed service times with a mean of 4 minutes, standard deviation 1 minute, and minimum of 0.01. (Hint: Change the mean arrival and service rates in cells B3 and B4, and change simulated run length to 8 in cell B5. To obtain the average number in the system (the number in the queue plus the number in service), create a new column measuring cumulative time in the system, and calculate the average in the same manner as that used to find the average time in queue. Then, create a new assumption cell similar to cell B14.)
4. Explain the concept of *transient state*. Why is this important in analyzing the output from simulation models?
5. Use file *Queueing Simulation—Terminating Model.xls* to simulate 1,000 customers, and graph the average waiting time and average percent idle time. Replicate the simulation five times (by pressing the F9 key), and use these results to estimate when steady-state begins.
6. Describe the method of batch means. What advantages does this approach have?
7. Apply the method of batch means to the following data, obtained from running a queueing model for 5,000 hours. Determine a 90 percent confidence interval for the mean queue length.

Batch Interval	Average Queue Length
(0, 1000)	5.62
(1000, 2000)	5.22
(2000, 3000)	4.16
(3000, 4000)	11.90
(4000, 5000)	7.84

8. A simulation model is used to compare the performance of two manufacturing system configurations. The following table shows the results for 10 independent trials of each system. Develop a 95 percent confidence interval for the differences in mean performance. Would you conclude that the systems are significantly different?

Trial	System 1	System 2
1	29.56	49.82
2	34.49	52.91
3	25.47	45.72
4	40.90	31.58
5	33.48	57.19
6	38.67	28.28
7	38.04	40.34
8	40.02	73.60
9	60.28	22.59
10	44.39	29.00

9. The following table shows the results of 10 trials for the same situation in problem 8, except that common random numbers were used. Find a 95 percent confidence interval for the difference in mean performance. Does your conclusion change? Compare your confidence interval with that constructed in problem 8. How do they differ?

Trial	System 1	System 2
1	56.74	29.44
2	33.43	24.62
3	35.28	26.30
4	34.92	42.46
5	38.87	32.54
6	32.70	37.19
7	51.46	36.84
8	42.84	41.42
9	48.39	60.95
10	23.18	40.94

10. For the data in problem 9, estimate the probability that system 1 is better than system 2 (where "better" refers to higher levels of performance). Assume the trials are independent. How does this compare with your conclusions in problem 9?

11. We wish to compare two machines that make keys in a hardware store. Keys sell for $1 each. The first machine costs $1,000 per year to own, and it costs $0.50 to make one key. Therefore, profit per year for machine 1 is ($1 − $0.50) × Demand − $1,000. Machine 2 costs $2,050 per year to own, but is more efficient, making keys at a variable cost of $0.30. Profit per year for machine 2 is ($1 − $0.30) × Demand − $2,050. Assume that demand is uniformly distributed between 3,000 and 7,000 keys per year.
   a. Determine the best system analytically.
   b. Given the following simulated results, develop a 90 percent confidence interval for the profit from each machine.

Run	Volume	Machine 1	Machine 2	Best	Difference
1	3,000	500	50	1	450
2	3,500	750	400	1	350
3	4,000	1,000	750	1	250
4	4,500	1,250	1,100	1	150
5	5,000	1,500	1,450	1	50
6	5,500	1,750	1,800	2	−50
7	6,000	2,000	2,150	2	−150
8	6,500	2,250	2,500	2	−250
9	7,000	2,500	2,850	2	−350

   c. Based on correlated sampling, develop a 90 percent confidence interval.
   d. Model the problem using Crystal Ball, and identify the 0.9 probability interval for the profit for each machine, as well as for the difference in profit. Use an overlay chart of machine 1 profit and machine 2 profit to compare results. Identify the probability of machine 1 having a higher profit than machine 2.

12. A company has a printer for legal documents. The printer has been a major bottleneck in operations. Demand has grown to an average of 5 jobs per hour, exponentially distributed. You have been requested to conduct a cost analysis of the

current system. Service time is normally distributed with mean 6 minutes (0.1 hours), standard deviation 0.03 hours, and minimum 0.001 hours. Two alternatives are available: a laser printer with normally-distributed service time having mean 3 minutes (0.05 hours), standard deviation 0.02 hours, and minimum 0.001 hours; and two letter-quality printers, both with normally-distributed service time having mean 0.1 hours, standard deviation 0.03 hours, and minimum 0.001 hours. If a job arrives prior to closing time, it is important to finish the copying job by using overtime. Assume 200 working days per year (8 hours each).

Management believes that operating-cost impact to the firm is $10 per hour in waiting time and $20 per hour in overtime. The investment cost per year of the laser printer is $4,000 more than the current system, and the cost of the two-server system is $3,000 more than the current system. Use *Queueing Simulation Terminating—Model.xls* to simulate 1,000 trials of 8 hours each for the single server models (changing service times). Cell B3 is already set up as the mean arrivals per hour for the exponential distribution (modeled in column D as =CB.Exponential($B$3)). Cell B4 is used for the mean processing time. Using cell B7 to record the standard deviation, column I needs to be adjusted to reflect normal distribution:

```
=MAX(0.001,CB.Normal($B$4,$B$7))
```

Create new assumption cells as follows:

```
Cell B18 Jobs=INDEX($C$3:$C$503,MATCH("Closed",$F$4:$F$503,0))
Cell B19 CumWaiting=INDEX($Q$3:$Q$503,MATCH("Closed",$F$4:$F$503,0))
Cell B20 OT=MAX(0,INDEX($J$3:$J$503,MATCH("Closed",$F$4:$F$503,0))-8)
Cell B22 OpCost=200*(10*B19+20*B20)
```

Modify this model to incorporate two servers. Compare the average number of jobs for each system, waiting time, overtime, and average operating cost. Find the probability that the sum of operating cost and investment for the current system is less than the corresponding numbers for each of the other two systems. Is the distribution of cost normal?

13. For the situation in problem 12, rework the problem assuming a heavier workload, with arrivals following the exponential distribution at the rate of 8 per hour. Compare average jobs per day, cumulative waiting time per day, overtime per day, and operating costs per year. Identify the probability that the sum of operating cost and investment for the current system is less than the corresponding numbers for each of the other two systems. Is the distribution of cost normal?

14. An entrepreneur sells premier corsages for the crowd at local football games. It costs $5 for materials for each corsage; corsages are sold for $10 each. Demand follows an exponential distribution with a mean of 20. Environmental protection regulations cover some of the materials used, and all corsages that are not sold must be destroyed at an additional cost of $1. Four policies being considered are to buy 10, 15, 20, or 25 corsages. Build a spreadsheet model to simulate this situation.
   a. For each simulated day, the quantity of corsages sold is the minimum of the quantity produced and the demand. Replicate each policy 30 times using correlated sampling, and compare the policies statistically.
   b. Use Crystal Ball to generate 1,000 trials of the four policies, and identify the 90 percent probability intervals for the policies of 10, 15, 20, and 25 corsages. Compare means and standard deviations with those obtained in part a. Use an overlay chart to compare results.

15. A manufacturing firm has signed a contract to produce a product to demand, receiving $500 per unit. The firm could either produce the product itself or outsource the production. Demand is exponentially distributed with a mean of 10,000 units per year. In-house production has normally-distributed fixed costs with mean $500,000 (standard deviation $100,000) and normally-distributed variable costs with mean $200 and standard deviation $10. Outsourcing has no fixed

cost, but variable costs are $250. Develop a forecast cell for the profit of each alternative, as well as for the difference in profit. Identify the mean of these three statistics, and construct an overlay chart for the two profit distributions. Finally, identify the probability that each alternative achieves at least $0 profit, $100,000 profit, and $200,000 profit.

16. Compare the alternatives from problem 15 in separate models, using the same and a different seed for generating demand (1,000 trials). Find the means and standard deviations, and the probability of achieving $0 profit, $100,000 profit, and $200,000 profit. Test the hypothesis that there is no significant difference between the alternatives.

17. How is experimental design used in simulation analysis? How might it be used in a waiting-line problem comparing one or two servers, as well as constant versus exponential service time distribution?

18. (This problem assumes prior knowledge of regression analysis and analysis of variance.) The following data were generated using the Mantel Manufacturing model. Analyze a simple factorial experimental design for the order quantity factor at two levels: 275 and 325, and the reorder point factor at two levels: 300 and 400. Use the multiple-regression procedure in Excel to perform the multiple regression and explain the results.

Q	ROP	Cost
275	300	215,300
275	400	60,964
325	300	83,253
325	400	68,888
275	300	82,250
275	400	31,966
325	300	81,905
325	400	22,054
275	300	254,829
275	400	227,430
325	300	48,510
325	400	211,343

19. Given the following simulation output for a queueing problem, analyze the factorial experimental design to determine the impact of each factor. The problem involves exponential arrivals at the rate of 15 per hour and services at the rate of 20 per hour. Four systems are being compared. The two factors are (1) the number of servers, either one or two and (2) the service distribution, either normal with a mean of 0.05 hours and a standard deviation of 0.01, or constant. The data are

Servers	Distribution Average	Waiting Time
1	Normal	0.088
2	Normal	0.066
1	Constant	0.049
2	Constant	0.065
1	Normal	0.077
2	Normal	0.119
1	Constant	0.0
2	Constant	0.123
1	Normal	0.041

(continued)

Servers	Distribution Average	Waiting Time
2	Normal	0.063
1	Constant	0.0
2	Constant	0.071
1	Normal	0.086
2	Normal	0.147
1	Constant	0.0
2	Constant	0.163
1	Normal	0.042
2	Normal	0.064
1	Constant	0.0
2	Constant	0.146

References

Ceric, V., and V. Hlupic. "Modeling a Solid Waste-Processing System by Discrete Event Simulation," *Journal of the Operational Research Society,* Vol. 44, no. 2, 1993, pp. 107–114.

Lentner, M., and T. Bishop. *Experimental Design and Analysis,* 2d ed. Blacksburg, VA: Valley Book Company, 1993.

Sherali, H. D., A. G. Hobeika, A. A. Trani, and B. J. Kim. "An Integrated Simulation and Dynamic Programming Approach for Determining Optimal Runway Exit Locations," *Management Science,* Vol. 38, no. 7, 1992, pp. 1049–1062.

CHAPTER 8

Systems Simulation Using ProcessModel

Chapter Outline

Developing the logic for how a system operates is the essence of simulation. Once this is done, a simulation model can be written in general-purpose programming languages such as BASIC, C++, FORTRAN, and PASCAL. General-purpose languages provide a maximum of modeling and output flexibility and were the first approaches to implementing simulation models. However, a major drawback to general-purpose languages is that many potential users of simulation may not be sufficiently fluent in the advanced programming techniques necessary for developing an efficient model. Even if you are comfortable with programming in a general-purpose language, specifying the detailed computations, file and data manipulation, and statistical processing necessary to implement a simulation can be quite tedious. Many of the

271

procedures required to implement a simulation, such as time advance, storing events in a calendar file, computing statistics, and so on, are common to many simulation models. To avoid having to recreate these for different applications, **simulation software** has been developed to ease the burden of developing and running simulation models.

Simulation software makes it easy to model complex problems and create useful reports for analysis and decision making much faster than general-purpose languages. Most modern systems use visual graphics, allowing the user to develop a model by simply clicking and dragging logic elements and making it unnecessary for the user to write any code. These languages also automatically perform many useful logic operations, such as generating entities in a random fashion, storing and retrieving entities in queues, collecting statistical information, and manipulating resources within a simulation. They automatically compute and report statistical information such as waiting times, queue lengths, and resource utilizations. The primary benefit to these systems is that they save enormous amounts of time and programming effort. Potential drawbacks include a loss of some flexibility in certain systems, the time required to learn to use them (although modern graphical software has made this less of a concern), and, of course, the price tag that may be associated with them.

GPSS and SIMSCRIPT were two of the earliest commercial simulation packages. GPSS is a process-oriented language, developed in 1961, that models a system using a block diagram that represents a sequence of activities through which *transactions* (entities) flow. As a transaction moves from one block to another, it triggers various logic operations. SIMSCRIPT was introduced by the RAND Corporation in 1963. SIMSCRIPT is not a transaction-based approach like GPSS but is a true programming language with its own syntax. SIMSCRIPT statements are English-like, thus facilitating the development of complex models. A SIMSCRIPT simulation model requires that the logic described by the event routines (such as the flowcharts in Figures 6-29 and 6-30) be translated into the programming language. This permits much more flexibility than GPSS but at a higher degree of difficulty in learning the language. Many other products have been developed over the years, including SLAM (Simulation Language for Alternative Modeling), SIMAN, and ARENA. In addition to these, special-purpose simulators have been developed for specific applications such as communication network applications, equipment maintenance and repair, and cellular manufacturing environments, to name just a few.

The software we will use in this and the following chapter is ProcessModel.[1] A student version of ProcessModel is included on a CD-ROM with this text. Designed specifically for managers and planners, ProcessModel combines simple flowcharting technology with powerful simulation capability to bring flowcharts to life through graphical animation. Typical ProcessModel applications include:

- Staff scheduling and shift planning
- Task prioritization and interruption
- Method selection
- Capacity planning
- Lot sizing
- Appointment scheduling
- Job sequencing
- Production scheduling
- Productivity improvement

---

[1]ProcessModel, Inc., 32 West Center, Suite 209, Provo, Utah 84601. Telephone: (801) 356-7165. Much of this chapter is adapted from the *Online User's Manual* accompanying the software on the CD-ROM. We will refer you to the manual as necessary for descriptions of more advanced capabilities of ProcessModel.

- Cycle-time reduction
- Cost reduction
- Quality management
- Bottleneck analysis
- Activity and resource-based costing
- Resource scheduling for breaks and downtime

ProcessModel allows you to create hierarchical models to better organize and manage large modeling projects. For example, teams can define different parts of a complex model and then put them together to simulate the entire process. Other special features include LIVE Animation™, which allows visualization of people, paperwork, and other objects flowing through the system, making it easy to recognize bottlenecks; OneStep modeling approach, which is designed to make modeling common processes easy to incorporate into a model; Visual Staffing™, which includes visual icons of people and resources in a model, thus helping to improve the face validity of the model and communication with users; and (in the Professional Package only) Process Optimization, which uses sophisticated optimization algorithms to optimize multiple factors simultaneously.

ProcessModel Overview

This section describes the basic features of ProcessModel. We encourage you to read this first to get a feel for the software and then study the example in the next section and work the SkillBuilder Exercises to gain some hands-on experience before tackling any of the problems in this chapter.

A **process model** is a flow diagram with associated operational information for simulating a process. A process might be a common business process such as order processing, a servicing process such as customer support, or a manufacturing process such as product assembly. A process flow diagram consists of *objects* (the graphic shapes in the flowchart) and *connections* (the lines connecting the graphic shapes). Objects represent the elements of the process; connections depict element relationships. A Properties Dialog displays the *operational information* for each object and connection.

MODEL ELEMENTS

ProcessModel provides a variety of *model elements* to build simulation models. The basic elements are described here. Chapter 3 in the *Online User's Manual* provides more detail about each of these.

- *Entities*—The items or people being processed; e.g., products, documents, or customers.
- *Activities*—The tasks performed on entities, such as assembly, document approval, or customer checkout. An activity is defined in terms of the activity time as well as any resource requirements. User-defined action logic may also be used to describe an activity. Activities have a processing capacity and may have an input and output queue associated with them.
- *Resources*—The agents used to perform activities and move entities, such as service personnel, operators, or equipment. Resources may be shared between several activities. If no resource is assigned to an activity, it is assumed that no resource is required.
- *Connections*—Several types of connections are used in ProcessModel:

  ENTITY ARRIVALS—Define where, when, and in what quantities entities enter the system to begin processing. Entity arrivals are defined by connecting an entity to the activity or storage where it begins processing. Multiple arrival connections can be created from an entity to one or more activities/storages.

ENTITY ROUTING—Defines the processing flow for entities. An entity routing is defined by connecting an activity or storage to the next activity or storage in the processing sequence. An activity or storage may have multiple input routings and multiple output routings. Entities do not move to the next activity or storage until there is available capacity and the condition or rule for routing the entity has been satisfied.

RESOURCE ASSIGNMENTS—Define the use of resources in performing activities or moving entities. A resource assignment is defined by connecting a resource to either an activity or a routing depending on whether it is used for an activity or to make a move. Resources may be captured before any activity or routing and freed after any activity or routing. Multiple or alternative resources may be used for an activity or routing.

ORDER SIGNALS—A connection between an activity or storage and an arrival or routing that signals the release of additional entities. The signal is triggered by a drop in inventory level at either the storage or the activity's input queue from which the signal originates.

- *Storages*—Waiting areas where entities can wait or be held until further processing. Storages are useful when controlling the order in which entities are allowed to move on through the model. Because an activity provides the option to have built-in input and output queues, a storage is primarily useful only for visual purposes or to model special queuing situations.
- *Labels*—A method to display dynamic variables during the simulation. Any information that is critical can be positioned on the screen for viewing during the model run.

In addition to these basic elements, several advanced elements provide further capability:

- *Schedules: Shifts and Breaks*—Elements to indicate when activities and resources are available. These allow you to schedule when activities and resources are available.
- *Action logic*—Simple logic statements to control the behavior of the model. Definable for any activity, arrival, and routing, action logic allows you to enter simple but powerful logic statements to control the behavior of the model that may otherwise be difficult or impossible to do using the properties provided.
- *Expressions*—Expressions fall into one of the following categories:

  DISTRIBUTIONS—ProcessModel allows you to use standard and user-defined distribution functions in defining times and other numeric values.

  ATTRIBUTES—Values or placeholders associated with individual entities such as size or condition, attributes can be changed in any arrival, activity, or routing's action logic. There are certain predefined attributes for each entity in addition to those defined by the user.

  VARIABLES—Used to store either values (real or integer) or single-word description called *descriptors*. Unassociated with a specific entity, variables allow you to track activities anywhere in the system, such as the number of entities that have completed a particular activity, or monitor the work in progress for a section or group of activities in the model.

- *Links*—Connections to complementary files for process improvement, such as spreadsheets, text documents, organizational charts, or business diagrams. The link will not show up in the output report or be recognized by the simulation. This is a convenient method of connecting files to the process definition.
- *Scenarios and Scenario Parameters*—Placeholders for values used when setting up multiple scenarios for a model. Using a scenario parameter in a model in place of a value allows you to experiment with different parameter values at the time you run the simulation. (Scenarios are discussed in section 4.2 of the *Online User's Manual.*)

The following steps describe the basic process for creating a running ProcessModel simulation:

1. Start ProcessModel. The opening screen will display the *Simulation* menu, Layout Window, Toolbox, and Gallery, and the Properties Dialog (described further in the next section).
2. Diagram the process flow by selecting or creating objects and connecting them on the layout to the activities and routings where they will be used.
3. In the Properties Dialog, define the behavior of each object and connection.
4. Connect resources to the activities and routings.
5. Add variability to the model by using distribution functions to represent properties.

After the process flow diagram is set up and the properties for the elements of the model are defined, you are ready to run a simulation. You may choose to view the results from your simulation on the screen immediately, or save the results to an output file to view later. You can compile the results in various forms: general summary statistics, selected statistics showing averages over a series of replications, and detailed graphs and charts.

THE MODELING ENVIRONMENT

The modeling environment, shown in Figure 8-1, provides all the tools and functions necessary to build and run process models. The modeling environment has several features:

Layout Window   The layout window for drawing the flow diagram is a scrollable drawing area that is divided into pages. Pages correspond to what you would see if the diagram were printed. You may start the diagram anywhere on the layout, although it is usually best to start in the upper-left corner. The diagram can be easily moved later if needed.

Toolbox   The Toolbox is the column of buttons displayed on the left side of the layout window. (The Toolbox can be opened and closed by selecting *Toolbars* from the *View* menu and checking or unchecking *Toolbox* in the dialog.) The primary buttons in the Toolbox are the *pointer* button, used to select objects; the *shape* button, used to place new shapes on the layout; and the *line* button, used to connect the shapes. The other buttons are for adding text, zooming, and changing the line style.

The Gallery   The Gallery is a formatting command center. From the Gallery, you can select the shape for your model and change color, font, line style, line ends, and shadowing. The Gallery is positioned on the right of the screen for easy access, but it can be positioned anywhere as a floating toolbar. The Gallery provides the ability to change the following:

- *Fill*—Change the color, pattern, or type of gradient used to fill graphics that are selected.
- *Font*—Modify the font type, style, and size for all text or graphics that contain text, for items that are currently selected.

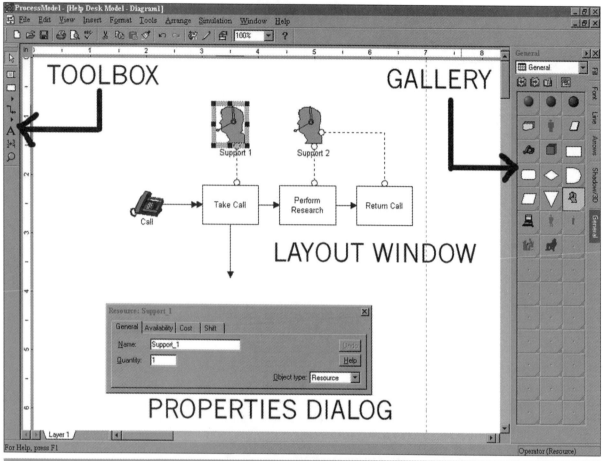

FIGURE 8-1 ProcessModel Modeling Environment

- *Line*—Change the line style, color, or weight for all routings, arrivals, resource connections, and graphic perimeters for all items that are currently selected.
- *Arrows*—Change the source line end or destination line end to one of 56 styles or create a custom line end.
- *Shadow/3D*—Add shadows or 3-dimensional appearance to entities, activities, and resources. Customize the direction, depth, and color of shadows and the depth and direction of 3-dimensional projections for all selected items.
- *General*—Default ProcessModel shape palette. Nine predefined ProcessModel palettes are provided. Custom palettes can be assembled from your company graphics, clip art, or digital pictures. A shape palette allows you to select each shape used to represent each object in the model. The default shape palette is called *general* and contains shapes that are commonly used for process diagramming. Other palettes and shapes may also be used if desired. All graphics in ProcessModel's shape palette have a default name and object type displayed in parentheses after the name of the graphic. For example, the *Customer Entity* graphic has been predefined to be an entity object. To display the default name and object type for a particular shape, simply move the pointer over the graphic in the shape palette. The type of object, as well as the name, may be changed once a shape is placed on the layout.

Properties Dialog   The Properties dialog allows you to define simulation information for each object and connection in the model. It stays open when you select a different object or connection, and if you close it, double-clicking on an object will open it again, or you may right-click on the object and select *Object Properties.* The Properties dialog changes context when you select another object. This allows you to view model element information instantly and edit all elements without having to open and close dialog windows each time you select another element. In the Properties dialog, you may define information for each element, such as its name, type, cost information, processing time, and capacity. You can also define action logic, create links to another flowchart, and assign schedules to resources and activities. The Properties dialog, like many other dialogs, is a tabbed dialog. Simply click on a tab to define information related to the tab's label.

DIAGRAMMING THE PROCESS FLOW

The best way to diagram a process is to create an entity object and then each activity or storage through which the entity will pass. Once the flow has been defined, resources can be placed on the layout and connected to the activities and routings where they are used. Finally, the appropriate information must be entered into the Properties dialog for each object and connection to define their behavior. If no output connection is defined from an activity or storage, or if an output connection does not attach to another activity or storage, it is assumed that the entities exit the system at that point.

To create an object on the layout:

1. Click on the shape button in the Toolbox to select the shape tool.
2. Click on a shape or symbol in the shape palette (in the Gallery) to represent the object you wish to create.
3. Click in the layout window to place the object shape.
4. Enter any necessary information for the object in the Properties dialog.

To place and connect graphics at the same time:

1. Click on the shape button in the Toolbox to select the shape tool.
2. Click on a shape or symbol in the shape palette (in the Gallery) to represent the object you wish to create.
3. Move the mouse over an existing object on the layout.
4. Click and hold while dragging the mouse. Notice that a line between the two objects automatically forms. Release the mouse when the second object is at the desired position.

INPUT DISTRIBUTIONS

ProcessModel provides the ability to randomly generate values from a number of common probability distribution functions. These functions may be used to represent any value in the Properties dialogs, including action logic. The distributions available in ProcessModel are

- *Continuous*—Uniform, normal, lognormal, triangular, exponential, Erlang, beta, gamma, Weibull, Pearson5, Pearson6, and Inverse Gaussian
- *Discrete*—Binomial, Poisson, geometric, and user defined

Examples of the syntax for the most common distributions are

Normal—$N(a, b, s)$, where a = mean, b = standard deviation, s = random number stream (optional)
Triangular—$T(a, b, c, s)$, where a = minimum, b = mode, c = maximum, s = random number stream (optional)

Uniform—$U(a, b, s)$, where a = mean, b = half range, s = random number stream (optional)

User defined—$D_n(\%_1, x_1, \ldots \%_n, x_n)$, where % = percentage (entries must total 100 percent), x = value (numeric or predefined descriptor), and n = number of $(\%, x)$ entries between two and five.

If you are unsure of which distribution to use, *Stat::Fit* is a tool to help you select the correct distribution. (See "Using Stat::Fit to Choose Distributions" in the *Online User's Manual,* page 243.)

DEFINING SPECIAL LOGIC

When you have completed your process flow diagram and begun to specify the properties for each object and connection, you may need to define variables and attributes to track, control, or respond to events within your model. Once you have defined the variables and attributes, you can use action logic to perform tests and control the flow of entities. Additionally, you might want to define schedules for resources and activities and perhaps even link to other charts. These advanced features are covered in Chapters 3 through 6 of the *Online User's Manual.*

RUNNING A SIMULATION

Once you have completed your process flow diagram and defined the properties for each of the elements in your model, you are ready to run the model and do any necessary fine-tuning. You can also set up scenarios for running a set of experiments for comparison. Scenarios are described in detail in Chapter 4 of the *Online User's Manual.*

Simulation Options　　To specify options for running the simulation, select *Options . . .* from the *Simulation* menu. The Options dialog will be displayed with *Run, Files, Graphics, Output Summary,* and *Package* tabs. Under the *Run* tab, you may specify how long your simulation should run, how long to allow the system to warm up to a steady state of operation, and the number of replications you want to run in order to collect multiple statistical samples. In addition, you are given several check-box options that allow you to choose animation and submodel options (see section 4.1, "Simulation Options," on page 265 of the *Online User's Manual*). Under the *Files* tab, you may opt to include shift information. The *Graphics* tab allows you to set up custom icons for your entities. Under the *Output Summary* tab, you may choose to show standard deviation in the cycle-time calculation and to set the size of the sample cycle. Under the *Package* tab, you can set the number of files to link together in your model package.

Simulate Chart　　When you select *Simulate Chart* from the *Simulation* menu, the model is first translated and checked for any rule violations. Then, the simulator is loaded and the model begins to run. As the simulation runs, you will see an animated representation of the model unless you have not checked the option to show animation in the Options dialog.

The Scoreboard　　If the *Scoreboard* option is selected in the Options dialog under the *Simulation* menu, the following performance measures will be displayed at the top of the simulation window:

SCOREBOARD	EntA	EntB	...
Qty Processed	data	data	...
Cycle Time (min)	data	data	...
VA Time (min)	data	data	...
Cost per Unit	data	data	...

The default time unit displayed following Cycle Time (time spent in the model) and VA Time (value-added time or the time spent in actual processing) is selected in the Options dialog, which is found under the *Simulation* menu.

The Simulation Window   The simulation window has a menu of its own, with selections for controlling many simulation parameters such as the run speed. You can also control the animation through panning, zooming, and pausing. The simulation window contains a speed control bar, along with the clock selection button for controlling the format of the clock readout. In addition, each resource has a status light that changes throughout the simulation to reflect the current operational state of each resource.

The *Simulation* menu in the simulation window allows you to pause or end the simulation. In the *Options* menu, you have the following selections:

- *Debug, Trace, Trace Output*—These three options are included for advanced debugging purposes.
- *Animation Off*—Temporarily disables animation to make the simulation run faster.
- *Zoom*—Allows you to zoom in or out on the animation.
- *Views*—Allows you to quickly and easily view any submodels defined. (Available only when hierarchical modeling is used.)
- *User Pause*—Allows you to specify the time of the next simulation pause. At the defined time, the simulation will pause to allow you to examine the layout and what is happening in the model.

The *Information* menu contains selections for obtaining system information during the run. The four selections allow inspection of queues, activities, and variables during the simulation. The *Window* menu allows you to rearrange windows and icons and select the active window. These options are standard to all Windows applications.

ProcessModel Example: Simulating a Phone Support Help Desk

This section provides step-by-step instructions for creating a process model of a familiar business process (a phone support help desk), running the simulation, and viewing the output reports and graphs. In the phone support center, incoming calls arrive about every 5 minutes, and a support representative evaluates the nature of each problem. The representative is able to resolve 75 percent of the calls immediately. However, 25 percent of the calls require that other support representatives do research and make a return call to the customer. The research itself combined with the return call requires 20 minutes. To simulate this situation, we need to identify the essential elements in the process (people and objects), the connections between them, and the rules that govern the operation of the process. This situation is more complex than the simple queueing models we illustrated in Chapters 6 and 7, and it would be virtually impossible to implement on a spreadsheet; however, the model is very easy to build and simulate using ProcessModel.

DEFINE PROCESS FLOW

The first step to take in building the model is to define and connect the entity and each activity of the process. This is done by placing shapes on the layout to represent the entity and each activity in the process sequence. Each shape is connected to the previous shape as it is placed on the layout. The shape tool is used to place shapes on the layout as well as to simultaneously place and connect shapes. You may also connect existing shapes on the layout using the connection (line) tool. All shapes are selected from the shape palette, each of which has a default name and object type.

In the ProcessModel screen, click on the telephone symbol (to represent the calls) in the shape palette to select it. Then move to the left of the layout and click to place the shape. ProcessModel can use default names in any shape-saving you do. In this case, it labels the telephone symbol *Call*. Next, select the rectangle called *Process* from the shape palette. Click on the *Call* shape and drag to the right. (A new shape is placed on the layout with a connection between it and *Call*. If you do not drag the shape, the connection will not be made, and you will have to create it manually or delete the new shape and try again.) With the shape selected, type *Take Call,* then click on the shape tool again. Click on the shape tool again and repeat these steps to create the *Perform Research* and *Return Call* activities and connections, as shown in Figure 8-2. In ProcessModel, if you double click on a shape from the left toolbar, the shape becomes a stamp. You can then insert the shape repeatedly in your model, until you click on a different shape or button.

Now, create a routing to exit for the 75 percent of the calls that the level 1 representative can handle immediately. To create this routing, first find the connection (line) tool on the left toolbar. Click on the connection (line) tool and drag a line from *Take Call* to a point somewhere below the *Take Call* box. The exit from *Take Call* will route the 75 percent of level 1 answered calls out of the system; the level 2 calls continue through to the *Return Call* activity before exiting. (The actual percentages will be entered later.) You could also define another route to exit from the *Return Call* activity, although it is assumed that entities exit if no further routings are defined.

SPECIFY RESOURCE REQUIREMENTS

The next step is to define resource requirements. We need to define two customer support representatives: one to answer calls and one to do research and return calls. First, we will define the support representative that takes calls. Select the person wearing the headset (to represent the support rep) from the shape palette. Move the mouse above *Take Call* and click to place the shape on the layout, and then type *Support 1*. Do the same thing above *Perform Research* and type *Support 2*. Now, both resources need to be connected to the activities. Click on the Line tool in the Toolbox. Drag a connection from *Support 1* to *Take Call*, and then drag a connection from *Support 2* to *Perform Re-*

FIGURE 8-2 Process Flow for Help Desk Problem

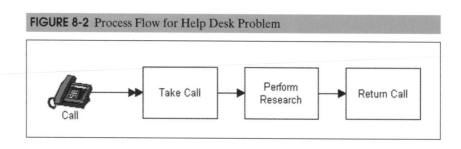

search. Support 2 also returns calls and has to be connected to that activity. Draw a connection from *Support 2* to *Return Call.* Your model should now look like that in Figure 8-3. The dashed lines indicate resource connections. If your model does not have this connection, then the symbol you chose was not a resource shape. This can be corrected easily by either deleting the symbol and choosing the resource symbol or by changing the object type to *Resource* in the object's Properties dialog.

SPECIFY PROCESS INFORMATION

The next step is to complete the process information for the model. The Properties dialog on the screen contains the information pertaining to the activities and connection in the model. When an activity is selected, the Properties dialog reflects the process information for that activity. We will define the frequency of arrivals, enter activity times for the *Take Call* and *Perform Research* activities, and define the percentage of calls that go to *Perform Research* and that exit the system. Finally, we will add cost information for the resources.

We will assume that calls arrive according to an exponential distribution with a mean interarrival time of 5 minutes. To specify the arrival distribution, select the arrival connection between the *Call* entity and the *Take Call* activity. (To select any object or connection, click on the Selector tool, the arrow at the top of the Toolbox, then click on the object). In the Properties dialog, select the *General* tab, and choose *Periodic* for the entity arrival type. In the *Repeat Every* field, type E(5), indicating an exponential distribution with a mean of 5 as the interarrival rate.

We will assume that the duration of phone calls follows a triangular distribution with minimum of 1/2 minute, most likely time of 2 minutes, and maximum time of 4 minutes. To specify this in ProcessModel, select the *Take Call* activity. Click in the *Time* field of the Properties dialog, delete the default time, and type T(.5, 2, 4). Default times are in minutes. We also assume that the distribution for the time to perform research is normal, with a mean of 20 minutes and a standard deviation of 5. Select the *Perform Research* activity, click in the *Time* field in the Properties dialog, and type N(20, 5).

FIGURE 8-3 Resource Additions to Help Desk Model

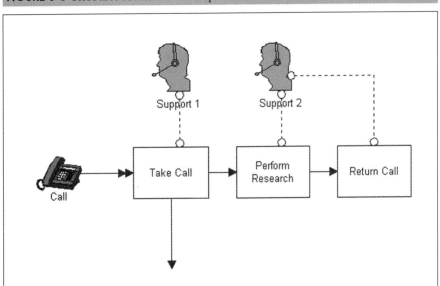

In the *Activity Dialog* general tab, there is an *Input Queue,* a *Capacity,* and an *Output Queue.* You can think of these as a desk containing an inbox, a work area, and an outbox. The default setting is to provide a large inbox, a work area for 1 entity, and no outbox. Because we will add staff at a later time, set the capacity, or the available work space, to 10.

Next, define the percentage of the calls that go to *Perform Research,* which is 25 percent. Click on the connection between *Take Call* and *Perform Research.* In the Properties Dialog, change the *Percent* field to 25. The percent field was automatically set previously to 50 (percent) because two branches were created earlier, balancing the percentages. While you are in the percentage dialog box, you can separate the statistics for calls requiring research from normal calls. This means that all of the easy call statistics (the ones that can be dealt with in less than 2 minutes) will not be lumped together with hard calls. This allows you to see what is happening to customers requiring advanced help. In the *New Name* field, type in HardCall. The percentage routing to exit is automatically updated to 75 percent because there are only two percentage routings from *Take Call.*

The *Return Call* activity takes 3 minutes. To enter this time, click on the *Return Call* activity. In the Properties Dialog, click in the field labeled *Time* and enter 3. The other Properties Dialogs do not require any editing. The same goes for the routing connection between *Perform Research* and *Return Call.*

To enter cost information, select *Support 2.* Highlight the *Cost* tab in the Activity box. In the *Hourly Cost* field, enter 20 (for $20 per hour). Select *Support 1.* Notice that the *Hourly Cost* tab remains selected. In the *Hourly Cost* field, enter 12.

RUN THE SIMULATION

The model is now complete, and we are ready to run the simulation. Click on the simulation pull-down menu and select *Save & Simulate.* You will be prompted to save your model. Type in the name of the file, for example, *Help Desk.* After the file has been saved, the simulation will begin to run, and the simulation window appears. As you are watching the simulation, you may want to take note of the following items:

- Telephone calls moving through the flowchart provide visual feedback of calls flowing through the process.
- Resources have status lights associated with them, indicating when they are in operation. The status light is green when the resource is being utilized and is blue when it is idle.
- Counters located above and to the left of each activity represent the number of calls waiting to process.
- An on-screen scoreboard keeps track of system statistics such as quantity processed, cycle time, value-added time, and cost per unit.

The simulation will run for a default time of 40 simulation hours. (This time can be changed from the *Simulation \ Options* menu.) You can speed up or slow down the simulation by sliding the speed control bar.

VIEW OUTPUT REPORTS

There are two types of output. One is an *output summary,* which gives basic management overview information. The other provides *detailed output* in the form of graphs, charts, and other decision-making information. Use the *View* menu in the Output Module to create new reports and graphs. You choose any of the following types of reports:

- *General stats*—Creates a general summary report that is generated automatically when you launch the Output Module from the simulation window when the simulation is ended.

- *Selected stats*—Allows you to create a replication report (information averaged over all replications) with only the specific statistics and elements you want in the report. This report is only available if there are multiple replications.
- *State or utilization summary*—Allows you to create graphs representing the state or utilization summary of a particular element of the model. The graphs are expressed as a percentage of the time elapsed during the simulation that the activity or resource was in a particular state.

From the *Options* menu, the Output Module allows you to customize the look and feel of each graph that you create. You can change the titles, fonts, and colors of each graph. Once you have the reports and graphs you want, you can select *Keep Reports & Graphs* from the *File* menu so that the same reports and graphs are displayed automatically but with new information when you run the model again. To display the Output Module with the results file, click on *Yes* from the *Yes/No* question box that pops up on the screen, asking if you want to see the results of the simulation when the simulation run ends. You can then create specific reports as well as bar graphs and pie charts.

The Output Module appears with the general statistics report opened; this report is shown in Figure 8-4. In the *ACTIVITIES* section, we see that support problems waited in the *Return Call inQ* activity on average over 496 minutes (your results may differ somewhat), and as many as 51 calls were waiting at any one time. Thus, this activity should be identified as a problem area suitable for process engineering efforts to improve performance. Note that the average time to take calls was 2.17 minutes, and the average time to perform research was 19.92 minutes; these are close to the means of the distributions defined for these activities. In the *RESOURCES* section, we see that Support 1 was busy about half the time; Support 2 was busy nearly 100 percent of the time. This also suggests that better allocation of resources should improve performance.

To examine the utilization of the support representatives from a different perspective, click on the *View* menu and select *State or Utilization Summary* to display the dialog shown in Figure 8-5. Select *Resource States* and click "OK," which displays the bar chart in Figure 8-6 on page 286. Anytime human resource utilization is above 80 percent for extended periods, the system will most likely result in long waiting times and queue lengths, requiring more resources or changes in the assignment of resources. To examine the utilization in a pie chart, click on a bar in the *Resource State* chart of the desired resource. Clicking on the *Support 2* bar displays the pie chart shown in Figure 8-7 on page 286. From the pie chart, you can see the utilization from a different perspective. In order to generate these same charts or reports the next time you run your simulation without having to execute these steps, you can print these charts from the *File* menu or select *Keep Reports* from the File menu. To exit the Output Module after the output data have been reviewed and analyzed, click on the *File* menu and select *Exit*.

From the Modeling Environment window, you may view the Output Summary by clicking on *View* and selecting *Output Summary*. You may choose to view a variety of data by clicking on the appropriate tree diagram entry:

- *Total Cost*—Total cost represents the sum of all the costs to run the process plus the addition of all unused resource costs.
- *Entity Cost*—Entity cost provides you with a realistic picture of the cost to produce one entity. Unused resource costs are applied to each entity type so you get a realistic picture of all costs involved in producing an entity.
- *Resource Cost*—Resource cost provides a breakdown of the cost of the resource for time utilized and for the time that the resource was utilized. This table helps you to quantify the resource waste in your process.

```
--------------------------------------------------------------------------------
Scenario       : Normal Run
Replication    : 1 of 1
Simulation Time : 40 hr
--------------------------------------------------------------------------------
```

ACTIVITIES

Activity Name	Scheduled Hours	Capacity	Total Entries	Average Minutes Per Entry	Average Contents	Maximum Contents	Current Contents	% Util
Take Call inQ	40	999	504	1.01	0.21	5	0	0.02
Take Call	40	1	504	2.17	0.45	1	0	45.62
Perform Research inQ	40	999	114	112.50	5.34	11	4	0.53
Perform Research	40	10	110	19.92	0.91	1	1	9.13
Return Call inQ	40	999	109	496.78	22.56	51	51	2.26
Return Call	40	1	58	3.00	0.07	1	0	7.25

ACTIVITY STATES BY PERCENTAGE (Multiple Capacity)

Activity Name	Scheduled Hours	% Empty	% Partially Occupied	% Full
Take Call inQ	40	84.85	15.15	0.00
Perform Research inQ	40	10.50	89.50	0.00
Perform Research	40	8.67	91.33	0.00
Return Call inQ	40	2.06	97.94	0.00

ACTIVITY STATES BY PERCENTAGE (Single Capacity)

Activity Name	Scheduled Hours	% Operation	% Idle	% Waiting	% Blocked
Take Call	40	45.62	54.38	0.00	0.00
Return Call	40	7.25	92.75	0.00	0.00

RESOURCES

Resource Name	Units	Scheduled Hours	Number Of Times Used	Average Minutes Per Usage	% Util
Support 1	1	40	504	2.17	45.62
Support 2	1	40	168	14.08	98.58

(a)

FIGURE 8-4 ProcessModel General Statistics Report

- *Cycle Time*—Cycle time is the length of time that an entity remains in the simulation model. Cycle time is given as an average.
- *Value-Added Time*—Value-added time is broken into value added, nonvalue added, and book value added. These breakdowns allow you to determine what part of your process provides the values and what part does not.

Figure 8-8 shows the resource cost for the example. Note that Support 1 is paid $12 per hour; thus, over a 40-hour simulation, this individual costs the company $480. Figure 8-8 breaks this cost down by the percentage of busy and idle time; we see that over half of this

RESOURCE STATES BY PERCENTAGE

Resource Name	Scheduled Hours	% In Use	% Idle	% Down
Support 1	40	45.62	54.38	0.00
Support 2	40	98.58	1.42	0.00

ENTITY SUMMARY   (Times in Scoreboard time units)

Entity Name	Qty Processed	Average Cycle Time (Minutes)	Average VA Time (Minutes)	Average Cost
Call	390	4.19	2.18	0.43
HardCall	58	596.99	24.99	8.04

VARIABLES

Variable Name	Total Changes	Average Minutes Per Change	Minimum Value	Maximum Value	Current Value	Average Value
Avg BVA Time Entity	1	0.00	0	0	0	0
Avg BVA Time Call	391	6.10	0	0	0	0
Avg BVA Time HardCall	59	39.04	0	0	0	0

(b)

FIGURE 8-4 (*continued*)

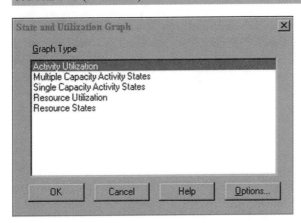

FIGURE 8-5 State and Utilization Graph Types

person's salary goes to nonproductive time. Consult Chapter 5 in the *Online User's Manual* for further information about output reports. At this point, you can return to the model and change activity times, the number of resources available, and branching percentages. You can then run the simulation and review the output again to evaluate the changes to the model.

SKILLBUILDER EXERCISE

Open ProcessModel. Follow the previous discussion and build and simulate the Help Desk model. What is the average cycle time for each type of call?

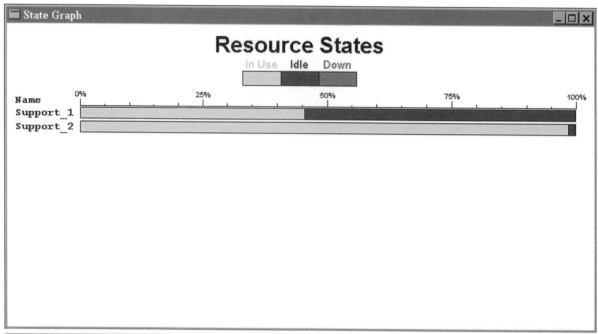

FIGURE 8-6 Resource State Bar Chart

FIGURE 8-7 Support_2 Utilization Pie Chart

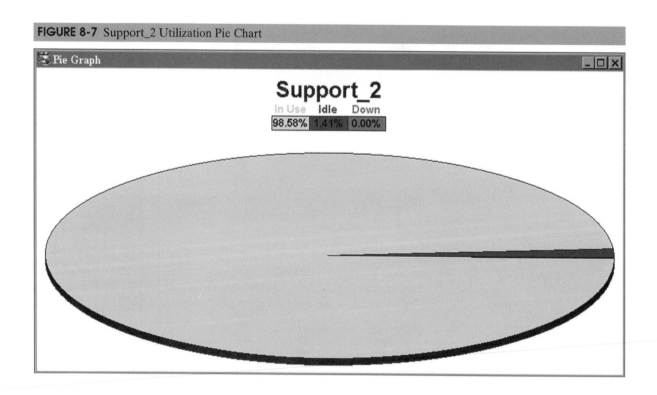

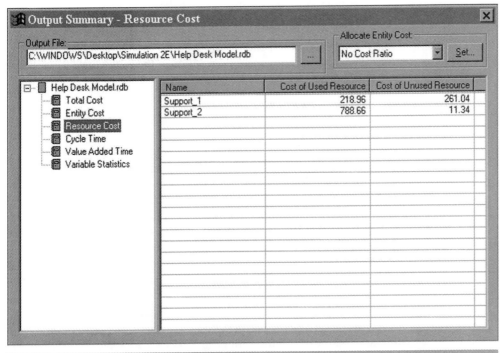

FIGURE 8-8 Viewing Resource Cost Summary Data

Enhancing the Help Desk Process Model

This section describes some additional features of ProcessModel that can be used in modeling other types of problems.

RENEGING

In queueing terminology, *reneging* refers to customers that leave the system after some time because of excessive waiting. For instance, callers in the Help Desk model might hang up after waiting for an average of 2 minutes with a standard deviation of 0.5 minutes in the hold queue (the input queue of *Take Call*). To model this, select the Line tool from the Toolbox and click and drag a routing from the *Take Call* activity. In the Properties Dialog for the connection, select *Renege* from the *Type* field. Enter a 0 for the *Move time* field and N(2, .5) in the *Renege after* field.

Call center managers are interested in tracking the number of calls that hang up. This may be done by giving the calls a new name as they hang up and keeping track of them on the scoreboard during the simulation. To do this, click on the renege routing connection to display its Properties Dialog. In the *New name* field, type *LostCall*. This will name each of the exiting entities *LostCall*.

SKILLBUILDER EXERCISE

Modify the Help Desk model to allow for reneging. What effect does this have on the simulation results?

VISUAL STAFFING

Currently, the two support resources are dedicated to performing specific tasks. One answers calls, and one performs research and returns calls. Running the model and looking at the statistics under this operating scenario resulted in the following observations:

- Customers requiring research wait, on average, nearly 10 hours to get a return call from support.
- One of the support representatives (Support 1) is underutilized; the other (Support 2) is overutilized.

To reduce the customer waiting time, you could add additional support representatives or you could cross-train and share the existing representatives. Because adding resources is seldom the first choice, let us concentrate on the second option, cross-training the support representative. You can model this by adding one line to the model. You can also structure the model so that the representative receiving calls may be interrupted while he or she performs research to take incoming calls.

Select the Connector Line tool. Draw a line from *Support 1* to the assignment line that connects *Support 2* to *Perform Research*. The result is shown in Figure 8-9. Notice that dialog box indicates that *Support 1* is an alternate to *Support 2*. If *Support 2* is busy, then *Support 1* will help perform research. To ensure that *Support 2* can be interrupted to answer incoming calls while performing research, select the connection between *Support 1* and *Take Call*. In the Properties Dialog, check the *Respond immediately* box.

SKILLBUILDER EXERCISE

Modify the Help Desk model to allow Support 2 to be interrupted to take incoming calls. What effect does this have on the response time and resource utilization?

CHANGING ARRIVAL PATTERNS

ProcessModel allows you to customize arrivals of entities to accurately reflect real-life situations such as nonstationary arrival patterns. For instance, suppose that the total daily call volume is 100 calls where 60 percent of the calls arrive between 8:00 A.M. and 12:00 P.M., and the remaining 40 percent arrive between 12:00 P.M. and 5:00 P.M. To model this pattern, click on the arrival connection between *Call* and *Take Call*. In the Properties Dialog, change the *Arrival Type* to *Daily Pattern*. Click on the *Define Pattern* button to display a dialog box. Click on the *New* button. A default arrival with a quantity of 1 appears in the list. Edit the start and end times to 8:00 A.M. to 12:00 P.M. and change the quantity to 60. (Make certain you change A.M. to P.M. for the 12:00 P.M. time.) Select the *New* button again. An entry for 12:00 P.M. to 4:00 P.M., with a quantity 60, appears. (This is a default to facilitate entering equal time periods.) Edit the start and end times to 12:00 P.M. to 5:00 P.M. and change the quantity to 40. To copy the same patterns to other days of the week, select *Monday* and click on the *Copy day* button. Then highlight *Tuesday* and click on the *Paste Monday* button. To paste this pattern to the rest of the days, click on the day of the week to which you want to paste this pat-

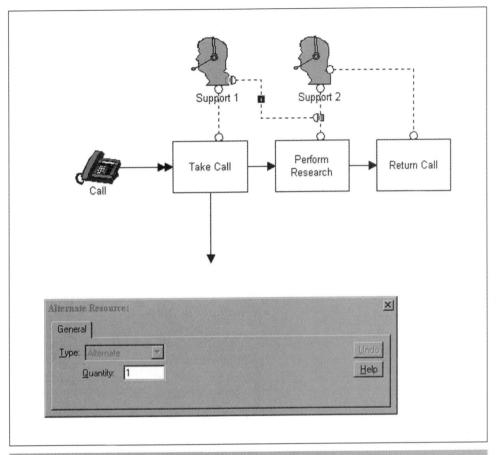

FIGURE 8-9 Linking Support_1 Resource to Support_2 Activity

tern, then click the *Paste Monday* button again. When completed, click on the *Close* button.

When you use the daily pattern arrival or scheduled arrival, you must understand that statistics, especially resource statistics, may be affected due to the way the clock works in ProcessModel. Each ProcessModel simulation begins at 12:00 A.M. (midnight) on Monday morning of the first week. Therefore, daily pattern or scheduled arrivals may skew statistical results, especially with regard to resource and input and output queue utilization. The solution is the use of shifts in conjunction with your daily pattern or scheduled arrivals. For more information see "Schedules—Shifts & Breaks" on page 205 of the *On-Line User's Manual.*

SKILLBUILDER EXERCISE

Change the arrival pattern of the basic Help Desk model to include the nonstationary arrival pat-tern. Summarize the results from the simulation output.

Simulation Options and Scenarios

Accessed from the *Simulation* menu, the Options dialog contains the settings and information used to run a model. Be sure to review these settings before running a simulation. The *Run* tab (see Figure 8-10) provides the basic settings as described here.

- *Run length*—The length of time for running the simulation. You may use any constant or expression to indicate the length of time the simulation will run.
- *Warmup length*—The length of time to run the simulation before collecting statistics. This option allows you to define that time of the transient period, as we discussed in Chapter 7. Again, any constant or expression may be used in this field.
- *Replications*—The number of replications to run for increasing the statistical significance of output results.
- *Report time units*—Allows you to select the default time units to use when you run the model.
- *Show Animation*—If checked, the process will be animated during the simulation. Disabling the animation enables the simulation to run faster and consume less memory.
- *Show Scoreboard*—Check to display basic performance measures during the simulation.
- *Scoreboard Time units*—Allows you to select the time unit displayed in the scoreboard at the time of simulation.
- *Disable subprocesses*—Causes the simulation to be run by the main model only (no submodels are executed). Individual submodels can be disabled in the Activity dialog.

Other simulation options are described in section 4.1 of the *Online User's Manual.*

Scenarios are experiments that are run with a model in which the value of one or more parameters varies for each run. Scenarios allow you to run and compare multiple test cases for a model all automatically. You create scenarios by defining scenario parameters to be used in your model, referencing them in your model, and then defining specific scenarios in which values are assigned to those parameters.

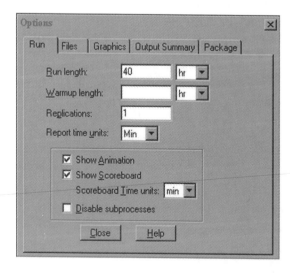

FIGURE 8-10 Run Options

Scenario parameters are placeholders like variables, only they are used to represent a value that changes from scenario to scenario. For example, if you want to run three model scenarios experimenting with different activity times, you would enter a scenario parameter as the activity time and then define three different scenarios with three different times in the *Scenarios* tab of the Scenarios & Parameters dialog.

To define a scenario parameter, select *Define Scenarios* . . . from the *Simulation* menu. Click on the *Scenario Parameters* tab in the dialog and then click on the *New* button. Enter the name of the parameter. You are free to adopt your own naming convention within the limits of the following rules:

- All letters are case insensitive ("A" is the same as "a").
- Only the letters A through Z (upper- or lowercase), the digits 0–9, and the underscores "_" may be used. No other symbols or characters may be used in ProcessModel names.
- Names must begin with a letter of the alphabet or an underscore "_". (e.g., Item5 or _Item5, but *not* 3_Item).
- Names must be single words (use underscores "_" for spaces).
- Do not use hyphens (e.g., the name high-color would be invalid).

Enter a default initial value. This could be a specific number (e.g., number of resources) or a distribution function. The *Scenario Parameters* tab has the following options:

- *New*—This button creates a new parameter that may then be edited.
- *Delete*—This button deletes the selected parameter.
- *Name*—The name of the parameter to be used in the model in place of actual values.
- *Default Value*—The value of the parameter when the *Save & Simulate Chart* menu item is selected rather than *Run Scenarios*.
- *Move Up*—Moves the selected parameter up in the list.
- *Move Down*—Moves the selected parameter down in the list.

To define scenarios, select *Define Scenarios* . . . from the *Simulation* menu. Click on the *Scenarios* tab in the dialog, then click on the *New* button. Enter the name of the scenario. Select parameter from the *Parameters* list box and enter a value in the *Value* field. Repeat this for each parameter you wish to change for the selected scenario. You may repeat this process to create additional scenarios.

For example, suppose we wish to define a scenario for the Help Desk model in which the mean interarrival time is 15 instead of its current value of 5. Figure 8-11 shows the *Scenario Parameters* tab, and Figure 8-12 shows the *Scenarios* tab. In the dialog in Figure 8-11, click on *New,* and enter mean_arrival_time as a parameter in the *Name* field, with 5 in the *Default Value* field. Click on the *Scenarios* tab (see Figure 8-12) and define a new scenario with a value of 15 for the parameter. To define the interarrival time as a parameter in the model itself, double click on the connector from *Call* to *Take Call* in the model. In the Properties dialog box, replace the expression E(5) by the expression E(mean_arrival_time). Then select *Run Scenarios* from the *Simulation* menu. This will cause ProcessModel to run the simulation twice—once for the mean interarrival time of 5 and a second run with a value of 15. With a longer average time between arrivals, we would expect less waiting and higher idle time for the resources (support 1 and support 2). In fact, the simulation results show that the average time for *Return Call inQ* drops to about 7 minutes, and the percentage of time that Support 2 is busy drops to under 30 percent. Using scenarios is much easier than changing and running each combination of parameters one at a time.

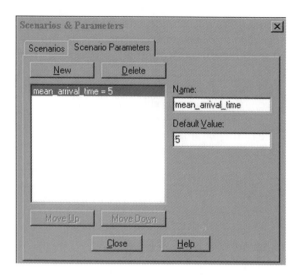

FIGURE 8-11 Scenario Parameters Dialog

SKILLBUILDER EXERCISE

Modify and run the Help Desk model for these five scenarios in which the mean interarrival time is varied as 5, 8, 11, 15, and 18 minutes. Examine the changes in waiting times, queue lengths, and resource utilization over this range of values.

Attributes and Variables

We introduced the notion of attributes in Chapter 6 when we discussed event-driven simulation models. In ProcessModel, **attributes** are values or placeholders for either values (real or integer) or descriptors (single-word descriptions) associated with individual entities that may, for example, indicate the entity's size or condition. An attribute's value can only be assigned, incremented, decremented, and examined by the entity to which the attribute belongs. For example, while activity A processes an entity, the attribute

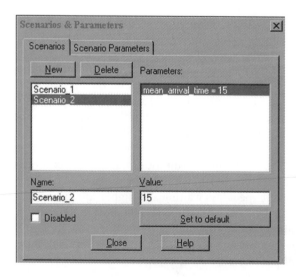

FIGURE 8-12 Scenarios Dialog

called *Color* could be tested for only that particular entity in the activity's action logic. Entities have the following predefined attributes:

- *Name*—The name of the entity. *Do not* use an assignment statement to assign this attribute; e.g., Name = BadCall will not work. Use the *NewName* statement or the available fields in the routing properties dialog.
- *Cost*—The current accumulated cost for an entity. To learn more about costs, see "ProcessModel and Activity-Based Costing" on page 529 of the *Online User's Manual.*
- *ID*—Unique identifying number assigned to each entity. (Created entities have the same ID as the entity that created them, so they can be reunited later, if desired.)
- *VATime*—Cumulative value-added time (in minutes).
- *CycleStart*—Time (in minutes) entity entered system.

Predefined attributes are manipulated automatically by the system, so they should only be manipulated manually with careful forethought.

User-defined attributes may be defined in the Attributes & Variables dialog that is accessed from the *Insert* menu (see Figure 8-13). User-defined attributes may be given an initial value using action statements defined for the entity arrival. Details about the Attributes & Variables dialog for attributes are given here:

- *New*—This button creates a new attribute.
- *Delete*—This button deletes the selected attribute. (Note that the predefined attribute may not be deleted.)
- *Name*—The name of the attribute. Letters, numbers, and the underscore "_" character are allowed in the name.
- *Type*—The type of attribute. This can be set to integer, real, or descriptive.

    INTEGER—Any whole number (no digits to the right of the decimal) between −32,000 and 32,000.

    REAL—Any number including those with digits to the right of the decimal. Use this when a high level of accuracy and detail is needed.

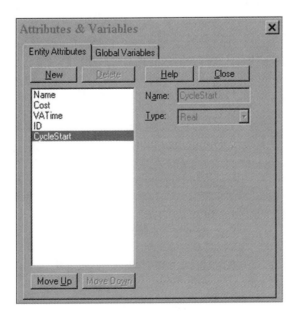

FIGURE 8-13 Attributes & Variables: Entity Attributes Tab

DESCRIPTIVE—Defined with a list of adjectives or descriptors that may be assigned to the attribute.

- *Descriptor list*—The list of adjectives or descriptors that may be assigned to the descriptive attribute. Only available for attributes whose type is descriptive.

To define an attribute, select *Attributes & Variables* from the *Simulation* menu, click the *New* button, and enter the name of the attribute. Select the type: real, integer, or descriptive. If you select descriptive, you must enter the list of descriptors in the edit box provided. (Each descriptor should be entered on a separate line.) Note that if the last attribute defined is descriptive, ProcessModel assumes that the current attribute is descriptive; thus, you should define real and integer attributes first.

Like attributes, **variables** are placeholders for either values (real or integer) or descriptors (single-word descriptions). They may be used to describe or track activities and states in the system, such as the number of entities that have completed a particular activity. Variables are global in nature and can be set, incremented, decremented, and examined in the *Action* tab of the *Properties* dialog. (Most elements have an action tab.)

Variables are of two types: predefined and user-defined. Predefined variables that are set up automatically include the following:

- *Qty_Processed <entity name>* Number of entities processed for an entity type. For example, Qty_Processed_EntA is the number of EntA processed.
- *Avg_VA_Time_<entity name>* Average value-added time (time units) for an entity type. For example, Avg_VA_Time_Orders is the number of time units the entity type *Orders* has spent in value-added activity.
- *Avg_Cycle_Time_<entity name>* Average cycle time (time units) for an entity type. For example, Avg_Cycle_Time_AssemblyA is the average number of time units that the entity type *AssemblyA* spent in the model.
- *Avg_Cost_<entity name>* Average cost for an entity type. For example, Avg_Cost_Call is the average cost for the entity type *Call* in the model. Average cost applies only to completed entities. Variables may also be displayed during the simulation on the scoreboard or in a user-defined position.

Like attributes, you may create user-defined variables in the Attributes & Variables dialog accessed from the *Simulation* menu (see Figure 8-14). The dialog has the following options:

- *New*—This button creates a new variable.
- *Delete*—This button deletes the selected variable. (Note that the predefined variables may not be accessed from the Attributes & Variables dialog.)
- *Name*—The name of the variable. Letters, numbers, and the underscore "_" character are allowed in the name.
- *Type*—The type of variable. This can be set to integer, real, or descriptive.

    INTEGER—Any whole number (no digits to the right of the decimal) between −32,000 and 32,000.
    REAL—Any number, including those with digits to the right of the decimal.
    DESCRIPTIVE—Defined with a string of adjectives or descriptors that may be assigned to the variable.

- *Initial Value*—The value assigned to the variable at the beginning of the simulation. If you do not enter a value or descriptor, ProcessModel will use zero (0) or the first descriptor in the list as the initial value for the variable.
- *Stats*—Changes the type of statistics that are collected.

    NONE—No statistics will be collected.

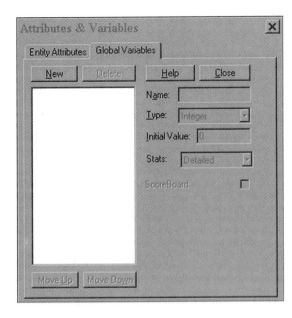

FIGURE 8-14 Attributes & Variables: Global Variables Tab

BASIC—Collects basic statistics such as total changes, average minutes per change, current value, and average value. Observation based.

BASIC TIME—Same information as basic, but the information is time-weighted.

DETAILED—Allows all of the same information as the basic option, plus standard deviation information.

DETAILED TIME—Same information as detailed, but the information is time-weighted.

OBSERVATION-BASED—Variable information is calculated based on a simple average.

TIME-WEIGHTED—Variable information is calculated based on the average of the products of the variable multiplied by the length of time it remained at that value.

- *Scoreboard*—The variable will be displayed on the simulation scoreboard when this object is checked.
- *Descriptor List*—The list of adjectives or descriptors that may be assigned to the descriptive variable. Only available for variables whose type is descriptive.

To define a variable, select *Attributes & Variables* from the *Insert* menu, click on the *Variables* tab, and then click on the *New* button. Enter the name of the variable. Select the type: real, integer, or descriptive. If you select descriptive, you must enter the list of descriptors in the box provided, with a separate descriptor entered on each line. To display a variable on the scoreboard, select *Attributes & Variables* from the *Insert* menu, click on the *Variables* tab, and select the Scoreboard check box.

ACTION LOGIC

ProcessModel allows you to design custom behavior in your model by using **action logic,** which allows you to define special logic that may not be easily defined using the normal property fields. Examples would include assigning values to attributes and variables or performing a test using an *IF . . . THEN* statement. Action logic can be defined for any activity, storage, arrival, or routing by clicking on the *Action* tab of the Properties dialog.

Depending on the object or connection for which the action is defined, only certain statements and other logic elements are meaningful and, therefore, valid. The valid statements and logic elements (variables, attributes, resources, distributions, operators, and scenario parameters) are displayed in the list box in the Action dialog. These statements and elements may be pasted from the list box into the action window to help you construct the desired action logic. If a larger work space is needed for action logic, click on the *Zoom* button to expand the window. To paste elements into the action window, select the type of element (including statements) you want from the pull-down box, select the specific element or statement from the list box, and then press the *Paste* button to insert the element at the position of the cursor in the action edit window, or simply double click the item.

Action logic often uses **expressions** (combinations of attributes, variables, numbers, and operators) in assigning values to attributes or variables, testing the value or state of a variable or attribute, and so forth. Expressions allow you to introduce variability into your model. They let you track, control, and respond to events. Some expressions provide a numeric value; these are called **numeric expressions** and consist of elements (attributes, variables, distributions, and constants) combined with mathematical operators $(+, -, \text{etc.})$ that result in a numeric value. Others provide a true/false value; these are called **Boolean expressions.**

To create an expression, you may use any combination of constants, probability distributions, attributes, and variables. You may also combine items to form a compound expression and use parentheses to set off parts of the expression to be evaluated first. Examples of numeric expressions include:

- `Attr1`
- `50.91`
- `Var 1 + 5`
- `Total_Pieces + 5 * Pkg_Qty`
- `(Weight + 5) * (Pkg_Qty / 2)`
- `N(25, 4.8) + Weight * (Total_Pieces - 10)`

Boolean expressions use logical operators to compare two numeric expressions, yielding a result of true or false. These expression may be used in *IF . . . THEN* statements and condition fields to make specific decisions in the model based on the values of two numeric expressions. You may use simple or compound numeric expressions on either side of the Boolean operator. Some examples are

- `IF Total_Pieces > 5 * Pkg_Qty THEN . . .`
- `IF (Weight + 5) <= (Pkg_Qty / 2) THEN . . .`
- `IF N(25, 4.8) + Weight = Total_Pieces - 10 THEN . . .`
- `IF Weight >= Total_Pieces AND Pkg_Qty > 20 THEN . . .`
- `IF Total_Pieces = Pkg_Qty OR Pkg_Qty > 35 THEN . . . Statements`

Statements are simply commands to be executed at particular stages in an entity's progress through the process. Some of the more common types of statements used in ProcessModel are described next; others are described in section 3.11.2 of the *Online User's Manual.*

( ) = ( )   This is the assignment statement, which allows you to assign a value (or descriptor) to a variable or to one of the attributes defined for your entities. The syntax is

`assignee = assignor`

where *assignee* is the variable or attribute to which the value is assigned, and *assignor* is the value assigned to the variable. This could be another variable or attribute, a predefined descriptor, or a mathematical expression. In the following example, the attribute Attr1 is assigned a value of 2. The second example assigns the value of PO_No to the attribute Invoice_No. Number three assigns the descriptor Red to the attribute Color. The last example assigns the product of 5 and the value of Base to the attribute Size.

1. Attr1 = 2
2. Invoice_No = PO_No
3. Color = Red
4. Size = 5 * Base

DEC   The decrement statement allows you to decrement a variable or attribute's value. It subtracts one (the default) or more from the value of the variable or attribute. The syntax is

```
DEC name [, expression]
```

where *name* is the name of the variable or attribute to be decremented. In the field [*expression*], you can optionally decrement the variable or attribute by more than one using an expression that can be a constant or a mathematical expression. The name and expression must be separated by a comma. (The square brackets illustrate only that this element is optional.) The following are several examples. The first decrements the value of Var1 by one. The second decrements the value of Attr1 by five. The third decrements the value of Number_in_System by the value of an attribute called Batch_Size.

1. DEC Var1
2. DEC Attr1, 5
3. DEC Number_in_System, Batch_Size

DISPLAY   This statement pauses the simulation and displays a message. The simulation will resume when the user selects "OK." The syntax is

```
DISPLAY <text string> {,<attribute / variable / function call>}
```

where *text string* is the message ProcessModel will display, and [*attribute / variable / function call*] is the text string or numeric value you wish to display. After the original set of information (i.e., text string, variable) the "$" character is used to add additional information (i.e., another text string or variable). You can force a carriage return by using the statement CHAR(13). Each new item that is appended to the statement must be prefaced with the "$" character.

The following example displays a message whenever a new order type begins processing at the current activity. A variable, Last_Order, stores the order type of the last entity processed at the activity. If the current entity's Order_Type attribute value is different from the previous order type, ProcessModel displays a message stating the new order's type.

```
IF Order_Type <> Last_Order THEN
BEGIN
DISPLAY "New Order Type:", Order_Type
Last_Order = Order_Type
END
```

The display statement is valuable for debugging complex models and for halting a model temporarily during a presentation to display information.

FREE The free statement allows you to free a resource (or resources) being used by the current entity. The syntax can be one of the following:

```
FREE [quantity] resource
FREE [quantity] resource, [quantity] resource, . . .
FREE ALL
```

where [*quantity*] is the number of units of the following resource to free. If no quantity is used, the quantity is assumed to be one. (The square brackets illustrate only that this element is optional); *resource* is the name of the resource or list of resource names to be freed. If any resource specified is not being used by the current entity, it is simply ignored. The keyword *ALL* is used with the *FREE* statement to free all captured resources. In the following example, an entity, which earlier captured the resource operator, frees the operator after a 3-minute activity time. This action is followed by an increment of the variable called TimesUsed.

```
TIME(3 min)
FREE Operator
INC TimesUsed
```

If no action statements follow the freeing of a resource, the resource can just as easily be freed by drawing a free resource assignment connection between the resource and the activity.

GET The GET statement enables an entity to obtain a resource. ProcessModel attempts to capture the resources in the order they are listed. If multiple resources are requested, but not available, those that are available will be captured and tied up until all are available. The syntax is

```
GET [quantity] resource, [priority]
GET [quantity] resource, [priority] AND [quantity] resource, [priority]
GET [quantity] resource, [priority] OR [quantity] resource, [priority]
```

where [*quantity*] allows you to optionally specify the number of resources to get if the resource has multiple units defined for it. (The square brackets illustrate only that this element is optional.) By default, *quantity* is equal to one unit of the resource. The field *resource* is the name of the resource to be captured. *AND* is used to capture more than one resource as each becomes available. To wait until all become available before capturing any of them, use the *JOINTLYGET* statement described on page 228 of the *Online User's Manual. OR* is used to capture one resource or the other. This is useful for situations where one of several resources could be used to accomplish the same thing. With [*priority*], you can optionally specify priority level to get the resource. The higher the number, the higher priority.

The following examples demonstrate the use of the GET statement. The first shows a simple request for a resource called *Operator*. The second tests the size attribute to determine whether or not the Operator and Helper are needed. The third requests three units of the resource called Operator.

1. GET Operator
2. IF Size > 10 THEN GET Operator AND Helper
3. GET 3 Operator

INC   The increment statement allows you to increment a variable or attribute's value. It adds one (the default) or more to the value of the variable or attribute. The syntax is

```
INC name [, expression]
```

where *name* is the name of the variable or attribute to be incremented, [*expression*] allows you to optionally increment the variable or attribute by more than one using an expression that can be a constant or a mathematical expression. The name and expression must be separated by a comma. (The square brackets illustrate only that this element is optional.) Three examples are shown here. The first increments the value of Var1 by one. The second increments the value of Attr1 by five. The third increments the value of Number_in_System by the value of Num_Processed plus one.

1. INC Var1
2. INC Attr1, 5
3. INC Number_in_System, Num_Processed +1

IF . . . THEN . . . ELSE   This statement allows you to test a variable or attribute's value and if the test is true to execute another statement. Optionally, an alternative statement can be executed if the test is not true. The syntax is

```
IF conditional expression THEN statement_1
IF conditional expression THEN statement_1 ELSE statement_2
```

where *conditional expression* is a comparative expression using comparison operators like the equals sign (=) and the less than/greater than symbols (<>). The result of this expression is either true or false (yes or no). Multiple or alternative conditions can be tested using the operators AND and OR. Parentheses may be used for nesting expressions. The statement *statement_1* is executed if the conditional expression is true. This can also be a block of statements started with a BEGIN keyword or symbol ({) and ended with the END keyword or symbol (}). The statement *statement_2,* preceded by the keyword *ELSE,* is executed if the conditional expression is false. This can also be a block of statements started with a BEGIN keyword or symbol ({) and ended with the END keyword or symbol (}). In the following examples, the *IF . . . THEN . . . ELSE* statement is used to make decisions about what happens in the model. In the first example, the variable *Calls* is incremented if the Name has the same descriptor value as Phone_Call. The second example shows a decision made based on the Patient_Type. If the patient is critical, both a *Nurse* and a *Doctor* resource are needed. Otherwise, only a Nurse is captured and the activity takes less time.

```
1. IF Name = Phone_Call THEN INC Calls
2. IF Patient_Type = Critical THEN
   BEGIN
   GET Nurse AND Doctor
   TIME(N(30, 5) min)
   FREE ALL
   END
   ELSE
   BEGIN
   GET Nurse
   TIME(N(8, 1) min)
   FREE ALL
   END
```

ProcessModel Example: Inventory Simulation with Back Orders

Figure 8-15 shows a ProcessModel diagram of a continuous review inventory system with back orders, similar to the situation we modeled in Chapter 6. The model follows the logic in Figure 6-7 almost exactly. This model makes use of the following variables, which are defined using the Attributes and Variables dialog from the *Insert* menu:

- Inventory_position
- Reorder_point
- Order_quantity
- Inventory_level
- Back_order_level
- Safety_stock

These variables are initialized by creating one entity (Item2 in Figure 8-15) at time 0 and setting the action logic of the arrival to set the values of the variables. In the dialog for the arrival connector, choose *Scheduled* as the type of arrival (see Figure 8-16). Click on the button *Define Schedule* to bring up the dialog in Figure 8-17. Set the time of the arrival to 12:00 A.M. on Monday of week 1; this is time "0" for the simulation. In the *Action* box, set the initial values. To facilitate this and eliminate typographical errors, select *Variables* in the *Filter* box, and then paste the names of the variables in the *Action* box to assist you in writing the proper expression. The arrival goes to the *Initialize* Variables activity (see Figure 8-18), which is used only for descriptive purposes.

FIGURE 8-15 Back-Order Inventory Simulation Model

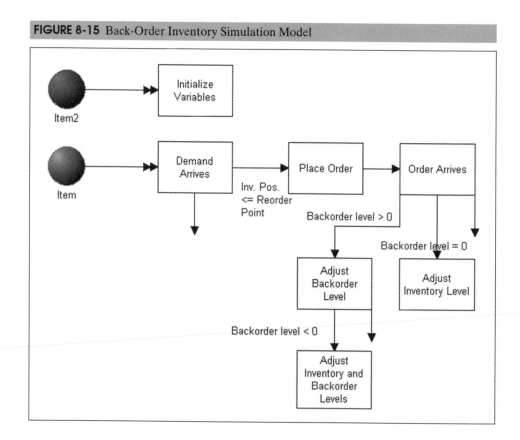

FIGURE 8-16 Scheduled Arrival Dialog

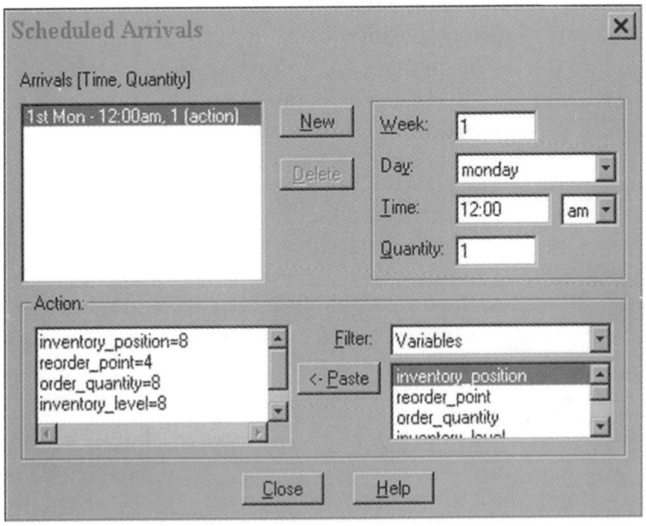

FIGURE 8-17 Define Schedule Dialog for Scheduled Arrivals

The main logic of the inventory simulation is defined by the model segment below the initialization segment. The *Item* icon represents a demand entity, each of which is created using a periodic arrival with an exponential time-between-demand distribution with a mean of 1. When the demand entity reaches the activity *Demand Arrives,* we do three things. First, the inventory position is decreased by 1; second, if the inventory level is positive, we decrease the value of the inventory level by 1; and third, if the inventory level is zero, increase the backorder level by 1. These are defined using statements in the *Action* tab of the Activity dialog (see Figure 8-19). We also need to set the activity time to zero in the dialog because the activity does not consume any simulated time. (This is true for all the activities in this particular model.)

FIGURE 8-18 Initialize Variables Activity

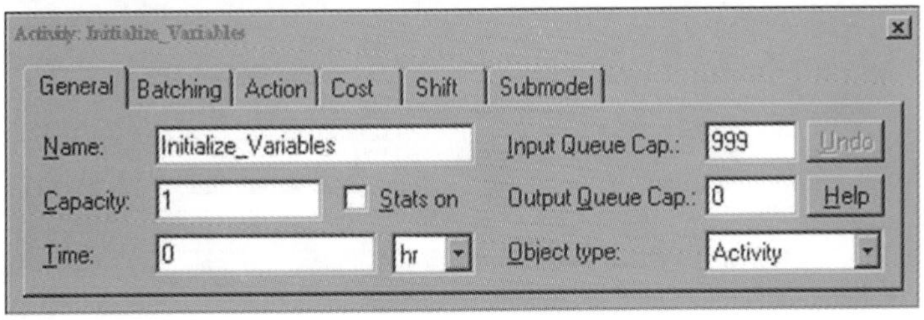

Next, we need to determine whether the inventory position is at or below the re-order point. This is done by using a conditional routing on the connector leading to the activity *Place Order,* as shown in Figure 8-20. We also must provide an alternative routing for the entity if the condition is not met. This is done with the connector directed down from the *Demand Arrives* activity; specify a conditional routing, but leave the condition blank. You may leave the condition blank as long as it is the last routing; however, we recommend that you define a condition. If an order is placed, the entity continues to activity *Order Arrives* after a delay that represents the lead time; this is specified as an exponential random variate with a mean of 3, as shown in Figure 8-21. When the order arrives, we first set the safety stock equal to the inventory level (this is the amount of inventory on hand when an order arrives), and then check if the number of back orders is zero or positive using conditional routings. If the back-order level is zero, we add the order quantity to the inventory level in the action logic of the *Adjust Inventory Level* activity (using the statement *INC inventory_level, order_quantity*). If the number of back orders is positive, the back-order level is decreased by the order quantity to satisfy as many back orders as possible. If the back-order level is less than zero, we will have satisfied all outstanding back orders, and the entity proceeds to the

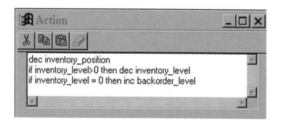

FIGURE 8-19 Action Logic for Demand Arrives Activity

```
dec inventory_position
if inventory_level>0 then dec inventory_level
if inventory_level = 0 then inc backorder_level
```

FIGURE 8-20 Conditional Routing Dialog

Conditional Routing:

| General | Cost | Action | Name |

Type: Conditional Move time: 0 hr Undo

Condition: inventory_position <= reorder_point Help

New name: (optional)

FIGURE 8-21 Setting Lead Time

Percentage Routing:

| General | Cost | Action | Name |

Type: Percentage Move time: E(3) hr Undo

Percent: 100 % Help

New name: (optional)

Adjust Inventory and Backorder Levels activity. At this point, we decrease the inventory level by the back-order level (which is negative) to give the correct level of inventory, and then set the back-order level to zero.

The simulation was run for 500 hours using $Q = 8$ and $r = 4$. A portion of the output is shown in Figure 8-22. The Total Entries column of the output provides information about the number of entities passing through each activity. We see that 512 demands occurred; this is about what we would expect with a mean interarrival time between demands of 1 hour. Sixty-four orders were placed; when the orders arrived, the back-order level was positive for 55 of the 64 times, or 86 percent. In the second portion of Figure 8-22, we have some statistics about the variables in the model. Of particular interest is the average inventory level (4.42), average back-order level (6.33), and average safety stock (0.26). These results suggest that the reorder quantity is too low or the reorder point is too high to prevent a high frequency of back-order accumulation. We would probably wish to define other scenarios to reduce the back orders that are incurred. We could also incorporate costs for holding inventory and incurring back orders to attempt to optimize the inventory system from a financial viewpoint.

FIGURE 8-22 Selected ProcessModel Results for Back-Order Inventory Simulation

Activity Name	Scheduled Hours	Capacity	Total Entries	Average Hours Per Entry	Average Contents	Maximum Contents	Current Contents	% Util
Demand Arrives inQ	500	999	512	0.00	0	1	0	0.00
Demand Arrives	500	1	512	0.00	0	1	0	0.00
Place Order inQ	500	999	64	0.00	0	1	0	0.00
Place Order	500	1	64	0.00	0	1	0	0.00
Order Arrives inQ	500	999	64	0.00	0	1	0	0.00
Order Arrives	500	1	64	0.00	0	1	0	0.00
Adjust Backorder Level inQ	500	999	55	0.00	0	1	0	0.00
Adjust Backorder Level	500	1	55	0.00	0	1	0	0.00
Adjust Inventory Level inQ	500	999	9	0.00	0	1	0	0.00
Adjust Inventory Level	500	1	9	0.00	0	1	0	0.00
Adjust Inventory and Backorder Levels inQ	500	999	11	0.00	0	1	0	0.00
Adjust Inventory and Backorder Levels	500	1	11	0.00	0	1	0	0.00

(a)

ENTITY SUMMARY (Times in Scoreboard time units)

Entity Name	Qty Processed	Average Cycle Time (Hours)	Average VA Time (Hours)	Average Cost
Item	512	0.35	0.00	0.00

VARIABLES (* indicates observation based variables)

Variable Name	Total Changes	Average Hours Per Change	Minimum Value	Maximum Value	Current Value	Average Value
Avg BVA Time Entity	1	0.00	0	0	0	0
Avg BVA Time Item	513	0.97	0	0	0	0
Avg BVA Time Item2	1	0.00	0	0	0	0
inventory position*	577	0.86	4	12	8	8
reorder point*	1	0.00	4	4	4	4
order quantity*	1	0.00	8	8	8	8
inventory level*	138	1.31	0	11	0	4.42
backorder level*	474	1.05	-6	25	4	6.33
safety stock*	64	7.76	0	3	0	0.26

(b)

SKILLBUILDER EXERCISE

Open the ProcessModel file *Backorder Inventory Model.* Use the Scenario option to evaluate the following combinations of order quantities and reorder points: (12, 4), (16, 4), (12, 8), and (16, 8). How do the results compare with the base case?

Simulation in Practice

A SIMULATION MODEL FOR AIR FORCE MAINTENANCE[2]

As one of the three U.S. Air Force maintenance depots, the Ogden Air Logistics Center Aircraft Directorate provides depot repair, modification, and maintenance support to major aircraft weapons systems. Currently, their workload mix includes the A-10 Warthog ground attack aircraft, C-130 Hercules transport, and the F-16 Falcon Fighter. Each year, the workload mix for the three different weapons systems is shifting. To meet customer deadlines at reasonable cost, program managers must compete for Aircraft Directorate facilities and resources. Before simulation modeling was introduced, planning and scheduling was a tedious manual process, and there was no analytical tool that integrated all three weapon systems at the top level. Numerous and diverse factors further complicated the goal of timely job completion. They included delinquent aircraft arrivals, varied requirements for depot maintenance, inclement weather, late delivery of repair materials, worker turnover, routing obstacles, and many other miscellaneous bottlenecks.

Because of the complexity of the project, discrete event simulation was selected as the most effective way to evaluate different scenarios. The Air Logistics Center contracted with PricewaterhouseCoopers (PwC) to design a simulation model that could be used to analyze strategic resource allocations and process improvements. The model also facilitates analysis of "what-if" scenarios by allowing the user to modify various components of the business. PwC developed the model using ProcessModel because it was user friendly, flexible, and cost effective, and it provided the detailed features needed to satisfy requirements.

After being implemented, the model correctly predicted bottlenecks arising from system variability, complexity, and resource constraints. The model has enabled the Aircraft Directorate to provide depot maintenance in a more timely, competitively-priced fashion. The Aircraft Directorate also believes that the simulation models will be useful for both developing long-term strategic plans and evaluating near-term tactical decisions. They plan to enhance the model and use it to analyze the most problematic shared resources and backshop routes. This will allow the Aircraft Directorate to dramatically improve resource allocation, thus creating a more sensible balance between the levels of cost and service.

Questions and Problems

1. What benefits does systems simulation software provide relative to computer programming languages?
2. Explain the differences among entities, activities, and resources.

---

[2]Adapted from Scott Sutherland and Ron Haltli, "Keeping Fighting Birds in Flight," *Simulation Success,* June 2000, a promotional publication of ProcessModel, Inc.

3. Explain the difference between an attribute and a global variable.
4. A firm has a copy machine that has become a bottleneck. A short-time study reveals that the arrival rate for work is one job every 5 minutes, following an exponential distribution. The average size of a job is 4 minutes and is normally distributed with a standard deviation of 1 minute. Develop a ProcessModel simulation to compare the results (time in queue, time in system, number in queue, number in system, printer utilization) with a new, faster copier capable of processing jobs in 3 minutes (normally distributed with standard deviation of 0.75 minutes). Use 100 replications of an 8-hour operating day.
5. Modify the copy machine model in problem 4 to represent the case in which jobs renege after waiting 5 minutes (use a renege connector). Compare the results with your answers to problem 3.
6. Modify the copy machine model in problem 4 to represent the case in which jobs balk if at least five other jobs are in line when a job arrives. (Hint: Change the input queue capacity from 999 to 4.) How do the results change?
7. Suppose that the copy machine in problem 4 is old enough that on average it jams once an hour (assume that running time is lognormally distributed with mean of 1 hour and standard deviation 0.2 hours), and it takes a constant 1 minute for a secretary to fix it. Incorporate this factor into your model, and compare your results with problem 4.
8. A help desk for computer users is open from noon until midnight, with steady demand throughout the day that is normally distributed with a rate of 10 minutes between arrivals and standard deviation of 3 minutes. Service varies substantially and is exponentially distributed with a mean of 30 minutes. There currently are three people at the help desk. Develop a ProcessModel simulation to compute relevant performance statistics.
9. An information systems project consists of the following activities and time distributions:

Delivery of database software	D3(10,16,70,24,20,32) hours
Install database	E(80) hours
Collect data	N(40,5) hours
Enter data	N(80,10) hours
Build interface	L(120,20) hours
Assemble hardware	N(40,5) hours
Testing	Constant 20 hours

Beginning installation of the database must wait for delivery of software and assembly of hardware. The data entry activity cannot begin until data is collected and the hardware assembled. Testing cannot begin until data is entered and the interface is built. Build a model, and simulate 100 replications to estimate the duration of each activity and completion of the project.
10. A car repair shop operation involves the following operations. Customers arrive at the front desk, spending a check-in time that is normal with mean = 5 minutes, standard deviation = 1 minute. Thirty percent of the customers are in for oil changes; 70 percent, for repair. Two servers are available to change the oil, with a time that is normal with mean = 15 minutes and standard deviation = 2 minutes. Four servers do repairs, with a time that is exponential with a mean of 90 minutes. When oil-changing and repair work is done, cars are cleaned. One server does the cleaning, with a normally-distributed time having a mean of 20 minutes and standard deviation of 3 minutes. Work arrives for oil changes at the rate of five per hour, and for repair at the rate of eight per hour. Simulate this system for 100 replications, and make recommendations for adding a server to each station.
11. Student registration at a large university is accomplished through 10 telephone lines. At the peak registration period, arrivals occur at the rate of 1,000 per hour,

exponentially distributed. Service is normally distributed with a mean of 2 minutes and a standard deviation of 0.5 minutes. A candidate for student office has proposed that student fees be increased $0.50 per student per semester to pay for a system with 30 lines. Evaluate the impact of the proposal over the current system by developing and running a ProcessModel simulation.

9

Applications of Systems Simulation

Chapter Outline

Systems simulation is used in a wide variety of organizations to address many different problems. We cannot even begin to scratch the surface of the possible applications of simulation in business and industry. In this chapter, we describe some practical applications of systems simulation in a variety of disciplines and provide simple examples using ProcessModel to show the flexibility of system simulation and its applicability in business. We also encourage you to study the ProcessModel examples in the *Demos* folder that is installed with the software. These provide other examples of practical problems that can be modeled using ProcessModel, and they also illustrate other features of the software.

Simulation in Practice: Designing Emergency Room Facilities Using Simulation[1]

Emergency rooms are important elements of medical systems, absorbing a high proportion of hospital resources. The emergency complex of Las Palmas de Gran Canaria in Spain involves a number of elements that must be coordinated. The complexity of the system precludes the use of analytic queueing models. Thus, simulation was used to analyze the N. S. del Pino hospital emergency department to evaluate the efficiency of possible modifications to personnel assignments, to evaluate expected system performance should demand increase, and to estimate the benefits of additional facilities in the way of an additional emergency laboratory or additional X-ray equipment.

The hospital serves over 400,000 people. Of all admissions in 1992, over 50 percent were to the emergency unit, which serviced almost 60,000 patients that year. Emergency room cost was much higher than that of office care; service quality was less than desired. Many of the visits to the emergency room could easily have been dealt with through normal office visits. Therefore, in 1987 a policy was implemented to assign a doctor to evaluate cases and send them to one of three care levels. These levels were to hospital care, ambulatory care, or no care.

Patients first had to go through an admission service. Some received diagnostic tests or therapies such as X rays, casts, first aid, laboratory tests, and other treatments. A sample of arrival rates and service times at each treatment station was taken to characterize the appropriate distribution. For example, lognormal distributions were determined for cast and first-aid arrivals, a Weibull distribution for laboratory arrivals, and Pearson type V distribution (related to the gamma distribution) for X-ray arrivals. The proportions of patients classified into medical levels, as well as tests and treatment proportions of each, were also identified.

The simulation experiment evaluated eight alternative scenarios, which varied by different numbers of service channels and servers. One alternative added a classifying doctor, while another decreased the original level of doctors by one. Different arrival rates were simulated to reflect the impact of general health policies. For instance, one scenario hypothesized a massive accident, increasing the arrival rate. New equipment acquisitions were simulated by reducing service time in the emergency laboratory and the X-ray department.

The model generated outputs, including average, maximum, and minimum times for waiting and service by location, as well as the number of patients waiting, in service, and in the system. The model also measured the utilization rate of each emergency room element. The simulation results suggested improvements in a number of reorganization options, specifically, identifying the benefits of purchasing new equipment and quantifying the cost-effectiveness of investing in additional doctors.

An Emergency Room Simulation Model

An emergency room consists of four stations for late-night operations (midnight until 8:00 A.M.). Arrivals to the emergency room occur at a rate of four patients per hour and are assumed to follow an exponential distribution. Incoming patients are initially screened to determine their level of severity. Past data indicate that five percent of incoming patients require hospital admission and leave the emergency room. Thirty percent of incoming patients require ambulatory care, after which they are released. Twenty percent of incoming patients are sent to the X-ray unit, and the last 45 percent are sent to the laboratory unit. Of

---

[1]Adapted from B. González López-Valcárcel and P. Barber Pérez, "Evaluation of Alternative Functional Designs in an Emergency Department by Means of Simulation," *Simulation*, Vol. 63, no. 1, 1994, pp. 20–28.

those going to the X-ray unit, 30 percent require admission to the hospital system, 10 percent are sent to the laboratory unit for additional testing, and 60 percent have no need of additional care and are thus released. Of patients entering the laboratory unit, 10 percent require hospitalization, and 90 percent are released. Current facilities are capable of keeping up with average traffic, although there is some concern that the existing laboratory facilities can become a bottleneck on particularly busy nights, especially as the community grows.

The basic emergency room system, shown in Figure 9-1, consists of the following activities, durations, and flows:

Activity	Duration	Routing
Arrivals	4/hour, exponential	Initial desk
Front desk	0.05 hour (constant)	0.30 ambulatory
		0.20 X ray
		0.45 lab
		0.05 hospital
Ambulatory care	Normal (0.25 hour, 0.1)	Released
X ray	Normal (0.25 hour, 0.05)	0.1 lab
		0.3 hospital
		0.6 released
Laboratory testing	Normal (0.5 hour, 0.1)	0.1 hospital
		0.9 released

Each station in the emergency facility is currently a single-server queueing system. Thus, the system consists of a network of queues, whose arrivals depend on the routings shown in the figure.

PROCESSMODEL IMPLEMENTATION

Figure 9-2 shows the ProcessModel flowchart of this situation, which corresponds closely to the system flow model in Figure 9-1. The patient entities arrive with a mean time between arrivals of 15 minutes, exponentially distributed. Patients spend a constant 3 minutes (0.05 hours) at the reception desk. Patients are then routed to one of the four activities that correspond to the ancillary departments or hospital admission.

Duration in the ambulatory activity is specified as N(0.25,0.1) hours in the *Time* field of the Properties dialog. The arrow emanating from this activity indicates that the patient is released. Duration in the X-ray activity is also normally distributed, with the time specified as N(0.25,0.05) hours. From this activity, patients are routed to the lab-testing activity with a probability of 0.1; to the hospital activity with a probability of 0.3; and are released with a probability of 0.6. The lab-testing activity has a duration of N(0.5,0.1) hours. Patients are routed to the hospital activity with probability of 0.1 or depart the system with probability 0.9. Once patients are admitted to the hospital, they are outside the boundary of the simulation model; thus, the duration is 0.

A summary of the activity icons and connectors in the model is as follows:

ACTIVITIES

Name	Capacity	Time	Input Q Cap.	Output Q Cap.	Object Type
Front desk	1	0.05 hours	999	0	Activity
Ambulatory	1	N(0.25, 0.1) hours	999	0	Activity
X ray	1	N(0.25, 0.05) hours	999	0	Activity
Hospital	1	0	999	0	Activity
Lab testing	1	N(0.5, 0.1) hours	999	0	Activity

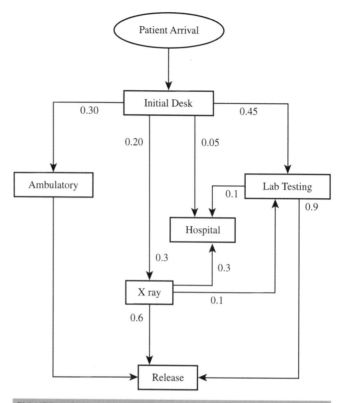

FIGURE 9-1 Emergency Room Model System Flow

FIGURE 9-2 ProcessModel Flowchart for Emergency Room Model

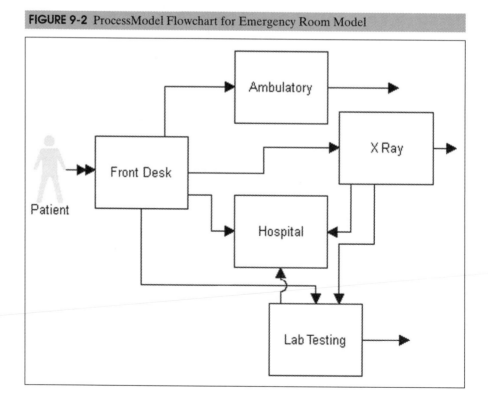

CONNECTORS

From	To	Type	Time	Quantity/ Arr.	First Time/ Renege After	Percentage
Patient	Front desk	Periodic	E(15) min.	1	0	
Front desk	Ambulatory	Percentage	1 min.			30
Ambulatory		Percentage	1 min.			100
Front desk	X ray	Percentage	1 min.			20
Front desk	Hospital	Percentage	1 min.			5
Front desk	Lab testing	Percentage	1 min.			45
X ray	Lab testing	Percentage	1 min.			10
X ray	Hospital	Percentage	1 min.			30
X ray		Percentage	1 min.			60
Lab testing	Hospital	Percentage	1 min.			10
Lab testing		Percentage	1 min.			90

Because we are interested in the nighttime operations, the simulation is run as a terminating model with a duration of 8 hours, and it was replicated 50 times. Results are shown here (with standard deviations in parentheses). The average number of patients processed is 33.02 (which we would expect at the arrival rate of four per hour for 8 hours). Waiting times are low, and the laboratory is the only facility working on average over 3 hours per shift, averaging 6.4 hours.

Station	Patients	Waiting Time	Work Time
Front desk	33.02 (5.40)	0.39 minutes (0.18)	1.646 hours (0.272)
Ambulatory	9.94 (2.69)	2.99 minutes (2.95)	2.430 hours (0.739)
X ray	6.26 (2.77)	1.36 minutes (2.15)	1.525 hours (0.652)
Laboratory	15.16 (3.44)	30.85 minutes (21.12)	6.400 hours (0.994)

SENSITIVITY ANALYSIS

Hospital management might certainly be interested in the effects that different arrival rates might have on the utilization of the facilities and patient waiting times. For example, suppose that demand increases to five patients per hour (every 12 minutes), exponentially distributed. This requires changing the time between arrivals from E(15) minutes to E(12) minutes. For this scenario, we obtained the following results:

Station	Patients	Waiting Time	Work Time
Front desk	41.52 (5.48)	0.51 minutes (0.22)	2.070 hours (0.276)
Ambulatory	12.48 (3.48)	4.47 minutes (4.51)	2.978 hours (0.878)
X ray	8.34 (2.32)	2.88 minutes (2.50)	2.047 hours (0.590)
Laboratory	19.34 (4.83)	55.62 minutes (30.46)	7.129 hours (1.030)

The system is slightly busier at the front desk, ambulatory, and X-ray processes, but it is still within reasonable levels. The laboratory, however, is becoming a serious bottleneck, working almost 90 percent of the 8-hour shift with patient waiting times of almost an hour.

A possible solution might be to increase the number of lab facilities to two. This is easy to do in the ProcessModel simulation; simply change the capacity in the lab-testing activity from 1 to 2. This was done for both arrival scenarios, yielding the following results:

Scenario	Four Arrivals per Hour		Five Arrivals per Hour	
	One Lab	Two Labs	One Lab	Two Labs
Patients	33.02	33.00	41.52	40.62
Waiting front desk	0.39 mins.	0.40 mins.	0.51 mins.	0.50 mins.
Waiting ambulatory	2.99 mins.	3.11 mins.	4.47 mins.	4.58 mins.
Waiting X ray	1.36 mins.	1.84 mins.	2.88 mins.	2.79 mins.
Waiting lab	30.85 mins.	3.58 mins.	55.62 mins.	6.09 mins.
Work front desk	1.646 hrs.	1.649 hrs.	2.070 hrs.	2.025 hrs.
Work ambulatory	2.430 hrs.	2.358 hrs.	2.978 hrs.	3.044 hrs.
Work X ray	1.525 hrs.	1.532 hrs.	2.047 hrs.	2.707 hrs.
Work lab	6.400 hrs.	7.486 hrs.	7.290 hrs.	8.821 hrs.

The difference in the number of patients simply reflects the change in average time between arrivals. The amount of waiting time at the reception desk, X-ray facility, and lab increases slightly when increasing the arrival rate, as expected. The waiting time at the lab increases more dramatically when the arrival rate increases and falls considerably when a second lab is added.

Work time at each station increases roughly proportional with the rate of arrivals in both systems for the desk, ambulatory, and X-ray systems. The work in the labs increases as the number of labs is increased by one. This is because the ability to process more patients increases (waiting time is less) in the two-laboratory case. Management would now need to make a cost-benefit analysis of adding a second laboratory. The simulation analysis clearly measures the expected change in work accomplished, as well as convenience to patients who have to wait less time.

SKILLBUILDER EXERCISE

This model assumed that the travel time between departments was constant at 1 minute, probably not a realistic assumption. Open the Process-Model file *Emergency Room Model* and examine the impact of changing the travel times to 5 minutes (constant) and, secondly, to normally-distributed times with a mean of 5 minutes and standard deviation of 1 minute.

Simulation in Practice: Simulation in Local Area Network Design[2]

The rapid growth in computers and computer-integrated manufacturing has resulted in increased emphasis on data communication systems that allows multiple-user access. For example, costly resources such as CPUs, printers, file servers, and other devices can be connected in a network, allowing shared use. However, many options exist for the network organization, and there often is a need to reconfigure networks to accommo-

---

[2]Adapted from R. Cobb, E. R. Mansfield, and J. M. Mellichamp, "Development of Design Guidelines for Local Area CSMA/CD Networks," *Simulation*, Vol. 58, no. 4, 1992, pp. 270–279.

date traffic growth, requests for new applications from users, and the opportunities to apply new and improved technology. Developing networks involves selecting transmission media, network configurations, and protocol options.

In one study, three transmission media—twisted-pair wire, coaxial cable, and fiber optic cable—were considered. Twisted-pair wire is less expensive but capable of handling only limited rates of data transmission. Coaxial cable is more expensive but has the capacity to handle higher transmission rates. Fiber optic cable has superior capacity but is even more expensive and involves installation restrictions.

The study investigated various network configuration options: tree, star, bus, and ring designs. Tree-structured designs had a network controller at a central node, with connecting nodes making up branches. Star structures had a central receiving and controlling node to which all other nodes were connected. Ring structures had a circular pattern, with each node connected to adjoining nodes. Bus structures had a length of cable to which all users were attached.

Protocol options available at the time of the study were token ring, token bus, and contention bus (Ethernet). The token ring protocol used a physical ring to pass control of the network. The token bus protocol used a physical bus design with token control based on logical design rather than physical design. The contention protocol used a physical bus design with token control based on logical design rather than physical design. The contention protocol used a bus design with all users attached on a length of cable, allowing users to transmit anytime the bus was not busy. With the contention protocol, signal collisions could occur when two or more messages were on the bus at the same time.

Network design requires balancing cost and performance. Performance is measured by the number of bits of information transmitted per time unit and by response time. From an analysis perspective, these networks can be modeled as waiting lines; thus, simulation offers a very useful means to compare alternatives.

The simulation model was developed using the GPSS simulation language. Key input data required by the simulation model used in the study were the number and size of messages, message routing, mean interarrival time, message size, and message origin and destination. Mean interarrival time was assumed to be exponential. Message sizes were also varied following the exponential distribution. Message origins and destinations were randomly generated.

A carrier-sensed multiple access with collision detection (CSMA/CD) protocol (contention bus, or Ethernet) was selected as the information flow control mechanism for the simulation model. Transmission logic sent a message if the network was idle; waited if the network was busy; and, in case of collision, detected the collision, stopped sending, and resent the message after a wait of random duration. If the message did not get through after 16 attempts, the system was notified of failure to send. The design variables in the simulation model were bus length, number of users, and user distribution. Bus length could be varied from 200 meters to 1,600 meters in 200-meter increments, representing the mean distance traveled by a signal in a microsecond. The number of users could be changed by the simulation analyst. User distribution (distance apart) influenced the time required for individuals on the system to sense activities of other users and, thus, affected system performance.

The simulation tracked events by sender, destination, size of message, collision retry attempts, delay, and transmission time. Collision resolution subroutines defined activity over each bus segment. Model output identified the number of collisions, bits of traffic transmitted, utilization of the bus, and delay ratio. The simulation was then used to generate data to study the relationship between the number of users per segment, user physical/geographical distribution, bus length, message size in bytes, and interarrival time in microseconds. The output from the simulation provided a means to develop a nonlinear regression to determine expected delay as a function of these variables. This regression model can be used by network designers to obtain a quick prediction of system delay should the simulation be available.

A Local Area Network Simulation Model

A local area network (LAN) consists of 120 workstations (organized into 12 nodes of about 10 users each) that share software. The workstations are connected by a bus to a network sharer. For most applications, there is no difficulty accessing software, but one database system receives especially high use; management is concerned that it might become a problem if the number of users increases significantly.

Typical groups of data, called *packets,* sent on the system are small user queries to the central processor. System responses are multiples of 64 kilobytes (kB). An analysis of historical data shows the following packet sizes and probabilities:

Packet Size	Probability
64 kB	0.118
256 kB	0.275
512 kB	0.239
1024 kB	0.368

The system currently uses Ethernet technology, restricting use on the bus network to one user at a time. When the user sends a packet, the computer checks the network to see if it is in use. The system is relatively old, with a rate of transmission of about 150 kB per second. If the network is not busy, the packet is sent. If the system is busy, the computer waits 1 second and rechecks to see if the system is still busy.

The current rate of use is an average of 4,200 milliseconds between submissions. The network currently involves transmission delays about half of the time, which is sometimes inconvenient, but bearable. However, a larger concern is that the system will become saturated over time. Thus, the key output measure of interest is the number of messages that might not get through within an hour of peak operation as a function of the number of users.

To develop a simulation model for this situation, we would need to generate arrival and service times for packets, check for possible collision, and calculate the finish time for each packet submission by time of arrival. Output measures would be the total milliseconds of transmission delay and average utilization of the system. To examine sensitivity of various arrival rates, we might conduct an experimental design by simulating mean arrival times of 4,800, 4,200, 3,600, 3,000, 2,400, and 1,800 milliseconds (all Erlang distributed with 120 sources) with five replications for each scenario.

PROCESSMODEL IMPLEMENTATION

Figure 9-3 shows a ProcessModel flowchart for the LAN model. Message entities arrive with periodic arrival times specified as ER(4.8,120) in the Properties dialog for the first run, with the 4.8 changed to the appropriate value for the other five models (4.2 seconds through 1.8 seconds). Entities are routed to the *Arrive* activity. The messages then move to a *Check Status* activity whose input queue capacity is set to 1. An arrow connector with 0 time duration then leads to the *SendPacket* activity. The *SendPacket* activity has a user-defined time duration reflecting the number of seconds and probability required to send each of the four packet sizes, as shown here.

Message Size	Seconds	Probability
64 kB	0.42	0.118
256 kB	1.68	0.275
512 kB	3.36	0.239
1024 kB	6.72	0.368

In ProcessModel, the user-defined distribution allows you to generate a random variate based on percentages. The syntax is $Dn(\%_1, x_1, \ldots \%_n, x_n)$, where $\%$ = percentage (entries must total 100 percent), x = value (numeric or predefined descriptor), and n = number of $(\%, x)$ entries between 2 and 5. As one example, D3(20, 35, 30, 37.5, 50, 45) will generate the outcome 35 twenty percent of the time, the outcome 37.5 thirty percent of the time, and the outcome 45 fifty percent of the time. In our example, we use the expression D4(11.8,0.42,27.5,1.68,23.9,3.36,36.8,6.72).

The connector arrow from the *SendPacket* block represents an exit from the system. If the *SendPacket* block is busy, messages are recycled through the *Process* block (after a delay of 1 second) using the renege option.

The ProcessModel components of the simulation model are summarized here.

ACTIVITIES

Name	Capacity	Time	Input Q Cap.	Output Q Cap.	Object Type
Message					Entity
Arrive	1	1 sec.	999	0	Activity
Check status	1	0	1	0	Activity
Send packet	1	D4 (11.8, 42, 27.5, 1.68, 23.9, 3.36, 36.8, 6.72) sec.	0	0	Activity
Process	1	0	999	0	Activity

FIGURE 9-3 ProcessModel Flowchart for Local Area Network Model

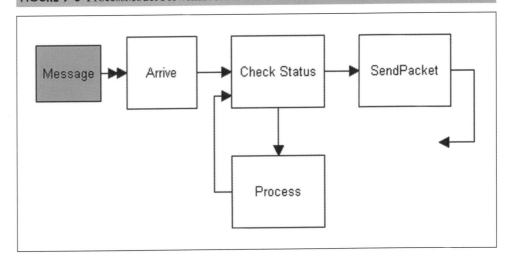

CONNECTORS

From	To	Type	Time	Quantity/ Arr.	First Time/ Renege After	Percentage
Message	Arrive	Periodic	ER(3,120) sec.	1	0	
Arrive	Check status	Percentage	0			100
Check status	Send packet	Percentage	0			100
Check status	Process	Renege	1 sec.		0	
Process	Check status	Percentage	0			100
Send packet		Percentage	0			100

SIMULATION RESULTS

Each of five runs is simulated for 15 minutes, and the results are given in the following table:

Arrival Rate (seconds between)	Messages Generated	Messages Processed	LAN Utilization	Ratio Processed/ Generated
1.8	16019.2	7641.6	0.999	0.477
2.4	12044.8	7705.6	0.999	0.640
3.0	9619.2	7699.2	0.999	0.800
3.6	7968.0	7456.0	0.987	0.936
4.2	6828.8	6803.2	0.882	0.997
4.8	5971.2	5945.6	0.799	0.996

Obviously, as the time between message arrivals decreases, the amount of waiting grows. The current situation, with an arrival rate of about 4.2 seconds between messages, shows a busy system, but one in which all of the packets get sent. The number of seconds of delay is fairly high. The system is going to perform better (obviously) for the slower rate of arrival (4.8 seconds between messages). As the rate of arrivals increases to one message every 3.6 seconds, delays increase; the system is almost constantly busy, and some packets do not get sent. At 3.0 seconds between messages, the system is no longer capable of doing the work. The ratio of messages processed (transmission completed) to messages generated drops to 0.640 for a rate of arrival of 2.4 seconds between messages, and to 0.477 at a rate of 1.8 seconds between messages.

SKILLBUILDER EXERCISE

Management of the network anticipates that the proportion of files sent by users will be larger in the near future. Modify the ProcessModel file *Lan Model* to use the following probability distribution for job sizes:

Job Size	Probability
64 kB	0.05
128 kB	0.15
256 kB	0.25
512 kB	0.55

Return the model for arrival rates of 3.6, 4.2, and 4.8 seconds between arrivals.

Simulation in Practice: Simulating Dispatching Policies in a Manufacturing Plant[3]

Jeffrey Division of Dresser Industries supplies underground mining equipment, mostly for coal mining. Underground coal mining has been a declining industry, and Jeffrey downsized its operations over 20 percent by the end of the 1980s. The company produced gears, large-parts machining, assembly, and major structural steel and repair work. Gear manufacturing was a high-volume activity with fairly predictable demand. Both quality and meeting schedules were critical; however, scheduling of gear manufacturing had historically been a problem, with a large proportion of orders expedited to finish work on time.

One approach considered to improve performance was to reorganize into smaller, more focused factories, each of which has its own mission and goals. Thus, gear manufacturing was split out from other operations and given its own building and supervision. Although just-in-time techniques were applied, performance did not immediately improve, and production planning and control problems continued.

The company commissioned a simulation study to better understand the dynamics of delivery performance in the gear manufacturing plant. This plant had a gear shop, a chain cell, and a heat treatment facility. The majority of the 2,900 parts produced required multiple operations with close tolerance machining with long setup times. About 90 percent of these parts were ordered in monthly batches. An MRP system was used to plan and release orders in batches. The plant had excess production capacity and hired labor as needed to meet increases in demand. The gear shop had 40 work centers, including five numerically-controlled machines. Most workers could operate more than one machine, and the company had training programs to increase worker capabilities, as well as an incentive plan.

The simulation model was developed using the simulation language SIMAN for the 204 gear parts with the heaviest workload over the prior 2 years. Each of these parts required one to 18 operations. Those work centers with heavy loads operated two 8-hour shifts per day; other work centers operated only a few hours per day. Dispatching followed the first-come-first-served rule. The simulation model generated interarrival times from an exponential distribution with means ranging between 1.8 and 3.3 hours. The variation in interarrival times created different shop load levels. The specific part associated with an order was then determined by sampling from a discrete probability distribution based on historical demand. Due dates were generated by multiplying the processing time by three and adding it to the order receipt time. The initial runs of the study applied the earliest due-date rule in an effort to improve performance.

The simulation began with an empty shop. The transient period was identified and eliminated, and statistics were collected over the next 6,000 simulated hours. Five runs were obtained for each combination of dispatching rule (FCFS and earliest due date first) and six shop load levels, resulting in a total of 60 different runs. The performance measures used were the percentage of tardy orders and the average flow time. The simulation model was validated by showing the results to supervisors, as no statistical data were available. For example, bottleneck centers identified in the simulation were found to be the same as experienced on the shop floor. Such results satisfied the company management that the model was indeed valid.

---

[3]Adapted from J. Hutchison, G. K. Leong, and P. T. Ward, "Improving Delivery Performance in Gear Manufacturing at Jeffrey Division of Dresser Industries," *Interfaces,* Vol. 23, no. 2, 1993, pp. 69–79.

TABLE 9-1   Average Queue Lengths by Work Center: Results of Dispatching Method Simulation

Job Interarrival Times:	3.0 hours		2.7 hours		2.4 hours		2.1 hours		1.8 hours	
Dispatching Rule:	*FCFS*	*EDD*	*FCFS*	*EDD*	*FCFS*	*EDD*	*FCFS*	*EDD*	*FCFS*	*EDD*
Work Center										
Bench	2	1	2	2	6	6	23	25	131	87
NC lathe 1	2	1	3	2	4	2	17	11	80	35
NC lathe 2	1	1	2	1	3	2	6	5	61	48
Grinder 1	1	1	2	2	3	2	8	6	35	56
NC lathe 3	1	1	1	1	3	2	5	3	27	19
Grinder 2	1	1	2	1	4	3	14	5	21	11
Gear Cutter 1	1	1	1	1	1	1	3	2	7	6
NC drill	1	1	1	1	1	2	2	3	6	7
Manual lathe 1	1	1	1	1	2	2	3	3	4	7
Gear cutter 2	1	[a]	1	[a]	2	1	2	1	3	2
Manual lathe 2	[a]	[a]	[a]	1	[a]	1	[a]	1	3	2
Gear cutter 2	1	[a]	1	1	1	1	1	2	3	4

[a] Average queue length < 0.5.

The simulation showed that the flow time in the shop increased as the shop loading increased (i.e., as interarrival times got smaller) for both the first-come-first-served and earliest-due-date rules, but at a faster rate with the FCFS rule. Similarly, the percentage of tardy orders increased for both rules with higher shop loads, but the earliest-due-date rule resulted in a smaller percentage of tardy orders than the FCFS rule.

For 12 work centers with the longest queues, Table 9-1 shows that the average queue lengths increased as jobs arrived more frequently for both rules. At high levels, flow time with the earliest due-date method was 50 percent less than with FCFS. The work center "bench" had the longest queues. Working overtime or adding another shift was suggested as a possible solution. Table 9-1 also shows that the penalties paid in longer waiting lines for heavier loads were much less severe with the EDD rule than with the FCFS rule. The difference between earliest-due-date and FCFS systems was found to be statistically significant for both flow time and percentage of tardy orders, using the paired *t*-test on 30 paired replications using common random numbers.

The study indicated that their reactive policy of allowing shop loads to increase without capacity adjustments led to severe penalties to catch up. Queues built up quickly at higher load levels at some workstations. Two improvement policies were identified as a result of the simulation experiments: (1) Use capacity planning techniques to adjust capacity based on forecasted shop loads and (2) Change dispatching practices to improve delivery performance and to reduce the effects of sudden spikes in demand.

A Job Shop Simulation Model

As the previous simulation shows, simulation can be used to analyze alternative factory layouts and rules for scheduling work. This is particularly important in job shops, where a variety of products are produced in a very dynamic demand environment. For example, in an automobile repair garage, as many as 40 different makes and models of automobiles, each with different service requirements, might arrive during a particular day.

In this section, we present an example of using simulation to evaluate different scheduling rules.

A small job shop specializes in the production of bronze and brass art works. The shop has two machines for bronze work but only one machine for brass work. Although each job is unique, all jobs fall into three major categories. Class I jobs require work only on the bronzing machine. Class II jobs require work only on the brass machine. Class III work requires processing on both types of machines. Many jobs coming into the shop are for one-of-a-kind pieces. However, a major customer is an art wholesaler, who brings in work in batch sizes of 12, 144, and 576 pieces each. Historical data reveal the following probabilities for the type of job and batch size:

Type	Probability	Batch Size	Probability
Class I	0.4	1	0.50
Class II	0.2	12	0.25
Class III	0.4	144	0.15
		576	0.10

Due dates, that is, the time customers would like the job completed, follow a uniform distribution between 0 and 100 days after arrival to the shop.

The firm currently schedules jobs on a first-in-first-out policy (FIFO). However, about 50 percent of their jobs are late. The firm wants to know if using either a policy of doing jobs in order of due date or a shortest-processing-time-first (SPT) policy will improve on-time performance.

A flowchart of the simulation model for this situation is shown in Figure 9-4. The first step is to generate jobs entering the system, including the type of job, due date, and batch size. The processing time for each unit is assumed to be normally distributed with a mean of 0.02 day and a standard deviation of 0.01 day (with a check assuring that duration is at least 0.008 day). The duration for the entire job is then obtained by multiplying this unit time by the batch size. Each of the three systems is simulated by sorting the data reflecting the scheduling rule. For the FIFO rule, the jobs are processed in the order generated. For the SPT rule, jobs are processed in order of duration, with shortest jobs first. For the due-date (DD) rule, jobs are processed in order of simulated due date.

The first processing action is to check for job type. If the job is class I, it is sent to a bronze machine. If the job is class II, it is sent to the brass machine. If the next job is class III, it is sent first to a bronze machine and then forwarded to the brass machine.

PROCESSMODEL IMPLEMENTATION

The problem involves generation of 100 jobs, with different routing, batch sizes, due dates, and unit production rates. The model is shown in Figure 9-5. The simulation begins by generating 100 job entities using scheduled arrivals beginning at 12:00 A.M. Monday (time "0" in ProcessModel). Several attributes are defined for each entity:

- Type (descriptive, with possible names of *Bronze*, *Brass*, or *Both*),
- Due_date (integer)
- Proctime (real)
- Finish (real)
- Late (real)

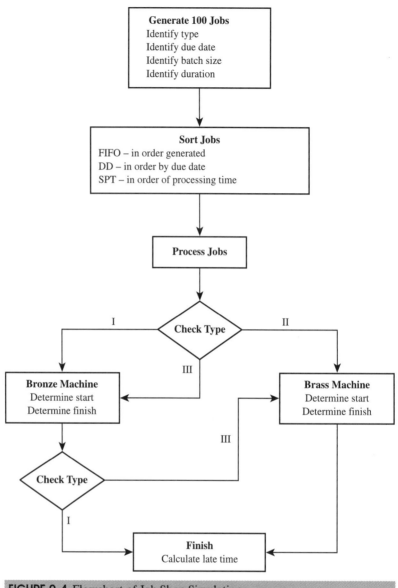

FIGURE 9-4 Flowchart of Job Shop Simulation

Two additional integer attributes are created for prioritization:

1. EDDpri (for earliest due date priority)
2. SPTpri (for shortest processing time priority)

The maximum due date is 800, with larger due dates having higher priorities. Thus, EDDpri = 800 − Due date, and SPTpri = (999 − Proctime)/10. Global variables are also created to measure total late time (Latebronze, Latebrass, and Lateboth). All of these variables are initially set equal to zero.

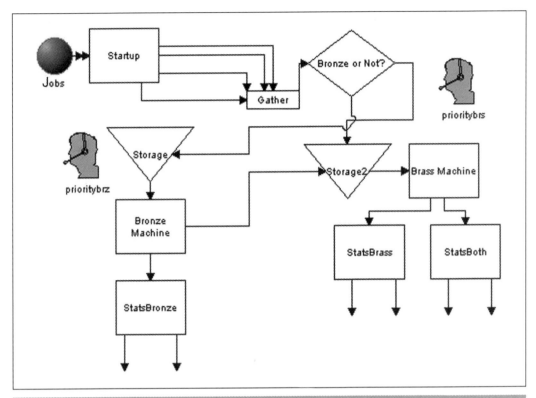

FIGURE 9-5 ProcessModel Flowchart for Job Shop Model

The arrival connector leads into the *Startup* activity from which entities are sorted by batch size using *Percentage* routing in the Properties dialog. For example, 50 percent of jobs leaving startup are routed to Storage1 and given a new name, Batch1; 25 percent are routed to Storage 2 and defined as Batch12, and so on. Action logic expressions are used to define the processing time. For example, for the top connector representing a batch size of one, we have

```
Proctime=N(0.16,0.08)
IF(Proctime<0.008, Proctime=0.008)
```

Each of the other three batch size paths is the same, except for a different mean for each batch size. The standard deviation is always one-half of the mean, and the *IF* statement placing a minimum time of 0.008 hours is included on all four paths.

The storage elements are connected to the *Gather* activity using a percentage connector. From the *Gather* activity, the entities proceed the *Bronze or Not?* decision element. This connector is a percentage connector, used to set the attributes:

```
Type=D3(40,BRONZE,20,BRASS,40,BOTH) and Due=U(400,400) using Action Logic.
```

The model logic now reflects the flow of entities. Two conditional connectors lead out of the *Bronze or Not?* decision element. The connector leading to the bronze machine has condition "Type=BRONZE" or "Type=BOTH." The connector to the brass machine has the condition "Type=BRASS." In order to apply a priority other than first-in-first-out, entities need to pass through a storage queue and, when entering the processing block, acquire a resource. Therefore, storage queues are created prior to entry into either the bronze machine or the brass machine. A resource *prioritybrz* is created to apply priority scheduling rules for jobs entering the bronze machine, along with a similar resource *prioritybrs* for the brass machine. Both resources have a capacity of one. (Although this is not needed for the FIFO model, we have included it so as to facilitate modifications for the other priority-based models.)

The bronze machine has a capacity of one, and time is defined by the attribute *Proctime*. On the two connectors leading out of this activity, action logic is used to calculate finish time and late time for the entity and release the resource *prioritybrz*. The expressions used for the action logic are

```
Finish=Clock(),
Late=Finish-Due, and
IF(Late<0), Then Late=0, and
Free prioritybrz.
```

From here, entities are routed to send both jobs back for processing on the brass machine and send bronze jobs out of the system using conditional connectors. Bronze jobs go to the *StatsBronze* block, where the action logic is used to calculate the total late time for bronze jobs using the formula: Latebronze=Latebronze+Late. Entities are then passed to another decision element to count bronze jobs that are late or on time.

Class III jobs leaving the bronze machine are routed back (using a condition connector) to the *Storage2* activity and on to the *Brass Machine* activity (this also enables the priority to be applied for the second and third models. The *Brass Machine* activity has the same components as the *Bronze Machine* activity. Calculating of late time for each job and freeing the resource *prioritybrs* occur on the connectors leading to the appropriate statistical calculation activity. Conditional routing connectors are used to separate brass and both jobs for calculation of late time and to sort through decision blocks to count the number of jobs late and on time.

Calculation of late time requires knowing when a process is finished. This calculation must be accomplished after the block where the process occurs, because Process-Model applies action logic calculations prior to general calculations. Therefore, calculation of finish = Clock() is done on the connector after the *Bronze Machine* and *Brass Machine* activities.

The activities in the model are summarized here:

Name	Capacity	Time	Input Q Cap.	Output Q Cap.	
Startup	1	0	999	0	
Gather	1	0	999	0	
Bronze or Not?	1	0	999	0	IF(Type=BOTH) THEN Proctime=Proctime/2)
Storage	999				
Bronze Machine	1	Proctime	999	0	
StatsBronze	1	0	999	0	Latebronze=Latebronze+Late
Storage2	999				
Brass Machine	1	Proctime	999	0	
StatsBrass	1	0	999	0	Latebrass=Latebrass+Late
StatsBoth	1	0	999	0	Lateboth=Lateboth+Late

The connectors in the model are summarized here:

From	To	Type	Time	%	Action	Condition
Jobs	Startup	Scheduled Arrival				
Startup	Gather	Percentage	0	50	Proctime=N(.16,.08)	
Startup	Gather	Percentage	0	25	Proctime=N(1.92,.96)	
Startup	Gather	Percentage	0	15	Proctime=N(23.04,11.52)	
Startup	Gather	Percentage	0	10	Proctime=N(92.16,46.08)	
Gather	Bronze or Not?	Percentage	0	100	Type=D3(40,BRONZE,20,BRASS, 40,BOTH); Due=U(400,400);EDDpri=(800-Due); SPTpri=(999-Proctime/10)	
Bronze or Not?	Storage	Conditional	0			Type=BRONZE or Type=BOTH
Storage	Bronze Machine	Percentage	0	100	Get prioritybrz	
Bronze or Not?	Brass Machine	Conditional	0			Type=BRASS
Storage2	Brass Machine	Percentage	0	100	Get prioritybrs	
Bronze Machine	StatsBronze	Conditional	0		Finish=Clock();Late=Finish-Due;IF(Late<0) Then Late=0	Type=BRASS
Bronze Machine	Storage2	Conditional	0		Finish=Clock();Late=Finish-Due;IF(Late<0) Then Late=0	Type=BOTH
Storage2	Brass Machine					
StatsBronze	Exit	Conditional	0			Late=0
StatsBronze	Exit	Conditional	0			Late>0
Brass Machine	StatsBrass	Conditional	0		Finish=Clock();Late=Finish-Due;IF(Late<0) Then Late=0	Type=BRASS
Brass Machine	StatsBoth	Conditional	0		Finish=Clock();Late=Finish-Due;IF(Late<0) Then Late=0	Type=BOTH
StatsBrass	Exit	Conditional	0			Late=0
StatsBrass	Exit	Conditional	0			Late>0
StatsBoth	Exit	Conditional	0			Late=0
StatsBoth	Exit	Conditional	0			Late>0

Earliest-Due-Date Scheduling Model   The model change to earliest due date simply applies priority for processing at the *Bronze Machine* and *Brass Machine* blocks. In the percentage connector leading into the *Bronze Machine* activity, the action *Get prioritybrz, EDDpri* is used, and *Get prioritybrs, EDDpri* in the connector leading to the brass machine.

Shortest-Processing-Time Model   The model is identical to the earliest-due-date model except that *SPTpri* is used instead of *EDDpri* as a basis for priority.

ANALYSIS OF RESULTS

Results for the FIFO model are given in the following table:

FIFO Model	Jobs Late	Jobs on Time	Total Days Late
Bronze	19.92	20.36	740.53
Brass	3.92	17.00	67.25
Both	25.22	13.58	1021.53
Total	49.06	50.94	1829.31

For the EDD model, we have

EDD Model	Jobs Late	Jobs on Time	Total Days Late
Bronze	19.36	20.92	334.66
Brass	2.76	18.16	33.80
Both	25.62	13.18	464.22
Total	47.74	52.26	832.68

For the SPT model, we have

SPT Model	Jobs Late	Jobs on Time	Total Days Late
Bronze	13.34	26.94	319.48
Brass	2.60	18.32	35.16
Both	16.98	21.82	461.99
Total	32.92	67.08	816.63

We see that in the earliest-due-date model, the number of jobs that are late is slightly smaller, although the total days late was dramatically reduced. The shortest-processing-time model resulted in the most jobs completed on time, matching theoretical expectations. All three models had roughly the same finishing times and, thus, similar waste times for the machines.

SKILLBUILDER EXERCISE

The job shop is expecting a number of changes in its business. The art wholesaler's work is expected to decline; however, single-batch-size work is expected to increase to match that decline. Management wants to know the impact of changes in proportions of work as described by the following probability distributions:

Type	Probability	Batch Size	Probability
Class I	0.3	1	0.70
Class II	0.1	12	0.15
Class III	0.6	144	0.10
		576	0.05

Open the ProcessModel file *Job Shop Model* and rerun the model for 100 jobs using each of the three priority systems and these new distributions. What differences do you see?

Simulation in Practice: Simulation in Supply Chain Management[4]

A supply chain is the total product-to-market system, from product source to consumer. Supply chains can be viewed as networks of organizations linked together by roles they play and are managed more effectively by considering them as a system. Coordination

[4]Adapted from J. G. A. J. van der Vorst, A. J. M. Beulens, and P. van Beek, "Modeling and Simulating Multi-echelon Food Systems," *European Journal of Operational Research,* Vol. 122, 2000, pp. 354–366.

of activities can lead to improved performance of the overall supply chain. Ways to improve operations include reduction of time consumption, improvement of quality, and increase in the availability of information for coordination. However, supply chains involve a high level of uncertainty, which is due to late deliveries, machine breakdowns, and order cancellations that lead to the use of higher inventory and excess capacity. As such, simulation can be of substantial benefit.

The focus of a supply chain is on meeting final customer requirements. For a food supply chain, for example, this could begin with government policy makers, through growers, basic food processors, producers of individual food items, manufacturers of packaged-food products, on through the distribution network to end users. The supply chain in this study involved producers, manufacturers, distribution centers, outlets, and consumers involved in a supply chain for chilled salads in the Netherlands. There was one producer and a retailer with one distribution center supplying 100 retail outlets. The outlets formed a working partnership, with the general objective to improve customer service at lower total chain costs, expecting to increase sales and, thus, profits. The producer wanted to create more space in retail outlets to expand his product assortment and improve product freshness. Simulation was used to evaluate alternative designs of supply chain infrastructure.

Both the configuration (design) level and operational (execution) level were of interest. At the configuration level, variables included the parties involved, their roles, constraints on how roles were executed, the manner of cooperation, information technology, and physical plants. Options at this level might include supplier relationships. At the operational level, variables related to timing, accuracy of information, quantities produced, and quality. Options here might include means of shortening lead time. Supply chain system performance was measured by the quality, availability, and timing of product delivery to consumers. Product quality related to shelf life, traceability, and health. Product availability referred to minimizing stock-outs and providing variety.

The supply chain simulation model developed measured system state, time, and place of each business entity after each transition. Orders were generated and processed, and turned into order pick lists, which were filled depending upon inventory levels. If shortages were encountered, a rationing policy was applied. Shipments were tracked by source to retail outlet by time, quantity, and batch number (which provided information on product description and remaining shelf life). Receipts were processed by employees who stocked inventory. Redesign options modeled were

- Diminishing time windows (waiting times, processing times)
- Synchronizing, eliminating, or reallocating business processes
- Shortening cycle times
- Coordinating decision policies
- Using of real-time information systems
- Sharing information, such as product demand and production schedules

Performance indicators related to supply chain costs and to service. Costs considered were holding costs (at both the distribution center and at the retail outlets), processing costs, and costs of product waste and necessary price reductions. Service was measured by stock-outs per hour, delivery reliability, average remaining product freshness, transport carrier utilization, and product assortment.

The supply chain model can be described as follows. Retail sales occurred during operating hours. Inventory levels were adjusted at each retail outlet at the end of the day. Twice a week, retail outlets placed orders with the distribution center following a base stock policy (to replenish available shelf space). There was a 1-day lead time, so there was no back-ordering. Order batch quantity was a fixed number of boxes with quantity that varied by product. During closing hours, stocked goods were checked for remaining shelf life and, if a product expired, it was discarded.

At the distribution center, outlet orders were processed at fixed times, determining what to ship each day to each retail outlet. Shipments from stock were made three times per week, with a lead time of 1 day. If shortages occurred, a rationing policy was implemented. The distribution center placed an order twice a week, based on sales forecasts derived from actual orders and historical sales patterns. The producer received these orders for 63 different product types, with an order lead time of 3 days. Another rationing policy was used if shortages were encountered.

The objective of the simulation analysis was to evaluate the consequences of different configurations and different control arrangements. At the configuration level, a computer-assisted ordering system at retail outlets would enable the producer to know outlet inventory positions in real time. At the operational level, three modifications to the system were considered: (1) decreasing the lead time from 3 days to less, (2) increasing ordering and delivery frequencies (producer to distribution center from two times a week to alternatives of three, four, or five times a week; from distribution center to outlets from three to four, five, or six times a week), and (3) adopting the distribution center ordering policy at the producer.

Five options were the focus of the simulation analysis:

1. The producers' lead time was decreased from 3 days to 1 day.
2. The order and delivery frequency between producer and distribution center was increased from two to three, four, or five times a week with lead time of 1 day, and frequency of orders from outlet to distribution center increased to three times a week.
3. Order and delivery frequency between distribution center and outlets was increased from three to four, five, or six times a week with lead time of 1 day and frequency of deliveries between producer and distribution center increased to three times a week.
4. Order frequency between producer and distribution center was increased to four times a week and delivery frequency to five times a week with lead time of 1 day and frequency of delivery from distribution center to outlets increased to six times a week.
5. Base stock ordering policy was changed to a system driven by order forecasts.

Total costs were calculated at each link, including costs of write-offs and price reductions. The simulation results suggested that implementation of a forecast-driven ordering system could reduce inventory levels. However, it was impossible to predict exact benefits because human interventions during peak demand and other disturbances could not be simulated. Nevertheless, the simulation model was found to be valuable in identifying policy improvements and predicting improvements in inventory and product freshness.

Simulating Alternative Supply Chain Systems

One of the most significant changes in manufacturing since World War II has been the implementation of just-in-time (JIT) approaches for supply chain management. The JIT concept attempts to reduce excess production and inventory by producing material as needed, ideally to arrive at the next production station no earlier (and no later) than it is needed, thus reducing inventory holding costs. Another benefit is that quality will increase, because the shop is not so cluttered. However, researchers have observed that a low variance in operations is required for the concept to work. We illustrate how simulation can be used to analyze supply chains using a simple example.

Eli Whitney was one of the first to attempt to develop mass-production capability for musket production using interchangeable parts (unfortunately, without much success). We may only wonder how the course of history may have changed had Eli had access to JIT principles and simulation tools! Suppose that the musket production process consists of four stages: produce the gun barrel, carve the stock, insert the flint and assemble, and inspect. We will assume that the times for each production step, initial inventory, and number of workers at each stage are

Process	*Distribution*	*Mean*	*Standard Deviation*	*Initial Inventory*	*Workers*
Produce gun barrel	Normal	6 days	0.6 day	2	2
Carve stock	Normal	3 days	0.6 day	Unlimited wood	1
Insert flint, assemble	Normal	1 day	0.2 day	Unlimited rocks	1
Inspect	Exponential	0.1 day			1

Stocks and flints are available in unlimited quantities. However, rough barrels have to be purchased from a supplier at a cost of $10 each and are refined into an accurate, rifled barrel. Gun barrels and stocks can be produced as long as material is available. One barrel and one stock are required before the flint can be inserted. Inspection does not occur until the flint is inserted. On average, 10 percent of the muskets are defective.

Currently, a reorder-point policy of ordering five barrels whenever the stock dips to two is used. Delivery time averages 5 days, normally distributed with a standard deviation of 3 days. The cost per order is $10. Muskets are sold for $100 each.

The reorder-point (ROP) system involves a series of queues, with an original two-server barrel production line connected with the reorder-point inventory system. The stock-carving center operates in parallel with the two barrel machines. These are followed by the flint insertion center and the inspection center. The inspection center rejects 10 percent of production. Profit is computed as

$$\text{Profit} = \text{guns produced} \times \$100 - \text{barrels} \times \$10 - \$10 \text{ per order}$$
$$- 20\%/\text{year} \times \text{average value of money tied up in barrels}$$

An alternative to the ROP model is a JIT ordering policy by which materials are ordered to arrive as needed. JIT requires low delivery-time variance, usually obtained by locating the supplier close to the plant. Therefore, we assume continuous delivery, with cost per order of $1. However, the price per barrel will increase to $17. We assume, as is borne out in practice, that the JIT system will improve quality, so that only 2 percent of the guns inspected will now be defective. However, the production process itself is not affected. Average inventory liability is incurred the beginning of the day that barrels are received, just after receipt. In the JIT system model, an order may include more than one barrel. Therefore, order costs are incurred each day barrels are ordered. The rejection rate is much lower, 2 percent of muskets produced. Profit in the JIT system is computed as

$$\text{Profit} = \text{guns produced} \times \$100 - \text{barrels} \times \$17 - \$1 \text{ per order}$$
$$- 20\%/\text{year} \times \text{average value of money tied up in barrels}$$

Figure 9-6 shows the flow of activity and differences for both approaches.

PROCESSMODEL IMPLEMENTATION

The basic framework of the simulation model for the ROP system is shown in Figure 9-7. The ProcessModel flowchart consists of an inventory module for raw metal barrels and workstations for barrel production (two stations), carving stocks, assembly and flint insertion, and inspection. The components of the model include entities for both raw barrels and trees (for carving). Raw barrels require reordering; trees are unlimited. The process and storage elements of the model are summarized here:

ACTIVITIES

Activity	Capacity	Time	Input Q Cap	Output Q Cap.	
Barrel storage	10				Queuing order=none
Barrel1	1	N(48,4.8) hr.	0	0	
Barrel2	1	N(48,4.8) hr.	0	0	
Completed barrels	999				Queuing order=none
Stocks	1	N(24,4.8) hr.	0	0	
Flint	1	N(8,1.6) hr.	999	0	
Inspect	1	E(0.8) hr.	999	0	

ROUTINGS

From	To	Type	Time	Quantity/Arrival		Percentage
Raw barrels	Barrel storage	Ordered	N(40,24) hr.			
Barrel storage	Order connector	Order signal		Order quantity = 5	Reorder level = 2	
Barrel storage	Barrel1	Percentage	0			100
Barrel storage	Barrel2	Else	0			
Barrel1	Completed barrels	Percentage	0			100
Barrel2	Completed barrels	Percentage	0			100
Trees	Stocks	Continuous				
Completed barrels	Flint	Percentage	0			100
Stocks	Flint	Attach	0	1		
Flint	Inspect	Percentage	0			100
Inspect	(accept)	Percentage				90
Inspect	(reject)	Percentage				10

A reorder-point system is modeled by using an *order signal* in ProcessModel, which is a connection between an activity or storage and an arrival or routing that signals the arrival or routing to order or release additional entities. The signal is triggered by a drop in inventory level at either the storage or the activity input queue. In the Order Signal dialog box, *Reorder Level* represents the level to which the storage or activity input queue must drop before signaling an order for more entities. Any expression is valid. *Order Quantity* is the quantity to order when an order is placed. Any expression is also valid. The *Place order at start* box is checked if an initial order is to be placed at the be-

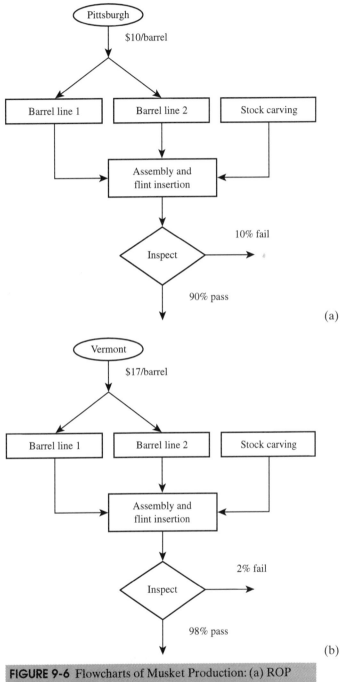

FIGURE 9-6 Flowcharts of Musket Production: (a) ROP System; (b) JIT System

ginning of the simulation run. If the initial order is insufficient to raise the queue or storage level above the reorder level, a *Periodic* arrival should be defined with its *First time* field set at time zero to initialize the inventory level. When an order signal is used to order entities at an arrival connection (the order signal is connected to an arrival connection), the arrival type must be *Ordered*. In this case, the connector for raw barrels to storage is specified as *Ordered*. Connecting the *Barrel storage* icon back to the *Ordered* arrow

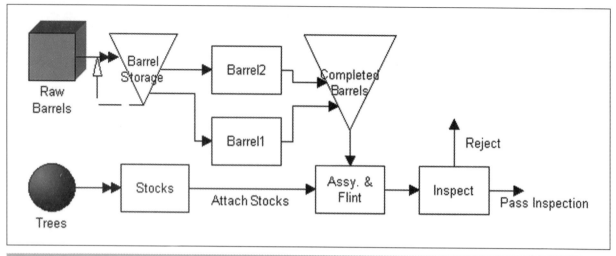

FIGURE 9-7 ProcessModel Flowchart for ROP System

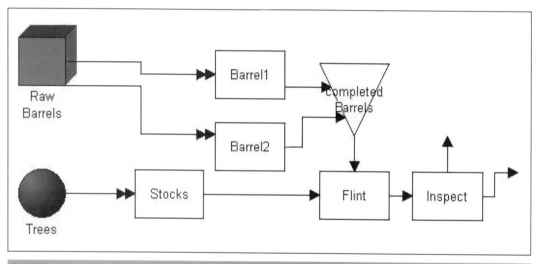

FIGURE 9-8 ProcessModel Flowchart for JIT System

creates an order signal. The *Barrel storage* icon is connected to each of the two barrel production processes with ordered routing arrows, each with 0 duration. The first connector is a percentage arrow, set at 100 percent. The second connector is an *Else type* arrow, so that if Barrel1 is busy, barrels are made available to the Barrel2 processor. The output of completed barrels is connected to the *Flint process* icon, using a percentage arrow set at 0 duration and 100 percent.

The simulation of carving stocks begins with the *Trees* entities feeding into the *Stocks* icon using a continuous arrival. Stocks output is connected to the *Flint process* icon, as an attach arrow with 0 time and quantity of 1. This arrangement tracks entity times based on barrels, with stocks attached as needed. Therefore, entity time in the system represents the inventory for barrels. The flint-processing operation passes entities on to the inspect process, where 90 percent of the entities pass inspection (and generate revenue), while 10 percent of the entities are rejected.

JUST-IN-TIME MODEL

The ProcessModel flowchart for this situation is shown in Figure 9-8. It is practically the same as the ROP model and, in fact, a little simpler. Raw barrels feed directly to each of the barrel process activities using continuous arrivals. The other key change to the model is that the output percentages from the inspect process are 98 percent to accepted output (generating revenue) and 2 percent to reject.

ANALYSIS OF RESULTS

Results for the two models based on 50 runs for each model by cost category are shown:

	Barrel1	*Barrel2*	*Stocks*	*Flint*	*Inspect*
ROP					
Entities	58.54	58.42	114.36	113.98	113.08
Utilization	0.994	0.993	0.975	0.322	0.033
JIT					
Entities	58.88	58.72	114.96	114.54	113.68
Utilization	1.000	1.000	0.973	0.324	0.033

There was a little waiting for barrel stock in the ROP system, but not much. The JIT system yielded no waiting for barrels. Because there was little waiting for barrels, the other processing components in the system had similar results. The profit results, shown in the following table, indicate that the JIT system is superior. This arises primarily from the higher acceptance rate. Differences in purchase-cost and ordering-cost balance are much better for the current ROP system, but this advantage is offset by the increased revenue due to quality.

	ROP	*Rate*	*Cash*	*JIT*	*Rate*	*Cash*
Revenue	102.24 muskets	$100	$10,224.00	111.54 muskets	$100	$11,154.00
Barrel cost	120.1 barrels	$10	−$1,201.00	117.6 barrels	$17	−$1,999.20
Order cost	24.02 orders	$10	−$240.20	117.6 orders	$1	−$117.60
Holding cost	5.985 bbl	$2	−$11.97	5.985 bbl	$3.4	−$11.66
Total			**$8,770.83**			**$9,025.54**

SKILLBUILDER EXERCISE

Eli's industrial engineer has identified a way to increase productivity at the flint machine. Instead of a mean time of 1 day and a standard deviation of 0.2 days, the new process would have a mean time of 0.8 days with a standard deviation of 0.1 days. Analyze the expected impact on profit of this change for the JIT model by modifying the ProcessModel file *Supply Chain JIT Model*.

Questions and Problems

1. Search the literature for additional examples of simulation applications other than those described in this chapter. Write a short application summary, similar to our Simulation in Practice cases, that describes the salient features of the application. Try to develop a small numerical example that you can model and simulate either on a computer or manually (using a spreadsheet to generate the appropriate random variates), and report your results.

2. Using the information in the emergency room simulation example, generate appropriate input data using a spreadsheet, and conduct a manual simulation of the system for 50 patients.

3. Using the information for the job shop simulation model, generate 10 jobs randomly, and manually conduct a simulation for each of the three rules described. Illustrate the differences in performance for applying each rule. Assume that class II jobs spend half their time on each machine in the sequence: bronze, brass.

4. Develop a flowchart similar to Figure 9-1 for a job shop simulation model that has an arbitrary number of work centers (up to 10) and jobs that can follow any routing among a subset of these work centers. (Hint: Maintain a list of work operations that need to be performed for each job during the course of the simulation, and route that job to the appropriate work center once it has completed the current operation.)

5. Interview a systems administrator at your school to understand the configuration of your school's computer system. Try to develop a conceptual simulation model of the system. How would you obtain the necessary data to simulate the model?

6. A company produces a component that involves the assembly of two manufactured parts (A and B). Parts A and B can be produced independently. Cellular manufacturing is used, so that there is no traveling within cells for each part. The work processes are as follows:
   - Get initial design: constant 2 minutes

   PART A
   - Gather materials: travel 5 minutes, constant delay of 20 minutes, travel 2 minutes
   - Lathe: normally distributed, mean 5 minutes, standard deviation 1 minute
   - Grind: normally distributed, mean 3 minutes, standard deviation 0.5 minutes
   - Polish: normally distributed, mean 2 minutes, standard deviation 0.5 minutes
   - Travel to assembly 1 minute

   PART B
   - Gather materials: travel 3 minutes, constant delay 8 minutes, travel 2 minutes
   - Lathe: normally distributed, mean 4 minutes, standard deviation 1 minute
   - Grind: normally distributed, mean 3 minutes, standard deviation 0.5 minutes
   - Travel to assembly 1 minute
   - Assemble: 4 minutes after both parts are available.

   Based on 100 trials, estimate average daily production (over 8 hours). Find the average utilization of each machine.

7. An army base provides physicals for troops. Every day, a particular company of recruits (uniformly distributed between 100 and 140 troops each) is processed. The first station involves filling out forms, 30 minutes in unison. Next come four stations that each troop must pass through, in no particular order.
   - Bloodwork: normally distributed (mean 10 minutes, standard deviation 2 minutes); 6 stations
   - Hearing: normally distributed (mean 8 minutes, standard deviation 2 minutes); 4 stations
   - Eyes: normally distributed (mean 4 minutes, standard deviation 1 minute); 2 stations
   - EKG: normally distributed (mean 10 minutes, standard deviation 2 minutes); 4 stations

   A bus transports the troops to a local hospital with an X-ray facility. When 30 troops have completed the four tests, the truck transports them (10-minute trip each way) to X ray. This process is normally distributed with mean 3 minutes and standard deviation 1 minute; there are two X-ray stations. After completing X ray, troops pass to a physician, where they have a final interview. This takes a constant 1 minute, and there is one physician. Develop a simulation model and, based on 50 trials, estimate the time required to process all troops. Find the utilization rates for each station.

8. A court system involves the following activities, each of which all those who are arraigned pass through. Arrivals occur around the clock, with an arrival rate of three per hour (exponentially distributed) between 12 A.M. and 9 A.M., five per hour (exponentially distributed) between 9 A.M. and 8 P.M., and 12 per hour (exponentially distributed) between 8 P.M. and 12 A.M.
   - Booking: one server; average time 4 minutes (normal, standard deviation 0.5 minute)
   - Processing: three servers; average time 15 minutes (normal, standard deviation 5 minutes)
   - Search: five servers; average time 5 minutes (normal, standard deviation 1 minute)
   - Cage: space for 200

   Three judges work from 8 A.M. until 12 P.M., and from 1 P.M. until 5 P.M. One night judge works from 6 P.M. until 12 A.M., and another from 12 A.M. until 7 A.M. Service time is exponentially distributed with an average of 10 minutes. Evaluate the system for bottlenecks, and suggest changes to the system.

9. A human resources office consists of the following activities, handling employment applications, employee terminations, and benefits complaints.
   - Front desk: average duration 30 seconds

   EMPLOYMENT APPLICATION INTERVIEWS
   - Fill out form: service time, normal 12 minutes (standard deviation 2 minutes); no capacity limit
   - Interviews: service time, normal 15 minutes, standard deviation 3 minutes; 3 interviewers available

   TERMINATIONS
   - Pull file: service time, normal 5 minutes, standard deviation 0.5 minutes; 2 servers
   - Forms: service time, normal 10 minutes, standard deviation 1 minute; no capacity limit

   BENEFITS COMPLAINTS
   - Pull file: service time, normal 5 minutes, standard deviation 0.5 minute; 1 server
   - Interview: service time, exponential 15 minutes; 1 server

   Simulate an 8-hour day for this office, which experiences a client every 4 minutes on average, exponentially distributed. Of these arrivals,
   - 60 percent are applying for work (have to travel 3 minutes after check-in)
   - 10 percent are terminations (1 minute travel after check-in)
   - 30 percent are for benefits (1 minute travel after check-in)

   Evaluate the system for bottlenecks, and suggest changes to the system.

10. Select a system or process that you are familiar with, such as a fast-food restaurant, gasoline station, grocery store or minimart, or college service office (e.g., financial aid, parking). Discuss the operation of the system with the manager and develop a systems simulation flowchart. Collect relevant data and implement a ProcessModel simulation. Design appropriate experiments to evaluate the system or improve its effectiveness. Summarize your model development, analysis, and findings in a formal report.

References

Cobb, R., E. R. Mansfield, and J. M. Mellichamp. "Development of Design Guidelines for Local Area CSMA/CD Networks," *Simulation,* Vol. 58, no. 4, 1992, pp. 270–279.

Costa, M. T., and J. S. Ferreira. "A Simulation Analysis of Sequencing Rules in a Flexible Flowline," *European Journal of Operational Research,* Vol. 119, 1999, pp. 440–450.

González López-Valcárcel, B., and P. Barber Pérez. "Evaluation of Alternative Functional Designs in an Emergency Department by Means of Simulation," *Simulation,* Vol. 63, no. 1, 1994, pp. 20–28.

Hutchison, J., G. K. Leong, and P. T. Ward. "Improving Delivery Performance in Gear Manufacturing at Jeffrey Division of Dresser Industries," *Interfaces,* Vol. 23, no. 2, 1993, pp. 69–79.

Van der Vorst, J. G. A. J., A. J. M. Beulens, and P. van Beek. "Modeling and Simulating Multiechelon Food Systems," *European Journal of Operational Research,* Vol. 122, 2000, pp. 354–366.

Simulation in Forecasting and Optimization

Chapter Outline

One of the major problems that managers face is forecasting future events in order to make good decisions. For example, forecasts of interest rates, energy prices, and other economic indicators are needed for financial planning; sales forecasts are needed to plan production and workforce capacity; and forecasts of trends in demographics, consumer behavior, and technological innovation are needed for long-term strategic planning. In addition to forecasting, managers need to make decisions that make the most effective use of resources. Although many decisions involve only a limited number of alternatives and can be addressed by examining a small number of alternatives, others have a very large or even an infinite number of possibilities. To identify the best decision in these situations, we often use optimization models that seek to minimize or maximize some quantity of interest—usually with constraints that limit the choices.

In this chapter, we discuss the role of simulation in forecasting and optimization. We first review some of the basic concepts of forecasting and optimization, with which we assume you have some prior familiarity. This chapter is not intended to be a thorough treatment of these topics; we refer you to the chapter references for more complete discussions.

Time Series Forecasting

Managers may choose from a wide range of forecasting techniques. Selecting the most appropriate method depends upon the time horizon of the variable being forecast, as well as available information upon which the forecast will be based. Many quantitative forecasting techniques are based on statistical time series models. A **time series** is a stream of historical data, such as weekly sales, which we denote as $A_1, A_2, \ldots A_t$, where A_t is the value of the time series in period t. We will use the notation F_t to denote the forecast for period t. Time series models assume that whatever forces have influenced sales in the recent past will continue into the near future; thus, forecasts are developed by extrapolating these data into the future. Among the most popular time series methods are moving averages and exponential smoothing.

Explanatory/causal models based on regression analysis are also used to identify factors that explain statistically the patterns observed in the variable being forecast. Simple time series models are used for short- and medium-range forecasts; regression analysis is the most popular method for long-range forecasting.

Time series forecasting assumes that historical data are a combination of a pattern and some random error. The error is the difference between the actual observation, A_t, and the forecast F_t. The underlying pattern may include some combination of a **trend**—a gradual shift in the value of the time series, and **seasonality**—short-term effects over a year, month, week, or even a day. At a neighborhood pharmacy, for instance, seasonal patterns may occur over a day, with the heaviest volume of customers in the morning and around the dinner hour. The pattern might even vary by day of the week, being higher on Friday, Saturday, and Monday than on other days. Understanding the pattern provides information to extrapolate the time series into the future, and statistical analysis of the error describes the accuracy of the forecasts.

Two principal time series techniques are

1. *Linear smoothing,* which estimates a smooth trend by removing extreme data and reducing data randomness, and
2. *Seasonal smoothing,* which combines smoothing data with an adjustment for seasonal behavior.

LINEAR SMOOTHING METHODS

Linear smoothing methods include the following:

Single Moving Average   This approach is best for volatile data with no trend or seasonality. Specifically, a simple moving average forecast for the next period is computed as the average of the most recent k observations:

$$F_{t+1} = (A_t + A_{t-1} + \ldots + A_{t+1-k})/k$$

The value of k is somewhat arbitrary, although its choice affects the accuracy of the forecast. The larger the value of k, the more the current forecast is dependent on older data; the smaller the value of k, the quicker the forecast responds to changes in the time series. The best value of k can be identified by analyzing forecast errors.

Excel has a tool for computing moving averages. From the *Tools* menu, select *Data Analysis,* and then *Moving Average.* Excel displays a dialog box in which you enter the *Input Range* of the data, the *Interval* (the value of k), and the first cell of the *Output Range.* You may also obtain a chart of the data and the moving averages, as well as a column of standard errors, by checking the appropriate boxes. However, we do not recommend using the chart or error options because the forecasts are not aligned correctly with the data (the forecast value aligned with a particular data point represents the forecast for the *next* month) and, thus, can be misleading. Rather, we recommend that you generate your own chart.

Double Moving Average   This applies the moving average technique twice, once to the last several periods of the original data, and then to the resulting single moving average data. This method then uses both sets of smoothed data to forecast into the future. Double moving averages are best for historical data with a trend but no seasonality.

Single Exponential Smoothing   Single exponential smoothing weights all of the past data with exponentially decreasing weights; the more recent the data, the higher the weight. The basic single exponential smoothing model is

$$\begin{aligned} F_{t+1} &= (1 - \alpha)F_t + \alpha A_t \\ &= F_t + \alpha(A_t - F_t) \end{aligned}$$

where F_{t+1} is the forecast for time period $t + 1$, F_t is the forecast for period t, A_t is the observed value in period t, and α is a constant between 0 and 1, called the **smoothing constant.**

Because the simple exponential smoothing model requires only the previous forecast and the current time series value, it is very easy to calculate; thus, it is highly suitable for environments such as inventory systems where many forecasts must be made. Different values of α affect how quickly the model responds to changes in the time series. For instance, a value of $\alpha = 0$ would simply repeat last period's forecast; $\alpha = 1$ would forecast last period's actual demand. The closer α is to 1, the quicker the model responds to changes in the time series because it puts more weight on the actual current observation than on the forecast. Like moving averages, this technique is best for volatile data with no trend or seasonality.

Excel also provides an exponential smoothing tool. From the *Tools* menu, select *Data Analysis,* and then *Exponential Smoothing.* Similar to the Moving Average dialog box, you must enter the *Input Range* of the time series data, the *Damping Factor* $(1 - \alpha)$—*not* the smoothing constant as we have defined it!—and the first cell of the *Output Range.* You also have options for labels, to chart output, and to obtain standard errors. As opposed to the Moving Average tool, the chart generated by this tool does correctly align the forecasts with the actual data.

Holt's Double Exponential Smoothing   Double exponential smoothing applies single exponential smoothing twice—once to the original data and then to the smoothed data. Holt's method can use a different parameter for the second smoothing calculations. This method is best for data with a trend but no seasonality. The formula is

$$F[2]_{t+1} = \beta F[1]_{t+1} + (1 - \beta)F[1]_t$$

where $F[2]_{t+1}$ is the doubly smoothed forecast for period $t + 1$, and $F[1]_{t+1}$ is the single exponential smoothing forecast for period $t + 1$ using a smoothing constant α as before.

SEASONAL SMOOTHING METHODS

Many time series exhibit trends and/or seasonality. Both simple moving average and simple exponential smoothing models tend to lag systematic changes in the data. Thus, it is beneficial to include trend and seasonality explicitly into the approach, making exponential smoothing more useful for many business forecasting situations. Seasonal exponential smoothing methods extend the simple exponential smoothing methods by adding an additional component to capture seasonality. These techniques include the following:

Seasonal, Additive Smoothing   This method calculates a seasonal index for historical data without a trend. It produces exponentially smoothed values for the level of the forecast (i.e., base value) and the seasonal adjustment. The seasonal adjustment is added to the base level to provide the forecast. This method is best for data without a trend, but with seasonality that does not change over time.

Seasonal, Multiplicative Smoothing   This approach calculates a seasonal index for historical data that do not have a trend. It produces exponentially smoothed values for the level of the forecast and the seasonal adjustment. This method is best for data without a trend, but with seasonality that changes over time; that is, where the seasonal effect may increase or decrease over time.

Holt-Winters' Additive Seasonal Smoothing   This technique is an extension of Holt's exponential smoothing that captures seasonality. The model produces exponentially smoothed values for the level, trend, and seasonal adjustment of the forecast. The seasonality factor is added to the trended forecast to produce the final estimate. This method is best for data with a trend and with seasonality that does not increase over time.

Holt-Winters' Multiplicative Seasonal Smoothing   This is similar to the additive method, but multiplies the trended forecast by the seasonality factor. This approach is best for data with a trend and seasonality that increases over time.

ERROR METRICS AND FORECAST ACCURACY

The quality of a forecast depends on how accurate it is in predicting future values of a time series. The error in a forecast is the difference between the forecast and the actual value of the time series once it is known, and it generally depends on the model parameters. In the simple moving average model, for example, different values for k will produce different forecasts, as will different values of α in the exponential smoothing model.

To analyze the accuracy of these models more precisely, we can define *error metrics*, which compare quantitatively the forecast with the actual observations. Three metrics that are commonly used are the *mean absolute deviation, mean square error,* and *mean absolute percentage error.* The **mean absolute deviation (MAD)** is the average difference between the actual value and the forecast, averaged over range of forecasted values:

$$\text{MAD} = \frac{\sum_{t=1}^{n} |A_t - F_t|}{n}$$

where A_t is the actual value of the time series at time t, F_t is the forecast value for time t, and n is the number of forecast values (*not* the number of data points because we do not have a forecast value associated with the first k data points). MAD is a fairly reliable error measure and is most accurate for normally distributed data.

Root mean square error (RMSE) is probably the most commonly used error metric. It penalizes larger errors because squaring larger numbers has a greater impact than squaring smaller numbers. The formula for RMSE is

$$\text{RMSE} = \frac{\sum_{t=1}^{n} (A_t - F_t)^2}{n}$$

Again, n represents the number of forecast values used in computing the average.

A third commonly used metric is the **mean absolute percentage error (MAPE).** MAPE is the average of absolute errors divided by actual observation values.

$$\text{MAPE} = \frac{\sum_{t=1}^{n} \frac{|A_t - F_t|}{A_t}}{n} \times 100$$

One advantage of MAPE is that it allows you to compare forecast accuracy between differently-scaled time series data.

REGRESSION MODELS

Multiple linear regression is used in forecasting when time series data may depend on other independent variables besides time, such as advertising, interest rates and other economic factors, or demographic data. For example, monthly energy consumption might depend on the temperature, the size of a house, and the number of residents. A multiple linear regression model has the form:

$$Y = \beta_0 + \beta_1 X_1 + \beta_2 X_2 + \ldots + \beta_k X_k + \varepsilon$$

where Y is the dependent variable
      $X_1 \ldots X_k$ are the independent (explanatory) variables
      β_0 is the intercept term
      $\beta_1 \ldots \beta_k$ are the regression coefficients for the independent variables
      ε is the error term

Multiple regression models for forecasting are used often in econometrics.

Forecasting with CB Predictor

CB Predictor is an Excel add-in like Crystal Ball and is part of the Crystal Ball Professional Edition package. CB Predictor may be accessed in Excel from the *CB Tools* menu. When CB Predictor is started, the dialog box shown in Figure 10-1 appears. The dialog box contains four tabs that query you for information one step at a time. *Input Data* allows you to enter the data on which to base your forecast; *Data Attributes* allows you to specify the type of data, whether or not seasonality is present, and regression option; *Method Gallery* allows you to select one of the eight time series methods we discussed earlier. The graphs shown in the *Method Gallery* (see Figure 10-2) suggest the method that is best suited for the data. However, CB Predictor will run each method you select and will recommend the one that best forecasts your data. The final tab, *Results,* allows you to specify a variety of reporting options.

To illustrate CB Predictor, we describe an Excel model for Monica's Bakery, a rapidly growing bakery in Albuquerque, New Mexico.[1] Monica has kept records of the

---

[1] Adapted from *CB Predictor User Manual,* Version 1.0, 1998–1999, Decisioneering, Inc.

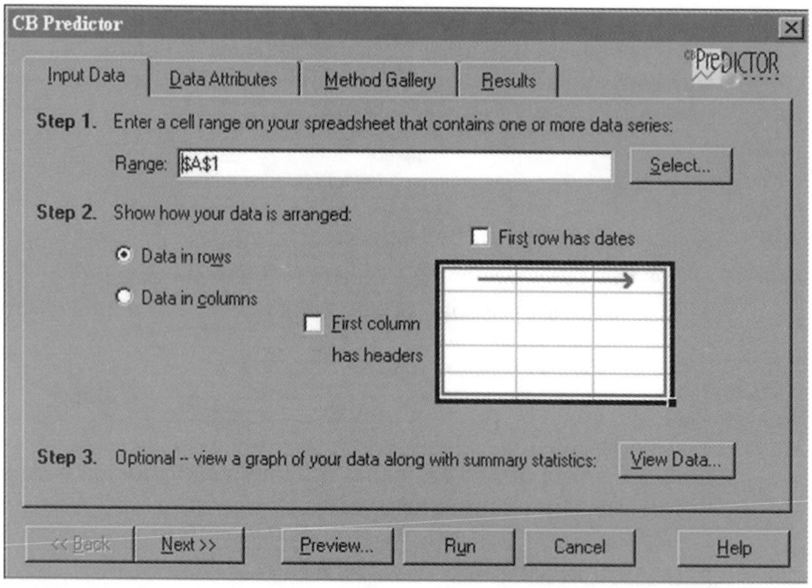

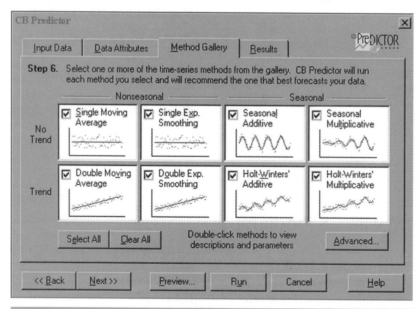

FIGURE 10-1 CB Predictor Dialog Box

FIGURE 10-2 CB Predictor Method Gallery

sales of her three main products: French bread, Italian bread, and pizza. Figure 10-3 shows the Excel workbook (*Monica's Bakery.xls*), which contains worksheets for sales data, operations, cash flow, and labor costs. Using forecasting techniques, Monica seeks to better predict sales, control inventory, market her products, and make strategic decisions. With better sales forecasts, for instance, Monica can better schedule deliveries to take advantage of bulk discounts while maintaining freshness of the ingredients.

Monica created an Excel PivotTable (beginning in row 41—see the actual Excel file) that summarizes the data for the three products in the Operations worksheet. This month, Monica must place an order that will be delivered at the end of this month for

	A	B	C	D	E	F
1	Monica's Bakery					
2				Daily Sales Data		
3						
4						
5		Date	French Bread	Italian Bread	Pizza	Total
6		05-Jun-96	$591.64	$852.96	$187.04	$1,631.64
7		06-Jun-96	582.36	1101.63	187.29	1871.28
8		07-Jun-96	591.63	1008.88	186.79	1787.31
9		08-Jun-96	589.60	1071.82	186.72	1848.14
10		09-Jun-96	582.48	1038.78	186.88	1808.14
11		10-Jun-96	579.34	1002.81	186.95	1769.10
12		11-Jun-96	606.50	995.83	187.13	1789.46
13		12-Jun-96	611.57	837.60	187.07	1636.25
14		13-Jun-96	605.11	1125.16	187.29	1917.55
15		14-Jun-96	617.31	1053.87	186.93	1858.11
16		15-Jun-96	630.87	1143.00	187.50	1961.37
17		16-Jun-96	656.43	1049.61	187.04	1893.08
18		17-Jun-96	667.51	1271.85	187.42	2126.78
19		18-Jun-96	658.08	1140.49	187.08	1985.65
20		19-Jun-96	670.71	1174.30	187.29	2032.30
21		20-Jun-96	658.58	1110.88	187.62	1957.08
22		21-Jun-96	660.71	1129.90	187.33	1977.95
23		22-Jun-96	652.45	1075.36	187.22	1915.04
24		23-Jun-96	671.35	1103.43	187.49	1962.27
25		24-Jun-96	670.21	1179.98	187.38	2037.57
26		25-Jun-96	679.56	1125.69	187.45	1992.69
27		26-Jun-96	685.69	1077.89	187.52	1951.10
28		27-Jun-96	688.59	978.21	187.47	1854.27
29		28-Jun-96	701.47	1060.75	187.48	1949.70
30		29-Jun-96	701.66	1072.89	187.66	1962.21
31		30-Jun-96	729.42	1029.71	187.24	1946.37
32		01-Jul-96	752.52	1109.18	187.43	2049.13

◄ ◄ ► ►◄ \ Sales Data ╱ Operations ╱ Cash Flow ╱ Labor Costs ╱

FIGURE 10-3 Excel Workbook for Monica's Bakery

the next month; therefore, she must forecast sales for the next 8 weeks (she is currently in week 173 of her business). The data for the last 4 weeks of this forecast (weeks 178–181) represent the month for which she is ordering.

From the Operations worksheet, select one cell in the PivotTable; CB Predictor will automatically select all the PivotTable data. Start CB Predictor, and ensure that the correct options are selected:

- Correct cell range, with headers, dates, and data in columns (*Input Data* tab)
- Time period in weeks with a seasonality of 52 weeks (*Data Attributes* tab)
- All time series methods selected (*Methods Gallery* tab)
- Number of periods to forecast is eight (*Results* tab)
- The only result selected is Paste Forecasts at the bottom of the PivotTable (*Results* tab).

After clicking on *Run,* the results paste to the end of the PivotTable, as shown in Figure 10-4. The last 4 weeks of forecast values for each data series are automatically summed and placed into the table at the top of the spreadsheet in the *Sales Forecast* column (see Figure 10-5). In the table, the monthly sales forecast is converted into the number of items sold and weight of each product. The second table takes the total weight of

	B	C	D	E
212	170	17,296.23	10,152.48	1,471.40
213	171	17,541.78	9,657.03	1,488.27
214	172	17,634.05	9,142.77	1,490.36
215	173	17,380.54	8,758.17	1,483.24
216	**174**	**17,386.18**	**8,443.90**	**1,479.81**
217	**175**	**17,340.74**	**8,338.27**	**1,473.22**
218	**176**	**17,294.74**	**8,520.45**	**1,460.56**
219	**177**	**17,395.24**	**8,642.11**	**1,445.23**
220	**178**	**17,483.23**	**8,333.47**	**1,436.11**
221	**179**	**17,406.77**	**8,372.50**	**1,430.78**
222	**180**	**17,396.03**	**8,881.09**	**1,423.12**
223	**181**	**17,500.20**	**8,909.12**	**1,414.73**

FIGURE 10-4 CB Predictor Forecasts for Monica's Bakery

each product and calculates how much of each ingredient is required to produce that much product. The ingredients for each are totaled into orders for the month in the third table. Thus, Monica should order 79,253 pounds of flour, 253 pounds of yeast, 358 pounds of salt, and 12 pounds of cheese.

SKILLBUILDER EXERCISE

Open the file *Shampoo.xls,* which shows weekly sales of a shampoo product for Tropical Cosmetics Co. Use CB Predictor to forecast weekly sales. Before running the model, click the *Preview* button in the *Results* tab to show the best-fitting method. Which method is best? Why?

FIGURE 10-5 Using Sales Forecasts to Compute Purchasing Requirements

	B	C	D	E	F	G
9						
10	Demand	Sales Forecast	Selling Price	Units	Unit Weight	Total Weight
11	French Bread	$ 69,786	$ 1.24	56,279.22	0.25	229,711.11
12	Italian Bread	$ 34,496	$ 2.08	16,584.70	1.05	15,794.95
13	Pizza	$ 5,705	$ 14.71	387.81	3.12	124.30
14						
15						
16	Ingredients	Forecast Weight	Per Pound	Pounds Required		
17	French Bread	229,711				
18	Flour		0.32	73,508		
19	Yeast		0.0010	228		
20	Salt		0.0015	342		
21	Italian Bread	15,795				
22	Flour		0.36	5,686		
23	Yeast		0.0015	24		
24	Salt		0.0010	16		
25	Pizza	124				
26	Flour		0.48	60		
27	Yeast		0.01	1		
28	Salt		0.00	0		
29	Tomato		0.10	12		
30						
31						
32	Consolidated Requirements		Pounds			
33	Flour		79,253			
34	Yeast		253		*Amounts to be Ordered*	
35	Salt		358			
36	Cheese		12			
37						

	B	C	D	E	F
49	Jun-99	13.71	122.7	500.0	9.0
50	Jul-99	14.19	127.3	500.1	8.1
51	Aug-99	14.12	128.1	500.1	7.6
52	Sep-99	14.27	127.9	500.5	7.4
53	Oct-99	$14.28	129.0	501.4	7.7
54	Nov-99	$14.37	129.2	502.5	7.4
55	Dec-99	$14.43	129.0	504.5	7.2
56	Jan-00	$14.44	129.5	505.5	7.4
57	Feb-00	$14.49	129.7	505.6	7.3
58	Mar-00	$14.52	129.9	505.5	7.1
59					
60	Coefficients for Monica's Average Wage	$0.00	0.1	0.0	(0.2)

FIGURE 10-6 Individual and Multiple Linear Regression Forecasts

	B	C	D	E	F	G
4						
5			Sep-99	Mar-00		
6	Wages		$ 14.27	$ 14.52	←	*Average labor costs from regression analysis*
7	Overhead		33%	33%		
8	Total Employees		18	18		
9	Total Labor Costs		$ 54,652.74	$ 55,627.87		
10	Labor Cost Change			2%		
11						

FIGURE 10-7 Calculation of Forecasted Labor Costs

CB Predictor also includes multiple regression forecasting. To illustrate this, we will examine the Labor Cost worksheet. Monica knows that a few key economic figures drive labor costs, such as the Industrial Production Index, local CPI, and local unemployment, all of which can be found from the Bureau of Labor Statistics Web site. The Labor Cost worksheet contains a PivotTable that lists her average hourly wage (dependent variable) for each month and the monthly numbers for these three indicators (independent variables).

In the worksheet, we select one cell in the Economic Variables for Regression Analysis table and start CB Predictor. Ensure that the cell range is selected correctly, monthly time periods are used with a seasonality of 12 months, the *Multiple Linear Regression* option is selected, the variables are specified correctly, all time series methods are selected (this brings up a new dialog showing the dependent and independent variables), the number of periods to forecast is six, and the only result selected is Paste Forecast. CB Predictor generates a regression equation between the variables, uses time series forecasting methods to forecast the independent variables individually, and then uses those forecasted values to calculate the dependent variable values using the regression equation. Figure 10-6 shows the forecasts pasted at the end of the table, and Figure 10-7 shows the forecasted labor costs. We see that the total payroll increase will only be 2 percent.

Integrating Forecasting and Simulation

CB Predictor may be used with Crystal Ball to analyze the effects of uncertainty in forecasts. For example, suppose that a manufacturing manager needs to forecast sales in order to make production decisions. By defining forecast values as Crystal Ball assumptions, you can generate a Crystal Ball forecast chart that shows the total amount of product that must be produced to meet demand with a specific level of certainty rather than simply using the expected forecast values.

You may set an option that automatically generates forecasted values as Crystal Ball assumptions. To do this, click on *Preferences* from the *Results* tab in CB Predictor. Select the *Paste* tab, and select *Paste Forecasts as Crystal Ball Assumptions,* then click "OK." Make sure you have selected the *Paste Forecasts At Cell* result. Then run the forecast procedure. For data series forecasted using time series methods, CB Predictor creates assumptions as normal distributions with mean equal to the forecasted value in the cell, and a standard deviation calculated using RMSE. For multiple linear regression, assumptions are created for the independent variable forecast values only. If you want to see the variability of the dependent variable, you can select the pasted formula cells and define them as Crystal Ball forecast cells. More likely, you might want to create one formula cell that represents the sum of the data in the dependent variable cells and define that formula cell as a Crystal Ball forecast.

Let us revisit Monica's Bakery. Monica is considering purchasing a flour silo and a delivery van. She needs to forecast when the bakery will be able to safely pay for these projects or whether she must finance them, and wants to incorporate uncertainty into her cash flow analysis. First, set up the run preferences in Crystal Ball to run 500 trials using Latin Hypercube sampling, an initial seed value of 999, and a sample size of 500. Open the Cash Flow worksheet (Figure 10-8) and select one cell in the PivotTable at

FIGURE 10-8 Cash Flow Worksheet for Monica's Bakery

	B	C	D	E	F	G	H	I
4								
5			Common-Sized	July	August	September	Forecasted from Monthly Data Collected from Pivot Table	
6	Revenue Forecast		100% $	123,853 $	125,502 $	117,662 ←		
7								
8								
9	Expenses		Common-Sized	July	August	September		
10	Cost of Goods							
11	Fixed		6,708	6,708	6,708	6,708		
12	Variable		23%	28,486	28,865	27,062		
13	Overhead							
14	Fixed		8,924	8,924	8,924	8,924		
15	Variable		18%	22,293	22,590	21,179		
16	Financing		5%	6,193	6,275	5,883		
17	Taxes		17%	21,389	21,673	20,319		
18	Total Expenses			$ 93,992 $	95,036 $	90,076		
19								
20								
21	Extraordinary Items			July	August	September		
22	Silo Construction			50,000.00	-	-		
23	New Van			-	35,000.00	-		
24								
25							Cash Flow based on Monthly Revenue Forecast	
26	Monthly Cash Flow			$ (20,140) $	(4,534) $	27,586 ←		
27								
28	Net Cash at Beginning of Month			$ 42,941 $	22,801 $	18,267		
29	Net Cash at End of Month			$ 22,801 $	18,267 $	45,854		
30								
31	Minimum Cash Target			20,000	20,000	20,000		

the bottom of the worksheet. Start CB Predictor, ensuring that the cell range is correct, the time periods are in months with a seasonality of 12 months, all time series methods are selected, the number of periods to forecast is three, the only result selected is Paste Forecast, and the *Paste Forecasts As Crystal Ball Assumptions* option is checked in the Preferences dialog under the *Results* tab. The results paste adjacent to the PivotTable as Crystal Ball assumptions and also are copied to the Revenue Forecast table at the top of the worksheet. Because she has committed to the silo construction, she wants to be at least 90 percent certain that she will not run out of cash and be unable to meet payroll and other critical bills. She has a minimum cash target each month. Therefore, define cells E29:G29, which represent the net cash at the end of each month, as Crystal Ball forecasts and run a simulation. The results are shown in Figure 10-9. From these results, we see that the probability of dropping below the reserve threshold of $20,000 is significant, so she decides to delay buying the van until September.

FORECASTING AND SIMULATING A NET INCOME MODEL

Suppose that a company uses a rolling 6-month average to estimate future sales. It developed a net income model (see the file *Net Income Model.xls* and Figure 10-10). For the next fiscal year, the model predicts $1.80 earnings per share (cell N22). Because the

FIGURE 10-9 Results of Simulating Net Cash Flow for Monica's Bakery

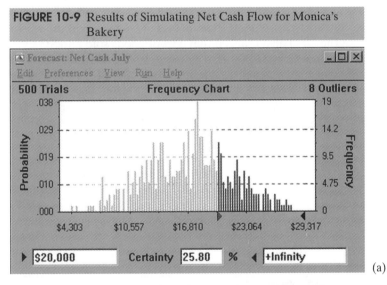

(a)

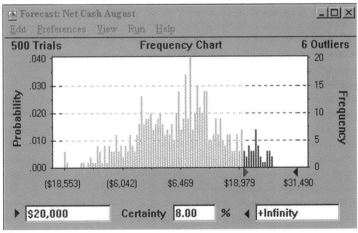

(b)

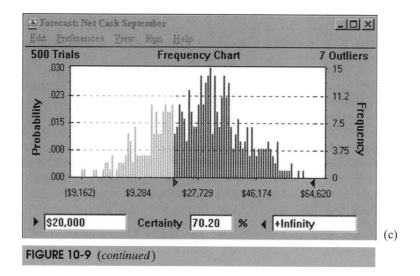

(c)

FIGURE 10-9 (*continued*)

penalty for missing earnings numbers is very steep, the CFO wanted to understand the certainty of reaching the forecasted sales. CB Predictor was used to forecast the sales for the next 12 months. (These are already entered in the worksheet.)

The CFO also realized that cost of goods sold and SG&A have considerable variability, so a Crystal Ball distribution was defined for these two variables. Total net income for the fiscal year and earnings per share (EPS) are defined as forecast cells. The results of a Crystal Ball simulation are shown in Figure 10-11. We see that there is only a 46.6 percent chance that earnings per share will be at least $1.80. What does the CFO tell Wall Street analysts?

FIGURE 10-10 Net Income Model

	A	B	C	D	E	F	G
1	Pro Forma Net Income Statement						
2							
3	Assumptions						
4							
5	COGS/Sales	61.0%	Taxes	38%			
6	SG&A/Sales	7.4%	Interest	8,000			
7	Depreciation	20,000	Shares Outst.	100,000			
8							
9							
10		July	August	September	October	November	December
11	Net Sales	$ 156,333	$ 161,556	$ 165,148	$ 167,006	$ 168,341	$ 167,897
12	COGS	95,363	98,549	100,740	101,874	102,688	102,417
13	Gross Profit	60,970	63,007	64,408	65,132	65,653	65,480
14	G&A Expenses	11,569	11,955	12,221	12,358	12,457	12,424
15	Depreciation	20,000	20,000	20,000	20,000	20,000	20,000
16	EBIT	29,401	31,052	32,187	32,774	33,196	33,056
17	Interest	8,000	8,000	8,000	8,000	8,000	8,000
18	EBT	21,401	23,052	24,187	24,774	25,196	25,056
19	Taxes	8,133	8,760	9,191	9,414	9,574	9,521
20	Net Income	13,269	14,292	14,996	15,360	15,621	15,534

(a)

	G	H	I	J	K	L	M	N
	December	**January**	**February**	**March**	**April**	**May**	**June**	**Total**
10	December	January	February	March	April	May	June	Total
11	$ 167,897	$ 164,380	$ 165,721	$ 166,416	$ 166,627	$ 166,564	$ 166,267	$ 1,982,256
12	102,417	100,272	101,090	101,514	101,642	101,604	101,423	1,209,176
13	65,480	64,108	64,631	64,902	64,984	64,960	64,844	773,080
14	12,424	12,164	12,263	12,315	12,330	12,326	12,304	146,687
15	20,000	20,000	20,000	20,000	20,000	20,000	20,000	240,000
16	33,056	31,944	32,368	32,587	32,654	32,634	32,541	386,393
17	8,000	8,000	8,000	8,000	8,000	8,000	8,000	96,000
18	25,056	23,944	24,368	24,587	24,654	24,634	24,541	290,393
19	9,521	9,099	9,260	9,343	9,369	9,361	9,325	110,349
20	15,534	14,845	15,108	15,244	15,286	15,273	15,215	180,044
21								
22								EPS 1.80

(b)

FIGURE 10-10 (*continued*)

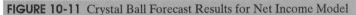

FIGURE 10-11 Crystal Ball Forecast Results for Net Income Model

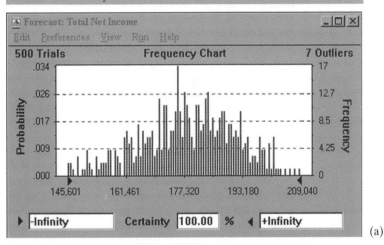

(a)

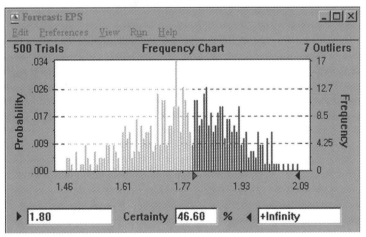

(b)

Optimization

An optimization model seeks to choose the best values for a set of decision variables to minimize or maximize an **objective function**—for instance, to minimize cost or maximize revenue. Most practical optimization problems have **constraints**—limitations or requirements that decision variables must satisfy. The presence of constraints usually makes identifying an optimal solution considerably more difficult. Some examples of constraints are as follows:

- The amount of material used to produce a set of products cannot exceed the available amount of 850 square feet.
- The amount of money spent on research and development projects cannot exceed the assigned budget of $300,000.
- Contractual requirements specify that at least 500 units of product must be produced.
- A mixture of fertilizer must contain exactly 30 percent nitrogen.
- We cannot produce a negative amount of product (*nonnegativity*).

Constraints are generally expressed mathematically as equations or inequalities. For the preceding examples, we might write:

- *Amount of material used $\leq$ 850 square feet.*
- *Amount spent on research and development $\leq$ $300,000.*
- *Number of units of product produced $\geq$ 500.*
- *Amount of nitrogen in mixture/total amount in mixture $= .30$.*
- *Amount of product produced $\geq$ 0.*

The left-hand side of each of these expressions is called a **constraint function.** When represented mathematically, a constraint function is a function of the decision variables. For example, suppose that in the first case the material requirements of three products are 3.0, 3.5, and 2.3 square feet per unit. If A, B, and C represent the number of units of each product to produce, then $3.0A$ represents the amount of material used to produce A units of product A; $3.5B$ represents the amount of material used to produce B units of product B; and so on. Note that dimensions of these terms are (square feet/unit)(units) = square feet. Therefore, the constraint that limits the amount of material that can be used can be expressed as

$$3.0A + 3.5B + 2.3C \leq 850$$

As another example, if two ingredients contain 20 percent and 33 percent nitrogen, respectively, then the fraction of nitrogen in a mixture of X pounds of the first ingredient and Y pounds of the second ingredient is expressed by the constraint function

$$(.20X + .33Y)/(X + Y)$$

If the fraction of nitrogen in the mixture must be 0.30, then we would have

$$(.20X + .33Y)/(X + Y) = 0.3$$

This can be rewritten as

$$(.20X + .33Y) = 0.3(X + Y)$$

or

$$-0.1X + 0.03Y = 0$$

Any solution that satisfies all constraints of a problem is called a **feasible solution.** An optimization problem that has no feasible solutions is called **infeasible.** The presence of constraints makes finding optimal solutions more difficult than for unconstrained problems. In fact, it may be very difficult to even identify a feasible solution, much less an optimal one. Thus, to solve constrained optimization problems, we generally rely on special solution procedures, such as the Excel Solver, which we briefly describe later.

TYPES OF OPTIMIZATION MODELS

The most common types of optimization problems are *linear, integer,* and *nonlinear.* A **linear optimization problem** (often called a **linear program**) has two basic properties. First, the objective function and all constraints are *linear functions* of the decision variables. This means that each function is simply a sum of terms, each of which is some constant multiplied by a decision variable. For example:

$$2*A + 3*B - 12*C$$

Second, all variables are *continuous,* meaning that they may assume any real value (typically nonnegative).

In an **integer (discrete) linear optimization problem,** some or all of the variables are restricted to be whole numbers. A special type of integer problem is one in which variables can only be zero or one. These *binary variables* help us to model logical, "yes or no" decisions. Integer linear optimization models are generally more difficult to solve than pure linear models. Finally, in a **nonlinear optimization problem,** the objective function and/or constraint functions are *nonlinear functions* of the decision variables; that is, terms cannot be written as a constant times a variable.

LINEAR OPTIMIZATION EXAMPLE: PORTFOLIO ALLOCATION

An investor has $100,000 to invest in four assets. The expected annual returns, and minimum and maximum amounts with which the investor will be comfortable allocating to each investment are shown here:

Investment	Annual Return	Minimum	Maximum
1. Life insurance	5%	$ 2,500	$5,000
2. Bond mutual funds	7%	$30,000	none
3. Stock mutual funds	11%	$15,000	none
4. Savings account	4%	none	none

The major source of uncertainty in this problem is the annual return of each asset. In addition, the decision maker faces other risks—for example, unanticipated changes in inflation or industrial production, the spread between high and low grade bonds, and the spread between long- and short-term interest rates. One approach to incorporating such risk factors in a decision model is arbitrate pricing theory (APT).[2] APT provides estimates of the sensitivity of a particular asset to these types of risk factors. Let

---

[2]See Schniederjans, M., T. Zorn, and R. Johnson, "Allocating Total Wealth: A Goal Programming Approach," *Computers and Operations Research,* Vol. 20, no. 7, 1993, pp. 679–685.

us assume that the risk factors per dollar allocated to each asset have been determined as follows:

Asset	Risk Factor/Dollar Invested
1. Life insurance	−0.5
2. Bond mutual funds	1.8
3. Stock mutual funds	2.1
4. Savings account	−0.3

The investor may specify a target level for the weighted risk factor, leading to a constraint that limits the risk to the desired level. For example, suppose that our investor will tolerate a weighted risk per dollar invested of at most 1.0. Thus, the weighted risk for a $100,000 total investment will be limited to 100,000. Then, if our investor allocates $5,000 in life insurance, $50,000 in bond mutual funds, $15,000 in stock mutual funds, and $30,000 in a savings account (which fall within the minimum and maximum amounts specified), the total expected annual return would be

$$.05(\$5,000) + .07(\$50,000) + .11(\$15,000) + .04(\$30,000) = \$6,600.$$

However, the total weighted risk associated with this solution is

$$-0.5(5,000) + 1.8(50,000) + 2.1(15,000) - 0.3(30,000) = 110,000$$

Because this is greater than the limit of 100,000, this solution could not be chosen. The decision problem, then, is to determine how much to invest in each asset to maximize the total expected annual return, remain within the minimum and maximum limits for each investment, and meet the limitation on the weighted risk.

A spreadsheet model for this problem is shown in Figure 10-12. Problem data are specified in rows 5 through 9. On the bottom half of the spreadsheet, we specify the model outputs, namely, the values of the decision variables (cells B13:B16), objective function (cell B17), and constraint limitations (the total weighted risk in cell E13 and total amount invested in cell E17). Note that the total weighted risk is computed as

FIGURE 10-12 Portfolio Allocation Spreadsheet Model

	A	B	C	D	E
1	Portfolio Allocation Model				
2					
3		Annual	Lower	Upper	Risk factor
4	Investments	return	bound	bound	per dollar
5	Life Insurance	5.0%	$2,500	$5,000	-0.5
6	Bond mutual funds	7.0%	$30,000	none	1.8
7	Stock mutual funds	11.0%	$15,000	none	2.1
8	Savings account	4.0%	none	none	-0.3
9	Total amount available	$100,000		Limit	100,000
10					
11		Amount		Constraints	Total weighted
12	Decision variables	invested			risk
13	Life Insurance	$5,000			146,000
14	Bond mutual funds	$50,000		Decision Variables	
15	Stock mutual funds	$30,000			Total amount
16	Savings account	$15,000			invested
17	Total expected return	$7,650		Objective	$100,000
18					

TABLE 10-1 Generic Examples of Linear Optimization Models

Type of Model	Decision Variables	Objective Function	Typical Constraints
Product mix	Quantities of products to produce and sell	Maximize contribution to profit	Resource limitations (e.g., production time, labor, material); maximum sales potential; contractual requirements
Process selection	Quantities of product to make using alternative processes	Minimize cost	Demand requirements; resource limitations
Blending	Quantity of materials to mix to produce one unit of product	Minimize cost	Specifications on acceptable mixture
Production planning	Quantities of product to produce in each of several periods; amount of inventory to hold between successive periods	Minimize production and inventory costs	Limited production rates; material balance equations (production + available inventory − inventory held to next period = demand)
Portfolio selection	Amounts to invest in different financial instruments	Maximize future expected return	Limit on available funds; sector requirements and restrictions (minimum and maximum amounts in different types of instruments)
Multiperiod investment planning	Amounts to invest in various instruments each year	Maximize return	Limit on available funds; cash balance equations between periods
Media selection	Number of advertisements in different media	Minimize cost	Budget limitation; requirements on number of customers reached; media requirements and restrictions

E5*B13 + E6*B14 + E7*B15 + E8*B16 and must be less than or equal to the value in cell E9. Similarly, the total amount invested, SUM(B13:B16), must be less than or equal to the value in cell E17. You can see that this particular solution is not feasible because the total weighted risk exceeds the limit of 100,000. Table 10-1 shows other generic examples of linear optimization models.

INTEGER OPTIMIZATION EXAMPLE: PROJECT SELECTION

A firm's R&D group has identified five potential new engineering and development projects; however, the firm is constrained by its available budget and human resources. The data are given in the spreadsheet in Figure 10-13. Each project is expected to generate

FIGURE 10-13 Project Selection Spreadsheet Model

	A	B	C	D	E	F	G
1	**Project Selection Model**						
2							
3							Available
4		Project 1	Project 2	Project 3	Project 4	Project 5	Resources
5	Expected Return (NPV)	$180,000	$220,000	$150,000	$140,000	$200,000	
6	Cash requirements	$ 55,000	$ 83,000	$ 24,000	$ 49,000	$ 61,000	$ 150,000
7	Personnel requirements	5	3	2	5	3	12
8							
9	Project selection decisions	0	1	0	1	0	
10	Cash expended	$132,000					
11	Personnel used	8					
12							
13	Total Return	$360,000					

a return (given by the net present value) but requires a fixed amount of cash and personnel. Because the resources are limited, all projects cannot be selected. Projects cannot be partially completed; thus, either the project must be undertaken completely or not at all.

To model this situation, we define the decision variables to be binary (that is, either zero or one), corresponding to either not selecting or selecting each project, respectively. These are defined in cells B9:F9. The objective function, computed in cell B13, is the total return, which can be expressed as the sum of the product of the return from each project and the binary decision variable:

$$\text{Total return} = B5*B9 + C5*C9 + D5*D9 + E5*E9 + F5*F9$$

To develop the constraints, note that if a project is selected, we must use both the cash and personnel requirements from the amounts available. These constraints can be written as

$$\text{Cash used} = B6*B9 + C6*C9 + D6*D9 + E6*E9 + F6*F9 \leq G6$$
$$\text{Personnel used} = B7*B9 + C7*C9 + D7*D9 + E7*E9 + F7*F9 \leq G7$$

The left-hand sides of these functions can be found in cells B10 and B11.

Constraint functions of binary variables can be used to model many different logical conditions. For instance, suppose that the R&D group has determined that, at most, one of projects 1 and 2 should be pursued. This can be modeled as

$$B9 + C9 \leq 1$$

Similarly, the constraint "If project 4 is chosen, then project 2 must also be chosen" can be modeled as

$$B9 \geq E9$$

or, equivalently,

$$B9 - E9 \geq 0$$

Note that if project 4 is chosen, then the value of cell E9 must be 1. From the preceding constraint, the value of B9 must, therefore, be greater than or equal to 1 also (i.e., project 2 must be chosen). On the other hand, if project 4 is not chosen (E9 = 0), then the constraint reduces to B9 ≥ 0, and project 2 can either be selected or not.

NONLINEAR OPTIMIZATION EXAMPLE: HOTEL PRICING

The Marquis Hotel is considering a major remodeling effort and needs to determine the best combination of rates and room sizes to maximize revenues. Currently, the hotel has 450 rooms with the following history:

Room Type	Rate	Daily Avg. No. Sold	Revenue
Standard	$ 85	250	$21,250
Gold	$ 98	100	$ 9,800
Platinum	$139	50	$ 6,950
		Total Revenue	$38,000

Each market segment has its own price/demand elasticity. Estimates are

Room Type	Price Elasticity of Demand
Standard	−1.5
Gold	−2.0
Platinum	−1.0

	A	B	C	D	E	F
1	**Marquis Hotel Problem**					
2						
3		Current	Average		Total Room	
4	Room type	Rate	Daily Sold	Elasticity	Capacity	
5	Standard	$ 85.00	250	-1.5	450	
6	Gold	$ 98.00	100	-2		
7	Platinum	$ 139.00	50	-1		
8						
9					Projected	
10					Rooms	Projected
11	Room type	New Price	Price Range		Sold	Revenue
12	Standard	$ 85.00	$ 70.00	$ 90.00	250	$21,250.00
13	Gold	$ 98.00	$ 90.00	$ 110.00	100	$ 9,800.00
14	Platinum	$ 139.00	$ 120.00	$ 149.00	50	$ 6,950.00
15				Totals	400	$38,000.00

FIGURE 10-14 Hotel Pricing Spreadsheet Model

This means, for example, that a *1 percent decrease* in the price of a standard room will *increase* the number of rooms sold by *1.5 percent.* Similarly, a 1 percent increase in the price will decrease the number of rooms sold by 1.5 percent. For any pricing structure (in $), the projected number of rooms of a given type sold can be found using the formula:

(Historical average number of rooms sold) + (Elasticity)*(New price − Current price)*(Historical average number of rooms sold)/(Current price)

The hotel owners want to keep the price of a standard room between $70 and $90; a gold room between $90 and $110; and a platinum room between $120 and $149. Although the rooms may be renovated, there are no plans to expand beyond the current 450-room capacity.

Figure 10-14 shows a spreadsheet model for this situation. The decision variables, the new prices to charge, are given in cells B12:B14. The projected numbers of rooms rented are computed in cells E12:E14 using the preceding formula. By multiplying the number of rooms rented by the new price for each room type, the projected revenue is calculated, as given in cells F12:F14. Note that because the projected number of rooms sold is a function of price, the revenue functions are nonlinear. The total revenue in cell F15 represents the objective function.

In addition, we have constraints: (1) The new price must fall within the allowable price range, and (2) The total projected number of rooms sold must not exceed 450. These can be expressed as

$$B12:B14 \geq C12:C14$$
$$B12:B14 \leq D12:D14$$

and

$$E15 \leq E5$$

Spreadsheet Optimization Using Excel Solver

Microsoft Excel contains an add-in called *Solver,* which allows you to find optimal solutions to constrained optimization problems formulated as spreadsheet models. (Check the list of available add-ins under *Tools/Add-Ins.* If Solver is not listed, you will have to reinstall Excel, using a custom installation.) Solver was developed and is maintained by Frontline Systems, Inc. (*www.frontsys.com*). Frontline Systems also sells a more powerful, industrial-strength version of Solver, called *Premium Solver.* We encourage you to visit this Web site for additional examples, tutorials, updates, and other information about the software.

To use Solver, you should design your spreadsheet to include the following:

1. A cell for the value of each decision variable
2. A cell that calculates the objective function value
3. Cells for the constraint functions

If you examine the examples we developed in the previous section, you will see that all of them include these cells. It is usually convenient to lay out your variables in rows or columns and provide descriptive labels either to the left of the columns or above the rows; this improves the readability and manageability of your models. You should also consider using different fonts, borders, or shading to enhance the readability of your spreadsheets.

In Solver, decision variables are called *adjustable cells,* or *changing cells;* the objective function cell is called the *target cell.* Solver identifies values of the changing cells that minimize or maximize the target cell value. Solver is easier to use if you define a cell for each of the constraint functions in your model (that is, the left-hand sides of the constraints).

To illustrate how to use Solver for linear programming models, we will solve the portfolio optimization model. Select Solver from the Excel *Tools* menu. In the Solver Parameters dialog box that appears (see Figure 10-15), set the target cell, optimization objective, and changing cells. Note that the changing cells are defined by the range B13:B16.

FIGURE 10-15 Solver Parameters Dialog Box

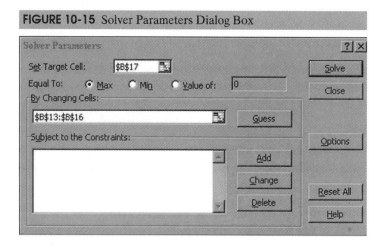

FIGURE 10-16 Add Constraint Dialog Box

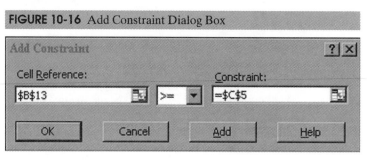

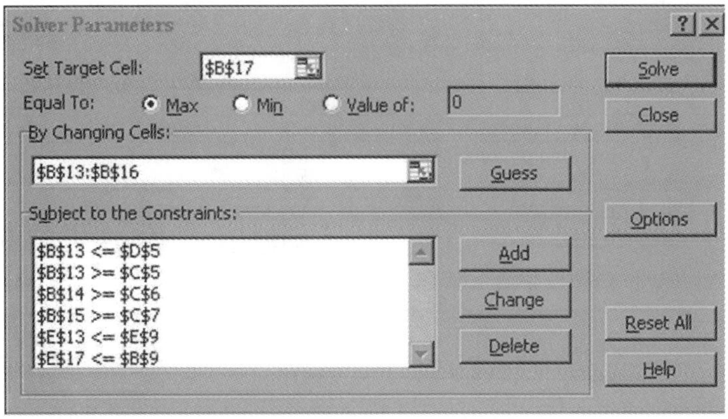

FIGURE 10-17 Portfolio Allocation Model in Solver

To add constraints, click the *Add* button. The Add Constraint dialog box (see Figure 10-16) will appear. *Cell Reference:* refers to the left-hand side of a constraint; *Constraint:* refers to the right-hand side. In either case, you may enter a single cell reference or a range of cells. The drop-down menu in the center of the dialog box allows you to choose the type of constraint: <=, =, >=, int, or binary. *Int* restricts the cell reference range to integers, and *binary* restricts it 0 or 1. Click the *Add* button to add new constraints. When you are finished adding constraints, click "OK." Figure 10-17 shows the Solver Parameters dialog box with the constraints entered. You may add, change, or delete these as necessary by clicking the appropriate buttons.

For linear models, you must select the *Options* button. This displays the Solver Options dialog box shown in Figure 10-18. Always check the boxes for *Assume Linear Model* and *Assume Non-Negative* when these are conditions of the problem. (You do not have to enter nonnegativity constraints explicitly in the model.) If you do not check *Assume Linear Model,* Solver will treat your model as nonlinear, and the output reports will not be in the proper form to interpret. Generally, you may leave the other options at their default values for linear models. Return to the Solver Parameters dialog box by clicking "OK."

To find the optimal solution, click the *Solve* button. The Solver Results dialog box will appear with the message "Solver found a solution." If a solution could not be found,

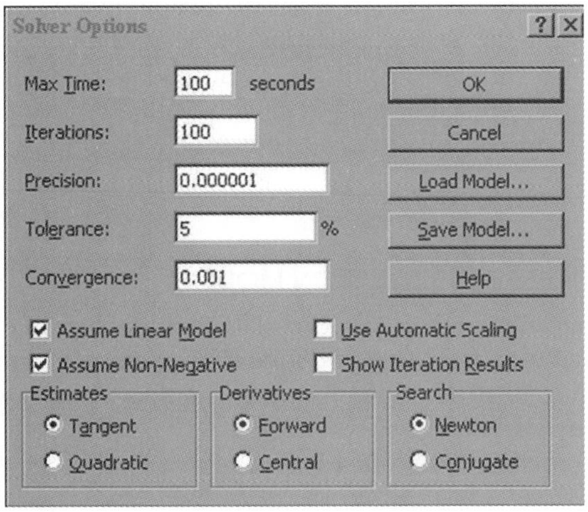

FIGURE 10-18 Solver Options

	A	B	C	D	E
1	**Portfolio Allocation Model**				
2					
3		**Annual**	**Lower**	**Upper**	**Risk factor**
4	**Investments**	**return**	**bound**	**bound**	**per dollar**
5	Life Insurance	5.0%	$2,500	$5,000	-0.5
6	Bond mutual funds	7.0%	$30,000	none	1.8
7	Stock mutual funds	11.0%	$15,000	none	2.1
8	Savings account	4.0%	none	none	-0.3
9	*Total amount available*	$100,000		*Limit*	100,000
10					
11		**Amount**		Constraints	*Total weighted*
12	**Decision variables**	**invested**			*risk*
13	Life Insurance	$5,000			100,000
14	Bond mutual funds	$30,000		Decision Variables	
15	Stock mutual funds	$28,333			*Total amount*
16	Savings account	$36,667			*invested*
17	*Total expected return*	$6,933		Objective	$100,000
18					

FIGURE 10-19 Solver Solution to the Portfolio Allocation Model

Solver will notify you with a message to this effect. This generally means that you have an error in your model, or you have included conflicting constraints that no single solution can satisfy. In such cases, you need to reexamine your model.

Solver generates three reports: Answer, Sensitivity, and Limits. To add them to your Excel workbook, hold the *Ctrl* key, click on each of them, and then click "OK." Solver places the solution into the spreadsheet model as shown in Figure 10-19. The maximum return is $6,933, obtained by the investment plan in cells B13:B16. We will not discuss how to interpret the reports here, but instead refer you to more complete descriptions found in textbooks on management science and decision models; for instance, James R. Evans and David L. Olson, *Statistics, Data Analysis,* and *Decision Modeling,* Prentice Hall, 2000.

SOLVING INTEGER OPTIMIZATION MODELS

Integer optimization models are set up in the same manner as linear models in Solver, except that any integer variables must be defined as such in the Add Constraint dialog box by using the *int* or *bin* options from the drop-down box. For example, to define all the variables for the Project Selection model as binary, we would select *bin* for the range of these variables. You should still choose *Assume Linear Model* in the Solver Options dialog box. For integer models, you also need to ensure that *the value of tolerance in the Solver Options dialog box is set to zero* to ensure finding an optimal solution.

SOLVING NONLINEAR OPTIMIZATION MODELS

Nonlinear optimization models are formulated with Solver in the same fashion as linear or integer models, except that you should *not* choose *Assume Linear Model* in the *Options* box. Figure 10-20 shows the Solver Parameters dialog box for the Marquis Hotel model. The optimal solution is shown in Figure 10-21. The optimal prices predict a demand for all 450 rooms with a total revenue of $39,380.65.

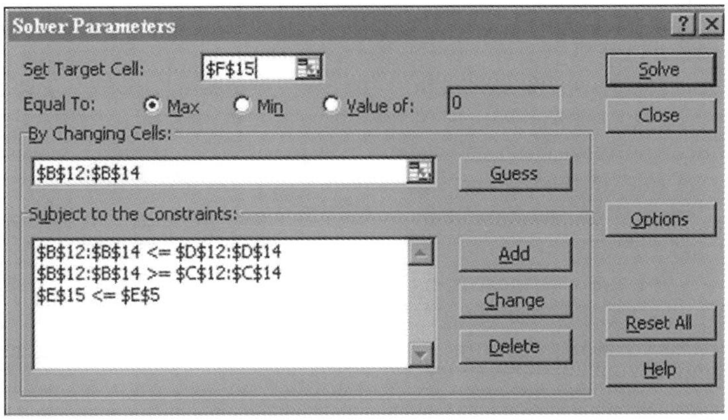

FIGURE 10-20 Solver Parameters for the Marquis Hotel Problem

	A	B	C	D	E	F
1	**Marquis Hotel Problem**					
2						
3		*Current*	*Average*		*Total Room*	
4	*Room type*	*Rate*	*Daily Sold*	*Elasticity*	*Capacity*	
5	Standard	$ 85.00	250	-1.5	450	
6	Gold	$ 98.00	100	-2		
7	Platinum	$ 139.00	50	-1		
8						
9					*Projected*	
10					*Rooms*	*Projected*
11	*Room type*	*New Price*	*Price Range*		*Sold*	*Revenue*
12	Standard	$ 76.87	$ 70.00	$ 90.00	286	$21,974.39
13	Gold	$ 90.00	$ 90.00	$ 110.00	116	$10,469.39
14	Platinum	$ 145.04	$ 120.00	$ 149.00	48	$ 6,936.87
15				*Totals*	450	$39,380.65

FIGURE 10-21 Optimal Solution to the Marquis Hotel Problem

Risk Analysis of Optimization Results

It is rare that any optimization model is completely deterministic; in most cases, some of the data will be uncertain. This implies that inherent risk exists in using the optimal solution obtained from a model. Using the capabilities of risk analysis software such as Crystal Ball, these risks can be better understood and mitigated. To illustrate this, we will use the Marquis Hotel pricing problem.

In the Marquis problem, the price-demand elasticities of demand are only estimates and most likely are quite uncertain. Because we probably will not know anything about their distributions, let us conservatively assume that the true values might vary from the estimates by plus or minus 25 percent. Thus, we model the elasticities by uniform distributions. Using the optimal prices identified by Solver earlier in this chapter, let us see what happens to the forecast of the number of rooms sold under this assumption using Crystal Ball.

In the spreadsheet in Figure 10-14, we select cells D5:D7 as assumption cells with uniform distributions having minimum and maximum values equal to 75 percent and 125 percent of the estimated values, respectively. The total rooms sold (E15) is defined as a forecast cell. The model was run for 2,500 trials, creating the forecasts in Figure 10-22. We see that the mean number of rooms sold under these prices is 450, which should be expected because the mean values of the elasticities were used to derive the optimal prices. However, because of the uncertainty associated with the elasticities, the probability that *more* than 450 rooms will be sold (demanded) is about 50 percent! This suggests that if the assumptions of the uncertain elasticities are true, the hotel might anticipate that demand will exceed its room capacity about half the time, resulting in many unhappy customers.

We could use these results, however, to identify the appropriate hotel capacity to ensure, for example, only a 10 percent chance exists that demand will exceed capacity. Figure 10-23 shows the forecast chart when the certainty level is set at 90 percent and the left grabber is anchored. We could interpret this as stating that if the hotel capacity were about 457 or 458 rooms, then demand will exceed capacity at most 10 percent of the time. If we shift the capacity constraint down by 7 rooms to 443 and use Solver to find the optimal prices associated with this constraint (which are $78.34, $90.00, and $146.51 for the three room types, respectively), we would expect demand to exceed 450 only about 10 percent of the time. Figure 10-24 shows the Solver results of a Crystal Ball run confirming that, with these prices, demand will exceed 450 less than 10 percent of the time.

FIGURE 10-22 Crystal Ball Forecast Charts for Marquis Hotel Problem

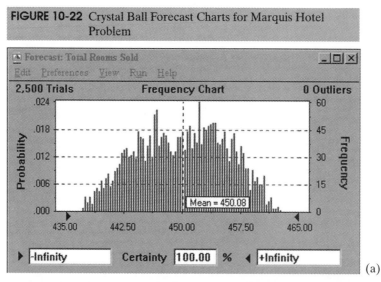

(a)

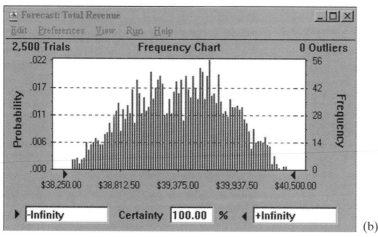

(b)

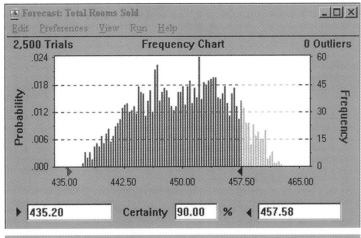

FIGURE 10-23 Evaluating Hotel Capacity with 90 percent Certainty

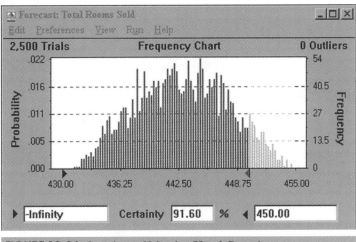

FIGURE 10-24 Certainty of Meeting Hotel Capacity

Combining Optimization and Simulation

To find an optimal set of decision variables for any simulation-based model, you generally need to search in a heuristic or ad hoc fashion. This usually involves running a simulation for an initial set of variables, analyzing the results, changing one or more variables, rerunning the simulation, and repeating this process until a satisfactory solution is obtained. This process can be very tedious and time-consuming, and often how to adjust the variables from one iteration to the next is not clear. We note that the commercial version of ProcessModel includes optimization; however, the student version does not. We refer you to Chapter 8 of the ProcessModel *Online User's Manual* for further information about its optimization capabilities.

For spreadsheet models, you may use the Crystal Ball Decision Table tool to help you optimize, but this is restricted to one or two variables. OptQuest, another of Decisioneering's Excel add-ins, overcomes these limitations by automatically searching for optimal solutions within Crystal Ball simulation model spreadsheets. Within OptQuest, you describe your optimization problem and search for values of decision variables that maximize or minimize a predefined objective. Additionally, OptQuest is designed to find solutions that satisfy a wide variety of constraints or a set of goals that you may define.

USING OPTQUEST

The basic process for using OptQuest is described as follows:

1. Create a Crystal Ball model of the decision problem.
2. Define the decision variables within Crystal Ball.
3. Invoke OptQuest from the Crystal Ball toolbar or the corresponding menu.
4. Create a new optimization file.
5. Select decision variables and set the bounds.
6. Specify constraints on decision variables.
7. Select an objective and any requirements.
8. Modify OptQuest options.
9. Solve the optimization problem.
10. Save the optimization files.
11. Exit OptQuest.

To illustrate the process, we will use the Portfolio Allocation example.

Create a Crystal Ball Model Using the basic spreadsheet model, first define the assumptions and forecast cells in Crystal Ball. We will assume that the annual returns for life insurance and mutual funds are uncertain, but that the rate for the savings account is constant. We will make the following assumptions in the Crystal Ball model:

- Cell B5: uniform distribution with minimum = 4 percent and maximum = 6 percent
- Cell B6: normal distribution with mean 7 percent and standard deviation 1 percent
- Cell B7: lognormal distribution with mean 11 percent and standard deviation 4 percent

We define the forecast cell to be the total expected return, cell B17. As would be the case with any Crystal Ball application, you would select *Run Preferences* from the *Run* menu and choose appropriate settings. Set the number of trials per simulation to 500.

Define Decision Variables The next step is to identify the decision variables in the model. This is accomplished using the *Define Decision Variables* option in the *Cell* menu. Position the cursor on cell B12. From the *Cell* menu, choose *Define Decision Variables*. Set the minimum and maximum values according to the problem data (i.e., columns C and D in the spreadsheet), as shown in Figure 10-19. Next, repeat the process of defining decision variables for cells B13, B14, and B15. When the maximum limit is *none,* you may use a value of $100,000 because this is the total amount available. You are now in a position to call OptQuest by selecting it from the *CB Tools* menu.

Creating a New Optimization File From the opening screen in OptQuest, select *New* from the *File* menu. This reads the model from the active Excel worksheet. This option allows you to create different optimization files for the same simulation. OptQuest will prompt you to select the subset of decision variables from your Crystal Ball model that will be used for optimization, the forecast cell and corresponding statistic that will be used as the objective to minimize or maximize, any forecast cells and corresponding statistics that will be used as goals, and any additional restrictions or constraints that you may wish to specify.

Select	Variable Name	Lower Bound	Suggested Value	Upper Bound	Type	WorkBook	WorkSheet	Cell
☑	Life Insurance	2500	5000	5000	Continuous	Portfolio Allocation Model.xls	Portfolio	B13
☑	Bond mutual funds	30000	50000	100000	Continuous	Portfolio Allocation Model.xls	Portfolio	B14
☑	Stock mutual funds	15000	30000	100000	Continuous	Portfolio Allocation Model.xls	Portfolio	B15
☑	Savings account	0	15000	100000	Continuous	Portfolio Allocation Model.xls	Portfolio	B16

Reorder   OK   Cancel   Help

FIGURE 10-25 OptQuest Decision Variable Selection Window

Select Decision Variables   Every decision variable in the Crystal Ball model appears in the *Decision Variable Selection* window shown in Figure 10-25. The first column indicates whether the variable has been selected for optimization. The check boxes show which variables are selected; initially, all decision variables are selected. For each selected variable, a lower and an upper bound must be given in the appropriate columns. If you would like to include a starting solution that OptQuest will improve upon, you can suggest the values of the selected variables in the *Suggested Value* column. The suggested value by default is the value that appears in the corresponding cell in your Crystal Ball model. If the suggested values are out of range or do not meet the problem's constraints, these values are ignored. The *Type* column indicates whether a variable is discrete or continuous. The variable type can be changed by clicking in this window or in the *Define Decision Variable* window of Crystal Ball. A step size is associated with discrete variables. A variable of the type Discrete_2, for example, has a step size of 2. Therefore, if the lower and upper bounds for this variable are 0 and 7, respectively, the only feasible values are 0, 2, 4, and 6. To change the step size, you must click on the *Discrete* item of the drop-down menu and enter the new step size in the dialog box. In Figure 10-25, we see that all decision variables are selected for the optimization model.

Specify Constraints   The next screen displayed allows you to specify any constraints (see Figure 10-26). A *constraint* is any limitation or requirement that restricts the possible solutions to the problem. In our example, we have two constraints. The first constraint limits the total weighted risk to 100,000, and the second ensures that we do not allocate more than $100,000 in total to all assets. In the OptQuest screen, a listing of all

FIGURE 10-26 OptQuest Constraint Window

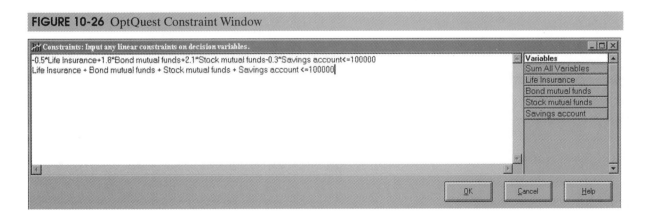

Constraints: Input any linear constraints on decision variables.

-0.5*Life Insurance+1.8*Bond mutual funds+2.1*Stock mutual funds-0.3*Savings account<=100000
Life Insurance + Bond mutual funds + Stock mutual funds + Savings account <=100000

Variables
Sum All Variables
Life Insurance
Bond mutual funds
Stock mutual funds
Savings account

OK   Cancel   Help

previously selected decision variables is displayed. Constraints may only use these variables. You then type the constraints one by one, placing a single constraint on each line. (To facilitate the process, you may click on the decision variable names in the right-hand column to move the name to where the cursor is.) Constraints should be one in each line. An asterisk must be used to indicate the product of a constant and a variable (e.g., $3*X$).

Thus, in our example, the risk constraint is

$$-0.5*\text{Life insurance} + 1.8*\text{Bond mutual funds} + 2.1*\text{Stock}$$
$$\text{mutual funds} - 0.3*\text{Savings account} <= 100000$$

and the total investment constraint is

$$\text{Life insurance} + \text{Bond mutual funds} + \text{Stock}$$
$$\text{mutual funds} + \text{Savings account} <= 100000$$

The newly-entered constraints are saved by clicking the *OK* button.

Select the Objective   Every OptQuest run requires the selection of a statistic for at least one forecast cell to act as the objective function to be minimized or maximized. You can select a forecast to be a *Maximize Objective* or a *Minimize Objective* from the drop-down menu in the *Select Objective/Requirements* column.

In addition to an objective, you may choose to set optimization *requirements*. Requirements are used to constrain forecast statistics to fall within specified lower and upper target values. This is done by choosing the *Requirement* option from the drop-down menu in the *Select Objective/Requirements* column and will be illustrated in other examples. In the Crystal Ball model, we have only defined one forecast, whose mean value we wish to maximize, as shown in Figure 10-27.

Modify OptQuest Options   Next, a window with the following three tabs appears:

- Time
- Settings
- Preferences

The *Time* tab allows you to specify the total time that the system is allowed to search for the best values for the optimization variables. You may either enter the total number of minutes or a date and time when the process must stop. Performance will depend on the speed of your microprocessor. The default optimization time is 10 minutes; however, you are able to choose any time limit you desire. Selecting a very long time limit does not present a problem, because you are always able to terminate the search by selecting *Stop* in the *Run* menu or pressing the *Esc* key. Additionally, you will be given the option to extend the search and carry the optimization process farther once the selected time has expired.

FIGURE 10-27 OptQuest Forecast Selection Window

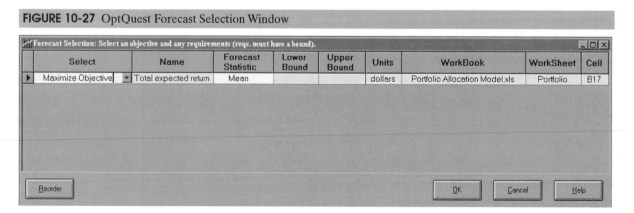

From the *Preferences* tab, you can select which Crystal Ball runs to save (default is *Only Best*) and change the name of the optimization model and font. You may also change the name of the log file. The log file records data related to the search, which can also be displayed by choosing the *Log* option of the *View* menu. The log file is particularly useful when the search abnormally terminates (e.g., due to a system crash), because a text editor can be used to read the information contained in this file, and the search does not result in a wasted effort.

Finally, in the *Advanced* tab, you can select a deterministic optimization (i.e., without simulation) instead of the default stochastic optimization. You are also able to change the settings for the optimization heuristic; however, we recommend leaving them at the default values.

Solve the Optimization Problem   The optimization process is initiated from the next dialog box or from the *Run* menu. As the simulation is running, you may also select three additional options from the *View* menu: *Performance Graph*, *Bar Graph*, or *Log*. The performance graph shows a plot of the value of the objective as a function of the number of simulations evaluated. The bar graph shows how the value of each decision variable changes during the optimization search procedure. Finally, the optimization log provides details of the sequence of best solutions generated during the search.

As the optimization progresses, the sequence of best solutions identified during the search is displayed on the *Status and Solutions* window. Each time a better solution is identified, a new line is added to the screen, showing the new objective value and values of the decision variables. In the upper-left corner of the *Status and Solutions* screen, you can monitor the time remaining and the simulation trial number currently under evaluation. (This information disappears when the time limit is reached.)

Figure 10-28 shows the *Status and Solutions* screen upon completion of the optimization. OptQuest identifies an initial feasible solution to begin the search process. It then searches intelligently within the limits defined for the decision variables, running Crystal Ball simulations for candidate solutions, and saves the best ones found during the search process.

Saving an Optimization File   The *Save* and *Save As . . .* options in the *File* menu allow you to save the current optimization model for future use. Note that the file that you save refers to the optimization problem and the OptQuest options only, and not to the Crystal Ball simulation model (which is saved in the Excel file). The optimization files are automatically given the extension name .OPT. The saved optimization file may be recalled by choosing *Open . . .* from the *File* menu.

FIGURE 10-28 OptQuest Status and Solutions Window

Simulation	Maximize Objective Total expected return Mean	Life Insurance	Bond mutual funds	Stock mutual funds	Savings account
1	5404.59	5000.00	41944.4	15000.0	15000.0
3	6023.94	2500.00	30000.0	15000.0	52500.0
5	6411.26	2500.00	45000.0	15000.0	37500.0
7	6948.80	5000.00	30000.0	28333.3	36666.7
32	6955.07	5000.00	30000.0	28300.7	36437.9
34	6975.18	5000.00	30000.0	28292.5	36380.7
Best: 36	6982.73	5000.00	30000.0	28268.0	36209.1

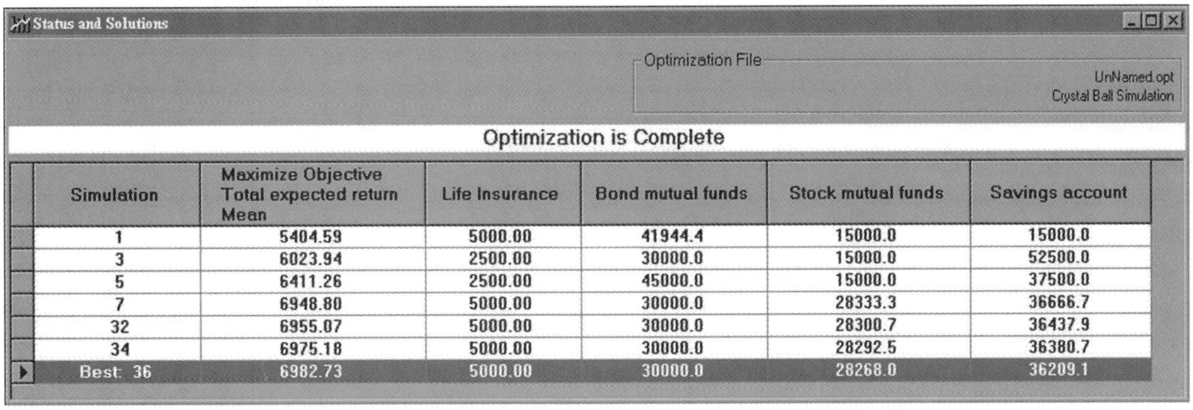

Status and Solutions — Optimization File — UnNamed.opt — Crystal Ball Simulation

Optimization is Complete

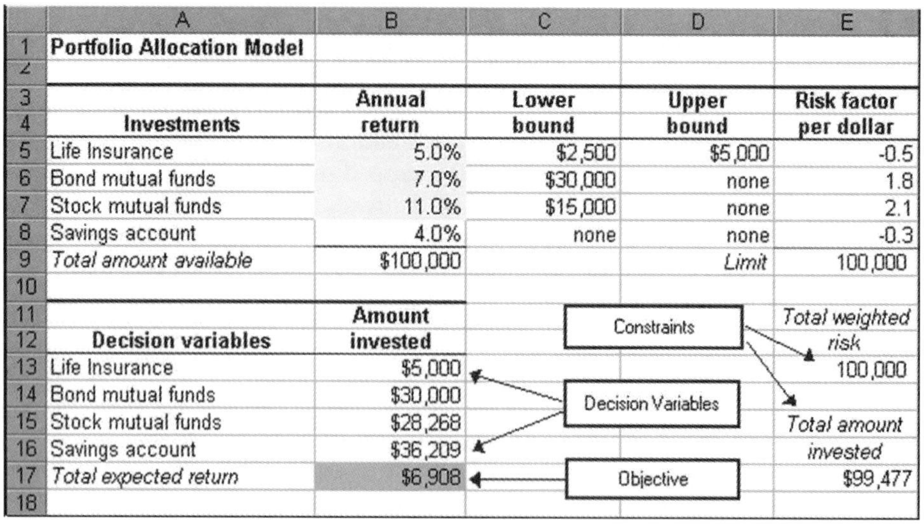

	A	B	C	D	E
1	**Portfolio Allocation Model**				
2					
3		**Annual**	**Lower**	**Upper**	**Risk factor**
4	**Investments**	**return**	**bound**	**bound**	**per dollar**
5	Life Insurance	5.0%	$2,500	$5,000	-0.5
6	Bond mutual funds	7.0%	$30,000	none	1.8
7	Stock mutual funds	11.0%	$15,000	none	2.1
8	Savings account	4.0%	none	none	-0.3
9	*Total amount available*	$100,000		*Limit*	100,000
10					
11		**Amount**		Constraints	*Total weighted*
12	**Decision variables**	**invested**			*risk*
13	Life Insurance	$5,000			100,000
14	Bond mutual funds	$30,000		Decision Variables	
15	Stock mutual funds	$28,268			*Total amount*
16	Savings account	$36,209			*invested*
17	*Total expected return*	$6,908		Objective	$99,477
18					

FIGURE 10-29 Best OptQuest Solution to the Portfolio Allocation Model

Exit OptQuest   To exit, choose *Exit* from the *File* menu. OptQuest will now save the best simulations for you and will restore the one you select when you exit. After choosing *Exit*, you will be given the opportunity to paste the best values found for the optimization variables in your Crystal Ball model. The results are shown in Figure 10-29. You can see that both constraints are satisfied. Alternatively, other values can be pasted by highlighting the corresponding row on the *Status and Solutions* window accessible from the *View* menu.

INTERPRETING RESULTS

You should note that the "best" OptQuest solution identified may not be the true optimal solution to the problem, but it will hopefully be close to the actual optimal solution. The quality of the results depends on the time limit you select for searching, the speed of the computer's microprocessor, the number of decision variables, and the complexity of the problem. With more of decision variables, you need a larger number of simulations and should run the problem for longer times.

After solving an optimization problem with OptQuest, you probably would want to examine the Crystal Ball simulation using the optimal values of the decision variables in order to assess the risks associated with the recommended solution. Figure 10-30 shows the Crystal Ball forecast chart associated with the best solution. Although the mean value was optimized, we see that a high amount of variability exists in the actual return because of the uncertainty in the returns of the individual investments. In fact, the total returns varied from about $4,000 to over $11,000.

ADDING A REQUIREMENT

A *requirement* is a forecast statistic that is restricted to fall within a specified lower and upper bound. The forecast statistic may be one of the following (which are the standard output statistics in Crystal Ball):

- Mean
- Median
- Mode
- Standard deviation

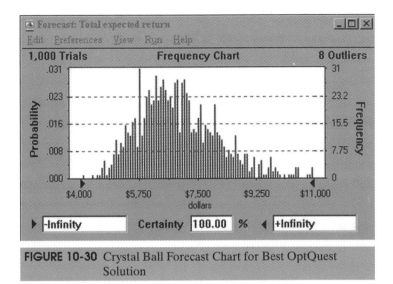

FIGURE 10-30 Crystal Ball Forecast Chart for Best OptQuest
Solution

- Variance
- Percentile (as specified by the user)
- Skewness
- Kurtosis
- Coefficient of variation
- Range (minimum, maximum, and width)
- Standard error

For example, to reduce the uncertainty of returns in the portfolio while also attempting to maximize the expected return, we might want to restrict the standard deviation to be no greater than 1,000. To add such a requirement in OptQuest, select *Forecast* from the *Tools* menu. This will bring up the *Forecast Selection* screen. Because we have only one forecast in the model, this row will be highlighted. The same cell may be simultaneously selected as an objective and as a goal. This can be achieved by highlighting the *Forecast Name* in the *Forecast Selection* window and choosing *Duplicate* from the *Edit* menu. This creates a new row, with the forecast named *Total Expected Return:2*. From the drop-down menu in the first column, select *Requirement*. Click on the *Forecast Statistic* cell and, using the drop-down menu, choose *Std_Dev*. Then set the upper bound to 1,000. Finally, click "OK."

You may now run the new model. The results are shown in Figure 10-31. The best solution among those with standard deviations less than or equal to 1,000 is identified.

FIGURE 10-31 OptQuest Solution with Standard Deviation Requirement

Simulation	Maximize Objective Total expected return:1	Requirement Total expected return:2 Std_Dev <= 1000	Life Insurance	Bond mutual funds	Stock mutual funds	Savings account
1	6905.89	1261.04 - Infeasible	5000.00	30000.0	28268.0	36209.1
2	3884.14	749.704	2500.00	30000.0	15000.0	0.00000
3	6021.33	768.067	2500.00	30000.0	15000.0	52500.0
5	6440.28	884.389	2500.00	45000.0	15000.0	37500.0
11	6443.30	879.451	2681.76	44908.8	15000.0	36649.7
Best: 13	6501.85	871.669	4251.40	45166.8	15000.0	35581.8

Simulation	Minimize Objective Total Cost Mean	Order Quantity	Reorder Point
21	3398.78	407	361
22	3208.96	385	351
23	3053.22	363	341
27	2925.82	348	334
28	2847.58	334	328
85	2839.91	332	328
Best: 101	2824.66	333	327

FIGURE 10-32 OptQuest Solution to Lost Sales Inventory Model

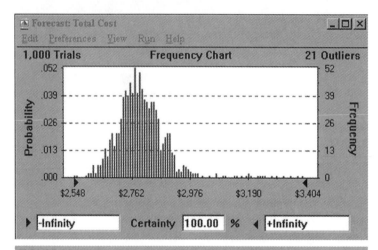

FIGURE 10-33 Total Inventory Cost Forecast Chart

FINDING AN OPTIMAL INVENTORY POLICY

In Chapter 6, we examined a lost sales inventory model and used the Crystal Ball Decision Table tool to converge toward an optimal solution (see Figure 6-5). OptQuest is designed for such problems. First, define the order quantity and reorder point as discrete decision variables with lower and upper bounds of 200 and 500, respectively, and a step size of 1. There are no constraints in the optimization model, and the objective is to minimize the average total cost.

Figure 10-32 shows the OptQuest results. The best solution identified was $Q = 333$ and $r = 327$ with a mean total cost of $2,825. Note that this solution is close to the point found in the decision table, but better. Figure 10-33 shows the Crystal Ball forecast chart for this solution, which provides a high degree of confidence that the total cost will be in the $2,500–$3,000 range.

Questions and Problems

1. Generate a forecast for Monica's Bakery using only the last 52 weeks of data (rows 164–215) in the worksheet *Monica's Bakery.xls*. Compare the RMSE with that of the forecast in the text, using all of the data. What do these relative results imply?
2. Using the data in the worksheet *Shampoo.xls*, find the mean and 95 percent forecast range of the unit sales of shampoo for 8 weeks into the future. Show the CB Predictor chart illustrating the forecast and range.
3. In the Excel workbook *Data.xls*, the worksheet Mower Unit Sales includes data on product sales by region for a company marketing lawn-care equipment. Use CB Predictor to select the best model to forecast 12 periods ahead for world mower unit

sales using data from January 1996 through December 2000. Compare the different models obtained, and find a 95 percent confidence range for the forecasts.

4. Compare the RSME, MAD, and MAPE for the eight CB Predictor models derived in problem 3. How much additional accuracy was gained by selecting a seasonal model?

5. In the workbook *Data.xls,* the worksheet Market Share Tractors includes data on lawn tractor sales by region for a company marketing lawn-care equipment. Use CB Predictor to select the best model to forecast 12 periods ahead for the North American (NA) region market share using data from January 1996 through December 2000. Predict market share for the world and by region.

6. The workbook *Data.xls* contains data on 1999 weekly prices for the Dow Jones Stock Index in the worksheet DowJones. Using the data for weeks 1 through 42, build a forecasting model to obtain forecasts for 10 weeks. If possible, update these data to the current week, and apply the model to forecast the next 10 weeks.

7. The worksheet GDP in the workbook *Data.xls* contains quarterly data on U.S. gross domestic product for 1999. Use CB Predictor to build a multiple regression forecasting model for four quarters for the variable *%Change in Disposable Personal Income,* using *%Change in Personal Consumption Expenditures, %Change in Private Domestic Investment,* and *%Change in Government Consumption Expenditures* as independent variables. If possible, update the data and apply the model to the current time.

8. The workbook *Data.xls* contains a worksheet Oil with 1999 daily prices of a barrel of West Texas intermediate crude. Use CB Predictor to forecast the price of oil 10 days into the future. Show the chart with the forecast and the 90 percent certainty interval. Comment on the forecast.

9. For the Portfolio Allocation model, add a fifth investment alternative, a penny mutual fund with an expected annual return of 16 percent (lognormally distributed with a standard deviation of 6 percent) and a risk factor of 3.0. Include the goal for the standard deviation to be less than or equal to $1,000, $1,500, and then $2,000 in turn, obtaining three solutions. Identify the investment amount for each option, as well as expected return and the probabilities of attaining $8,000, $9,000, $10,000, and $11,000 in profit, based on the best OptQuest solution you find. Use 500 trials per simulation.

10. In the Project Selection spreadsheet model (with constraints on budget and personnel included), assume the following assumptions for the expected returns:

    Project 1 Normally distributed; mean $180,000; std. dev. $18,000
    Project 2 Lognormally distributed; mean $220,000; std. dev. $40,000
    Project 3 Lognormally distributed; mean $150,000; std. dev. $15,000
    Project 4 Normally distributed; mean $140,000; std. dev. $28,000
    Project 5 Lognormally distributed; mean $200,000; std. dev. $50,000

    Assign cells B9:F9 as decision cells, and the total return (cell B13) as a forecast cell. Use OptQuest to find the set of projects that maximizes return, and then use Crystal Ball to identify the probability of attaining a profit of $400,000, $500,000, and $600,000.

11. A small canning company specializes in gourmet canned foods. They can five combinations of ham, lima beans, and jalapeno peppers.

Product	Maximum (includes contracts)	Contracts/Day (minimum demands)
Ham & Beans	10,000 cans/day	5,000 cans/day
Jalapeno Ham & Beans	4,000	1,000
Lima Beans	6,000	1,000
Jalapeno Lima Beans	4,000	2,000
Jalapeno Peppers	1,000	0 (new product)

The production department obtains input materials and fills 16-oz. cans. All quantities are in ounces; all costs and sales prices/can are in dollars. There is a maximum production limit of 24,000 cans/day. Canning costs are constant.

Product	Ham	Limas	Jalapenos	Water	Can Cost
Ham & Beans	4 oz.	9 oz.	None	3 oz.	$0.05
Jalapeno Ham & Beans	3 oz.	9 oz.	1 oz.	3 oz.	$0.05
Lima Beans	None	14 oz.	None	2 oz.	$0.05
Jalapeno Lima Beans	None	12 oz.	1 oz.	3 oz.	$0.05
Jalapeno Peppers	None	None	12 oz.	4 oz.	$0.05
Material cost	$0.40/oz.	$0.05/oz.	$0.10/oz.	free	

The company has a contract with a ham supplier for daily delivery of up to 30,000 oz. of ham at $.40/oz. They also have a contract with a lima bean supplier for up to 100,000 oz. of lima beans per day at $.05/oz. They do not have to pay for materials they do not use. They grow their own jalapenos, which cost $.10/oz to pick. There is more jalapeno supply than can be used. There is also an unlimited supply of tangy bayou water. The sales prices for these products are uncertain. Analysis of past data indicate different distributions for each product as follows:

Product	Distribution	Mean	Standard Deviation
Ham & Beans	Normal	$2.31	$0.25
Jalapeno Ham & Beans	Lognormal	$2.00	$0.40
Lima Beans	Normal	$0.85	$0.10
Jalapeno Lima Beans	Lognormal	$0.90	$0.30
Jalapeno Peppers	Exponential	$1.35	

a. Use Excel Solver to identify the optimal solution based on the mean sales prices per can. Then use Crystal Ball with 1,000 trials to identify the mean return, minimum return, maximum return, and probabilities for returns of at least 0, $1,000/day, $2,000/day, $3,000/day, and $5,000/day.
b. Use OptQuest to generate a solution minimizing the variance in expected profit. Then use Crystal Ball to identify the mean return, minimum return, maximum return, and probabilities for returns of at least 0, $1,000/day, $2,000/day, $3,000/day, and $5,000/day.
c. Discuss the relative tradeoffs between the solutions obtained in parts a. and b.
12. Air Transport Corp. produces airplanes for large smugglers, medium-sized revolutions, and small governments. They produce three models of airplane: the Raven, the Hawk, and the Falcon. This industry is highly dynamic, and Air Transport must set its production schedule for the next year prior to obtaining firm orders. Production data are shown here:

Model	Minimum Required by Contracts	Maximum Demand	Fuselages	Missile Launchers	Cannon
Raven	0	40	1	4	0
Hawk	15	30	1	2	2
Falcon	0	50	1	4	4

There are 50 fuselages available per year, 400 missile launchers, and 40 cannon. Management has assigned probabilities for three levels of profitability per plane.

Model	Low	Probability	Most Likely	Probability	High	Probability
Raven	$ 4 million	0.1	$ 5 million	0.7	$ 6 million	0.2
Hawk	$10 million	0.3	$20 million	0.4	$25 million	0.3
Falcon	$10 million	0.1	$20 million	0.3	$30 million	0.6

a. Use Solver to identify the solution with optimal expected profit. Then use Crystal Ball to identify mean annual profit, minimum, maximum, and probabilities of profit of at least $400 million and $600 million.

b. Use OptQuest to find a solution with the requirement that a profit of at least $400 million be achieved with 90 percent certainty.

APPENDIX A

Normal Distribution
(mean 0, standard deviation 1)

z	.00	.01	.02	.03	.04	.05	.06	.07	.08	.09
0.0	.5000	.5040	.5080	.5120	.5160	.5199	.5239	.5279	.5319	.5359
0.1	.5398	.5438	.5478	.5517	.5557	.5596	.5636	.5675	.5714	.5753
0.2	.5793	.5832	.5871	.5910	.5948	.5987	.6026	.6064	.6103	.6141
0.3	.6179	.6217	.6255	.6293	.6331	.6368	.6406	.6443	.6480	.6517
0.4	.6554	.6591	.6628	.6664	.6700	.6736	.6772	.6808	.6844	.6879
0.5	.6915	.6950	.6985	.7019	.7054	.7088	.7123	.7157	.7190	.7224
0.6	.7257	.7291	.7324	.7357	.7389	.7422	.7454	.7486	.7517	.7549
0.7	.7580	.7611	.7642	.7673	.7704	.7734	.7764	.7794	.7823	.7852
0.8	.7881	.7910	.7939	.7967	.7995	.8023	.8051	.8078	.8106	.8133
0.9	.8159	.8186	.8212	.8238	.8264	.8289	.8315	.8340	.8365	.8389
1.0	.8413	.8438	.8461	.8485	.8508	.8531	.8554	.8577	.8599	.8621
1.1	.8643	.8665	.8686	.8708	.8729	.8749	.8770	.8790	.8810	.8830
1.2	.8849	.8869	.8888	.8907	.8925	.8944	.8962	.8980	.8997	.9015
1.3	.9032	.9049	.9066	.9082	.9099	.9115	.9131	.9147	.9162	.9177
1.4	.9192	.9207	.9222	.9236	.9251	.9265	.9279	.9292	.9306	.9319
1.5	.9332	.9345	.9357	.9370	.9382	.9394	.9406	.9418	.9429	.9441
1.6	.9452	.9463	.9474	.9484	.9495	.9505	.9515	.9525	.9535	.9545
1.7	.9554	.9564	.9573	.9582	.9591	.9599	.9608	.9616	.9625	.9633
1.8	.9641	.9649	.9656	.9664	.9671	.9678	.9686	.9693	.9699	.9706
1.9	.9713	.9719	.9726	.9732	.9738	.9744	.9750	.9756	.9761	.9767
2.0	.9772	.9778	.9783	.9788	.9793	.9798	.9803	.9808	.9812	.9817
2.1	.9821	.9826	.9830	.9834	.9838	.9842	.9846	.9850	.9854	.9857
2.2	.9861	.9864	.9868	.9871	.9875	.9878	.9881	.9884	.9887	.9890
2.3	.9893	.9896	.9898	.9901	.9904	.9906	.9909	.9911	.9913	.9916
2.4	.9918	.9920	.9922	.9925	.9927	.9929	.9931	.9932	.9934	.9936
2.5	.9938	.9940	.9941	.9943	.9945	.9946	.9948	.9949	.9951	.9952
2.6	.9953	.9955	.9956	.9957	.9959	.9960	.9961	.9962	.9963	.9964
2.7	.9965	.9966	.9967	.9968	.9969	.9970	.9971	.9972	.9973	.9974
2.8	.9974	.9975	.9976	.9977	.9977	.9978	.9979	.9979	.9980	.9981
2.9	.9981	.9982	.9982	.9983	.9984	.9984	.9985	.9985	.9986	.9986
3.0	.9987	.9987	.9987	.9988	.9988	.9989	.9989	.9989	.9990	.9990
3.1	.9990	.9991	.9991	.9991	.9992	.9992	.9992	.9992	.9993	.9993
3.2	.9993	.9993	.9994	.9994	.9994	.9994	.9994	.9995	.9995	.9995
3.3	.9995	.9995	.9996	.9996	.9996	.9996	.9996	.9996	.9996	.9997
3.4	.9997	.9997	.9997	.9997	.9997	.9997	.9997	.9997	.9998	.9998
3.5	.9998	.9998	.9998	.9998	.9998	.9998	.9998	.9998	.9998	.9998
3.6	.9998	.9999	.9999	.9999	.9999	.9999	.9999	.9999	.9999	.9999

t Distribution

one tailed test

degrees of freedom	P = .400	.300	.200	.100	.050	.020	.010
1	1.376	1.963	3.078	6.314	12.706	31.821	63.657
2	1.061	1.386	1.886	2.920	4.303	6.965	9.925
3	0.978	1.250	1.638	2.353	3.182	4.541	5.841
4	0.941	1.190	1.533	2.132	2.776	3.747	4.604
5	0.920	1.156	1.476	2.015	2.571	3.365	4.032
6	0.906	1.134	1.440	1.943	2.447	3.143	3.707
7	0.896	1.119	1.415	1.895	2.365	2.998	3.499
8	0.889	1.108	1.397	1.860	2.306	2.896	3.355
9	0.883	1.100	1.383	1.833	2.262	2.821	3.250
10	0.879	1.093	1.372	1.812	2.228	2.764	3.169
11	0.876	1.088	1.363	1.796	2.201	2.718	3.106
12	0.873	1.083	1.356	1.782	2.179	2.681	3.055
13	0.870	1.079	1.350	1.771	2.160	2.650	3.012
14	0.868	1.076	1.345	1.761	2.145	2.624	2.977
15	0.866	1.074	1.341	1.753	2.131	2.602	2.947
16	0.865	1.071	1.337	1.746	2.120	2.583	2.921
17	0.863	1.069	1.333	1.740	2.110	2.567	2.898
18	0.862	1.067	1.330	1.734	2.101	2.552	2.878
19	0.861	1.066	1.328	1.729	2.093	2.539	2.861
20	0.860	1.064	1.325	1.725	2.086	2.528	2.845
21	0.859	1.063	1.323	1.721	2.080	2.518	2.831
22	0.858	1.061	1.321	1.717	2.074	2.508	2.819
23	0.858	1.060	1.319	1.714	2.069	2.500	2.807
24	0.857	1.059	1.318	1.711	2.064	2.492	2.797
25	0.856	1.058	1.316	1.708	2.060	2.485	2.787
26	0.856	1.058	1.315	1.706	2.056	2.479	2.779
27	0.855	1.057	1.314	1.703	2.052	2.473	2.771
28	0.855	1.056	1.313	1.701	2.048	2.467	2.763
29	0.854	1.055	1.311	1.699	2.045	2.462	2.756
30	0.854	1.055	1.310	1.697	2.042	2.457	2.750
∞	0.842	1.036	1.282	1.645	1.960	2.326	2.576

P = probability of having t this large or larger in size by chance

Chi-Square Limits

χ^2 scores greater than these limits fail with the given probability of rejecting a valid sample.

d.f.	$\chi^2_{.99}$	$\chi^2_{.98}$	$\chi^2_{.95}$	$\chi^2_{.10}$	$\chi^2_{.05}$	$\chi^2_{.02}$	$\chi^2_{.01}$
1	.000157	.000628	.00393	2.706	3.841	5.412	6.635
2	.0201	.0404	.103	4.605	5.991	7.824	9.210
3	.115	.185	.352	6.251	7.815	9.837	11.341
4	.297	.429	.711	7.779	9.488	11.668	13.277
5	.554	.752	1.145	9.236	11.070	13.388	15.086
6	.872	1.134	1.635	10.645	12.592	15.033	16.812
7	1.239	1.564	2.167	12.017	14.067	16.622	18.475
8	1.646	2.032	2.733	13.362	15.507	18.168	20.090
9	2.088	2.532	3.325	14.684	16.919	19.679	21.666
10	2.558	3.059	3.940	15.987	18.307	21.161	23.209
11	3.053	3.609	4.575	17.275	19.675	22.618	24.725
12	3.571	4.178	5.226	18.549	21.026	24.054	26.217
13	4.107	4.765	5.892	19.812	22.362	25.472	27.688
14	4.660	5.368	6.571	21.064	23.685	26.873	29.141
15	5.229	5.985	7.261	22.307	24.996	28.259	30.578
16	5.812	6.614	7.962	23.542	26.296	29.633	32.000
17	6.408	7.255	8.672	24.769	27.587	30.995	33.409
18	7.015	7.906	9.390	25.989	28.869	32.346	34.805
19	7.633	8.567	10.117	27.204	30.144	33.687	36.191
20	8.260	9.237	10.851	28.412	31.410	35.020	37.566
21	8.897	9.915	11.591	29.615	32.671	36.343	38.932
22	9.542	10.600	12.338	30.813	33.924	37.659	40.289
23	10.196	11.293	13.091	32.007	35.172	38.968	41.638
24	10.856	11.992	13.848	33.196	36.415	40.270	42.980
25	11.524	12.697	14.611	34.382	37.652	41.566	44.314
26	12.198	13.409	15.379	35.563	38.885	42.856	45.642
27	12.879	14.125	16.151	36.741	40.113	44.140	46.963
28	13.565	14.847	16.928	37.916	41.337	45.419	48.278
29	14.256	15.574	17.708	39.087	42.557	46.693	49.588
30	14.953	16.306	18.493	40.256	43.773	47.962	50.892

For larger values of n, m $\sqrt{2\chi^2} - \sqrt{2n-1}$ may be used as a normal deviate with unit variance (the probability of χ^2 corresponds to one tail of the normal curve).

Kolmogorov-Smirnov Limits

Calculated scores above these limits fail with the given probability of rejecting a valid sample.

One-sided test $p = .90$	*.95*	*.975*	*.99*	*.995*		$p = .90$	*.95*	*.975*	*.99*	*.995*
Two-sided test $p = .80$	*.90*	*.95*	*.98*	*.99*		$p = .80$	*.90*	*.95*	*.98*	*.99*
n						*n*				
1	.900	.950	.975	.990	.995	21	.226	.259	.287	.321	.344
2	.684	.776	.842	.900	.929	22	.221	.253	.281	.314	.337
3	.565	.636	.708	.785	.829	23	.216	.247	.275	.307	.330
4	.493	.565	.624	.689	.734	24	.212	.242	.269	.301	.323
5	.447	.509	.563	.627	.669	25	.208	.238	.264	.295	.317
6	.410	.468	.519	.577	.617	26	.204	.233	.259	.290	.311
7	.381	.436	.483	.538	.576	27	.200	.229	.254	.284	.305
8	.358	.410	.454	.507	.542	28	.197	.225	.250	.279	.300
9	.339	.387	.430	.480	.513	29	.193	.221	.246	.275	.295
10	.323	.369	.409	.457	.489	30	.190	.218	.242	.270	.290
11	.308	.352	.391	.437	.468	31	.187	.214	.238	.266	.285
12	.296	.338	.375	.419	.449	32	.184	.211	.234	.262	.281
13	.285	.325	.361	.404	.432	33	.182	.208	.231	.258	.277
14	.275	.314	.349	.390	.418	34	.179	.205	.227	.254	.273
15	.266	.304	.338	.377	.404	35	.177	.202	.224	.251	.269
16	.258	.295	.327	.366	.392	36	.174	.199	.221	.247	.265
17	.250	.286	.318	.355	.381	37	.172	.196	.218	.244	.262
18	.244	.279	.309	.346	.371	38	.170	.194	.215	.241	.258
19	.237	.271	.301	.337	.361	39	.168	.191	.213	.238	.255
20	.232	.265	.294	.329	.352	40	.165	.189	.210	.235	.252

for $n > 40$, approximated by $\dfrac{1.07}{\sqrt{n}}$ $\dfrac{1.22}{\sqrt{n}}$ $\dfrac{1.36}{\sqrt{n}}$ $\dfrac{1.52}{\sqrt{n}}$ $\dfrac{1.63}{\sqrt{n}}$

Bibliography

Optimization

Lim, J.–M., B.–J. Yum, H. Hwang, and K.–S. Kim. Determination of an optimal configuration of operating policies for direct–input–output manufacturing systems using the Taguchi method. *Computers & Industrial Engineering*, 31(3,4), 1996, 555–560.

White, T. P., R. Toland, J. A. Jackson, Jr., and J. M. Kloeber, Jr. Simulation and optimization of a new waste remediation process. *Omega*, 24(6), 1996, 705–714.

Penkuhn, T., T. Spengler, H. Puchert, and O. Rentz. Environmental integrated production planning for the ammonia synthesis. *European Journal of Operational Research*, 97(2), 1997, 327–336.

Rubinstein, R. Y. Optimization of computer simulation models with rare events. *European Journal of Operational Research*, 99(1), 1997, 89–112.

Boender, G. C. E. A hybrid simulation/optimisation scenario model for asset/liability management. *European Journal of Operational Research*, 99(1), 1997, 126–135.

Futschik, A., and G. C. Pflug. Optimal allocation of simulation experiments in discrete stochastic optimization and approximative algorithms. *European Journal of Operational Research*, 101(2), 1997, 245–260.

Ramasesh, R. V., and M. D. Jayakumar. Inclusion of flexibility benefits in discounted cash flow analyses for investment evaluation: A simulation/optimization model. *European Journal of Operational Research*, 102(1), 1997, 124–141.

Sotskov, Y., N. Y. Sotskova, and F. Werner. Stability of an optimal schedule in a job shop. *Omega*, 25(4), 1997, 397–414.

Milosevic, Z., and C. Ponhofer. Refiner improves steam system with custom simulation/optimization package. *Oil & Gas Journal*, 95(34), 1997, 90–94.

Novomestky, F. A dynamic, globally diversified, index neutral synthetic asset allocation strategy. *Management Science*, 43(7), 1997, 998–1016.

Consiglio, A., and S. A. Zenios. A model for designing callable bonds and its solution using tabu search. *Journal of Economic Dynamics & Control*, 21(8,9), 1997, 1445–1470.

Li, S., and D. Tirupati. Impact of product mix flexibility and allocation policies on technology. *Computers & Operations Research*, 24(7), 1997, 611–626.

Pierreval, H., and L. Tautou. Using evolutionary algorithms and simulation for the optimization of manufacturing systems. *IIE Transactions*, 29(3), 1997, 181–189.

Fu, M. C., and K. J. Healy. Techniques for optimization via simulation: An experimental study on an (s,S) inventory system. *IIE Transactions*, 29(3), 1997, 191–199.

Fu, M. C., and S. D. Hill. Optimization of discrete event systems via simultaneous perturbation stochastic approximation. *IIE Transactions*, 29(3), 1997, 233–243.

Coley, K., S. Wright, E. Park, and C. Ntuen. Optimizing the usability of automated teller machines for older adults. *Computers & Industrial Engineering*, 33(1,2), 1997, 209–212.

Switek, W., and T. Majewski. Dynamic modeling and optimization for technology management. *Computers & Industrial Engineering*, 33(1,2), 1997, 11–14.

Hurrion, R. D. An example of simulation optimisation using a neural network metamodel: Finding the optimum number of kanbans in a manufacturing system. *Journal of the Operational Research Society*, 48(11), 1997, 1105–1112.

Hueter, J., and W. Swart. An integrated labor-management system for Taco Bell. *Interfaces*, 28:1, 1998, 75–91.

Lin, C.–R., and J. Buongiorno. Tree diversity, landscape diversity, and economics of maple-birch forests: Implications of Markovian models. *Management Science*, 44(10), 1998, 1351–1366.

Orman, A. Optimization of stochastic models: The interface between simulation and optimization. *Journal of the Operational Research Society*, 49(6), 1998, 675.

Ng, M. K. Heuristics approach to printed circuit board insertion problem. *Journal of the Operational Research Society*, 49(10), 1998, 1051–1059.

Mason, A. J., D. M. Ryan, and D. M. Panton. Integrated simulation, heuristic and optimisation approaches to staff scheduling. *Operations Research*, 46(2), 1998, 161–175.

Kamrani, A. K., K. Hubbard, H. R. Parsaei, and H. R. Leep. Simulation-based methodology for machine cell design. *Computers & Industrial Engineering*, 34(1), 1998, 173–188.

Kim, K. H., T. G. Kim, and K. H. Park. Hierarchical partitioning algorithm for optimistic distributed simulation of DEVS models. *Journal of Systems Architecture*, 44(6,7), 1998, 433–455.

Sexton, R. S., B. Alidaee, R. E. Dorsey, and J. D. Johnson. Global optimization for artificial neural networks: A tabu search application. *European Journal of Operational Research*, 106(2,3), 1998, 570–584.

Chen, X., W. Wan, and X. Xu. Modeling rolling batch planning as vehicle routing problem with time windows. *Computers & Operations Research*, 25(12), 1998, 1127–1136.

Liu, L., and X. Liu. Dynamic and static job allocation for multiserver systems. *IIE Transactions*, 30(9), 1998, 845–854.

Bashyam, S., and M. C. Fu. Optimization of (s,S) inventory systems with random lead times and a service level constraint. *Management Science*, 44(12 (Part 2)), 1998, S243–S256.

Chien, C.–F., and W.–T. Wu. A recursive computational procedure for container loading. *Computers & Industrial Engineering*, 35(1,2), 1998, 319–322.

Rossetti, M. D., and G. Clark. Evaluating a queueing approximation for the machine interference problem with two types of stoppages via simulation optimization. *Computers & Industrial Engineering*, 34(3), 1998, 655–668.

Duarte, A. M., Jr. Optimal value at risk hedge using simulation methods. *Derivatives Quarterly*, 5(2), 1998, 67–75.

Xie, X. Stability analysis and optimization of an inventory system with bounded orders. *European Journal of Operational Research*, 110(1), 1998, 126–149.

Greenwood, A. G., L. P. Rees, and F. C. Siochi. An investigation of the behavior of simulation response surfaces. *European Journal of Operational Research*, 110(2), 1998, 282–313.

S, S.–K., and E. del Castillo. Calculation of an optimal region of operation for dual response systems fitted from experimental data. *Journal of the Operational Research Society*, 50(8), 1999, 826–836.

Cooper, R., J. Haltiwanger, and L. Power. Machine replacement and the business cycle: Lumps and bumps. *American Economic Review*, 89(4), 1999, 921–946.

Kulturel, S., N. E. Ozemirel, C. Sepil, and Z. Bozkurt. Experimental investigation of shared storage assignment policies in automated storage/ retrieval systems. *IIE Transactions*, 31(8), 1999, 739–749.

Iassinovski, S. I., C. Raczy, and A. Artiba. Intelligent simulation based decision support environment. *Computers & Industrial Engineering*, 37(1,2), 1999, 227–230.

Lopez-Garcia, L., and A. Posada-Bolivar. A simulator that uses tabu search to approach the optimal solution to stochastic inventory models. *Computers & Industrial Engineering*, 37(1,2), 1999, 215–218.

Debuse, J. C. W., V. J. Rayward-Smith, and G. D. Smith. Parameter optimisation for a discrete event simulator. *Computers & Industrial Engineering*, 37(1,2), 1999, 181–184.

Kurian, T. K., and C. V. K. Reddy. Online production control using a genetic algorithm. *Computers & Industrial Engineering*, 37(1,2), 1999, 101–104.

Albino, V., and A. C. Garavelli. Limited flexibility in cellular manufacturing systems: A simulation study. *International Journal of Production Economics*, 60–61, 1999, 447–455.

Consiglio, A., and S. A. Zenios. Designing portfolios of financial products via integrated simulation and optimization models. *Operations Research*, 47(2), 1999, 195–208.

Azadivar, F., and G. Tompkins. Simulation optimization with qualitative variables and structural model changes: A genetic algorithm approach. *European Journal of Operational Research*, 113(1), 1999, 169–182.

Wierzbicki, A. P., and J. Granat. Multiobjective modeling for engineering applications: DIDASN++ system. *European Journal of Operational Research*, 113(2), 1999, 374–389.

Simpson, N. C. Multiple level production planning in rolling horizon assembly environments. *European Journal of Operational Research*, 114(1), 1999, 15–28.

Bertrand, J. W. M., and W. G. M. M. Rutten. Evaluation of three production planning procedures for the use of recipe flexibility. *European Journal of Operational Research*, 115(1), 1999, 179–194.

Heidergott, B. Optimisation of a single-component maintenance system: A smoothed perturbation analysis approach. *European Journal of Operational Research*, 119(1), 1999, 181–190.

Homem-de-Mello, T., A. Shapiro, and M. L. Spearman. Finding optimal material release times using simulation-based optimization. *Management Science*, 45(1), 1999, 86–102.

Guldmann, J.–M., and F. Wang. Optimizing the natural gas supply mix of local distribution utilities. *European Journal of Operational Research*, 112(3), 1999, 598–612.

Ho, Y.–C., C. G. Cassandras, C.–H. Chen, and L. Dai. Ordinal optimisation and simulation. *Journal of the Operational Research Society*, 51(4), 2000, 490–500.

Theory Simulation Experimental Design

Sinuany-Stern, Z. , I. David, and S. Biran. An efficient heuristic for a partially observable Markov decision process of machine replacement. *Computers & Operations Research*, 24(2), 1997, 117–126.

Caprihan, R., and S. Wadhwa. Impact of routing flexibility on the performance of an FMS—A simulation study. *International Journal of Flexible Manufacturing Systems*, 9(3), 1997, 273–298.

Lambert, S., B. Cyr, G. Abdul-Nour, and J. Drolet. Comparison study of scheduling rules and set-up policies for a SMT production line. *Computers & Industrial Engineering*, 33(1,2), 1997, 369–372.

Cyr, B., S. Lambert, G. Abdul-Nour, and R. Rochette. Manufacturing flexibility: SMT factors study. *Computers & Industrial Engineering*, 33(1,2), 1997, 361–364.

Avramidis, A. N., and J. R. Wilson. Correlation-induction techniques for estimating quantiles in simulation experiments. *Operations Research*, 46(4), 1998, 574–591.

Koksal, G., and Y. Fathi. Design of economical noise array experiments for a partially controlled simulation environment. *Computers & Industrial Engineering*, 35(3,4), 1998, 555–558.

Glasserman, P., P. Heidelberger, P. Shahabuddin, and T. Zajic. Multilevel splitting for estimating rare event probabilities. *Operations Research*, 47(4), 1999, 585–600.

Cheng, R. C. H., and J. P. C. Kleijnen. Improved design of queueing simulation experiments with highly heteroscedastic responses. *Operations Research*, 47(5), 1999, 672–777.

Hurrion, R. D., and S. Birgil. A comparison of factorial and random experimental design methods for the development of regression and neural network simulation metamodels. *Journal of the Operational Research Society*, 50(10), 1999, 1018–1033.

Skeels, C. L., and F. Vella. A Monte-Carlo investigation of the sampling behavior of conditional moment tests in tobit and probit models. *Journal of Econometrics*, 92(2), 1999, 275–294.

Metters, R., and V. Vargas. A comparison of production scheduling policies on costs, service level, and schedule changes. *Production & Operations Management*, 8(1), 1999, 76–91.

Fox, B. L. Separability in optimal allocation. *Operations Research*, 48(1), 2000, 173–176.

Random Number Generation

Mata-Toledo, R. A., and M. A. Willis. Visualization of random sequences using the chaos game algorithm. *Journal of Systems & Software*, 39(1), 1997, 3–6.

Amir, A., and E. Dar. An improved deterministic algorithm for generating different many-element random samples. *Information Processing Letters*, 62(2), 1997, 95–101.

Dion, J. P., and N. M. Yanev. Limit theorems and estimation theory for branching processes with an increasing random number of ancestors. *Journal of Applied Probability*, 34(2), 1997, 309–327.

Artikis, T. P., A. P. Voudouri, and J. I. Moshakis. A stochastic present value model in selecting risk management processes. *Journal of Applied Business Research*, 13(4), 1997, 119–125.

Kao, C., and H. C. Tang. Systematic searches for good multiple recursive random number generators. *Computers & Operations Research*, 24(10), 1997, 899–905.

Lurie, P. M., and M. S. Goldberg. An approximate method for sampling correlated random variables from partially-specified distributions. *Management Science*, 44(2), 1998, 203–218.

Avramidis, A. N., and J. R. Wilson. Correlation-induction techniques for estimating quantiles in simulation experiments. *Operations Research*, 46(4), 1998, 574–591.

L'Ecuyer, P. Good parameters and implementations for combined multiple recursive random number generators. *Operations Research*, 47(1), 1999, 159–164.

McCullough, B. D. Econometric software reliability: EViews, LIMDEP, SHAZAM and TSP. *Journal of Applied Econometrics*, 14(2), 1999, 191–202.

Ruhkin, A. L. Approximate entropy for testing randomness. *Journal of Applied Probability*, 37(1), 2000, 88–100.

Validation

Kozan, E. Increasing the operational efficiency of container terminals in Australia. *Journal of the Operational Research Society*, 48(2), 1997, 151–161.

Jayjock, M. A. Uncertainty analysis in the estimation of exposure. *American Industrial Hygiene Association Journal*, 58(5), 1997, 380–382.

Umamageswaran, K., K. Subramani, P. A. Wilsey, and P. Alexander. Formal verification and empirical analysis of rollback relaxation. *Journal of Systems Architecture*, 44(6,7), 1998, 473–495.

Schmidt, D. C., J. Haddock, S. Marchandon, G. Runger, et al. A methodology for formulating, formalizing, validating, and evaluating a real-time process control advisor. *IIE Transactions*, 30(3), 1998, 235–245.

Ashley, R. A new technique for postsample model selection and validation. *Journal of Economic Dynamics & Control*, 22(5), 1998, 647–665.

Kleijnen, J. P. C., B. Bettonvil, and W. Van Groenendaal. Validation of trace-driven simulation models: A novel regression test. *Management Science*, 44(6), 1998, 812–819.

Martin, E. IV. Centralized bakery reduces distribution costs using simulation. *Interfaces*, 28(4), 1998, 38–46.

Kleindorfer, G. B., L. O Neill, and R. Ganeshan. Validation in simulation: Various positions in the philosophy of science. *Management Science*, 44(8), 1998, 1087–1099.

Zenios, S., M. R. Holmer, R. McKendall, and C. Vassiadou-Zeniou. Dynamic models for fixed-income portfolio management under uncertainty. *Journal of Economic Dynamics & Control*, 22(10), 1998, 1517–1541.

Odijk, M. A. Sensitivity analysis of a railway station track layout with respect to a given timetable. *European Journal of Operational Research*, 112(3), 1999, 517–530.

Jacobson, S. H., and E. Yucesan. On the complexity of verifying structural properties of discrete event simulation models. *Operations Research*, 47(3), 1999, 476–481.

Kleijnen, J. P. C., and R. G. Sargent. A methodology for fitting and validating metamodels in simulation. *European Journal of Operational Research*, 120(1), 2000, 14–29.

Garcia, D. F., A. M. Campos, and A. Aguilar. Automatic auditing of the quality of environmental simulation tests in the aerospace industry. *Computers in Industry*, 42(1), 2000, 1–12.

Dengiz, B., and K. S. Akbay. Computer simulation of a PCB production line: Metamodeling approach. *International Journal of Production Economics*, 63(2), 2000, 195–205.

Phillips, M. R., and Marsh, D. T. The validation of fast-time air traffic simulations in practice. *Journal of the Operational Research Society*, 51(4), 2000, 457–464.

Variance Reduction

Fishman, V., P. Fitton, and Y. Galperin. Hybrid low-discrepancy sequences: Effective path reduction for yield curve scenario generation. *Journal of Fixed Income*, 7(1), 1997, 75–84.

Boyle, P., M. Broadie, and P. Glasserman. Monte-Carlo methods for security pricing. *Journal of Economic Dynamics & Control*, 21(8,9), 1997, 1267–1321.

Willard, G. A. Calculating prices and sensitivities for path-independent derivative securities in multifactor models. *Journal of Derivatives*, 5(1), 1997, 45–61.

Wang, P., I. M. Cockburn, and M. L. Puterman. Analysis of patent data—a mixed-poisson-regression-model approach. *Journal of Business & Economic Statistics*, 16(1), 1998, 27–41.

Usabel, M. Applications to risk theory of a Monte-Carlo multiple integration method. *Insurance Mathematics & Economics*, 23(1), 1998, 71–83.

Wang, C.–L. On the transient delays of M/G/1 queues. *Journal of Applied Probability*, 36(3), 1999, 882–893.

Srikant, R., and W. Whitt. Variance reduction in simulations of loss models. *Operations Research*, 47(4), 1999, 509–523.

Kleinman, N. L., J. C. Spall, and D. Q. Naiman. Simulation-based optimization with stochastic approximation using common random numbers. *Management Science*, 45(11), 1999, 1570–1578.

Henderson, S. G., and P. W. Glynn. Derandomizing variance estimators. *Operations Research*, 47(6), 1999, 907–916.

Statistical Interpretation

Zietz, J. Aggregate consumption with heterogeneous agents and a changing income distribution. *Atlantic Economic Journal*, 24(4), 1996, 361–370.

Hajivassiliou, V. A., and D. L. McFadden. The method of simulated scores for the estimation of LDV models. *Econometrica*, 66(4), 1998, 863–896.

Kaufman, D. E., and R. L. Smith. Direction choice for accelerated convergence in hit-and-run sampling. *Operations Research*, 46(1), 1998, 84–95.

Glasserman, P., P. Heidelberger, and P. Shahabuddin. Asymptotically optimal importance sampling and stratification for pricing path-dependent options. *Mathematical Finance*, 9(2), 1999, 117–152.

Kleijnen, J. P. C., and R. G. Sargent. A methodology for fitting and validating metamodels in simulation. *European Journal of Operational Research*, 120(1), 2000, 14–29.

Bootstrapping

Brockman, P., and M. Chowdhury. Deterministic versus stochastic volatility: Implications for option pricing models. *Applied Financial Economics*, 7(5), 1997, 499–505.

Horowitz, J. L. Bootstrap methods for covariance structures. *Journal of Human Resources*, 33(1), 1998, 39–61.

Lothgren, M., and M. Tambour. Testing scale efficiency in DEA models: A bootstrapping approach. *Applied Economics*, 31(10), 1999, 1231–1237.

Ferrall, C., and A. A. Smith, Jr. A sequential game model of sports championship series: Theory and estimation. *Review of Economics & Statistics*, 81(4), 1999, 704–719.

Systems Dynamics

Mandal, P., A. Howell, and A. S. Sohal. A systemic approach to quality improvements: The interactions between the technical, human and quality systems. *Total Quality Management*, 9(1), 1998, 79–100.

Rus, I., J. Collofello, and P. Lakey. Software process simulation for reliability management. *Journal of Systems & Software*, 46(2,3), 1999, 173–182.

Visual Simulation

Derrick, E. J., and O. Balci. Domino: A multifaceted conceptual framework for visual simulation modeling. *Infor*, 35(2), 1997, 93–120.

Hicks, D. A. Simulation market forces can't be ignored. *IIE Solutions*, 30(5), 1998, 18–19.

Ledlow, G. R., D. M. Bradshaw, and M. J. Perry. Animated simulation: A valuable decision support tool for practice improvement. *Journal of Healthcare Management*, 44(2), 1999, 91–102.

Hurrion, R. D. A sequential method for the development of visual interactive meta-simulation models using neural networks. *Journal of the Operational Research Society*, 51(6), 2000, 712–719.

Applications Queueing

Banks, J., and J. G. Dai. Simulation studies of multiclass queueing networks. *IIE Transactions*, 29(3), 1997, 213–219.

Resnick, S., and G. Samorodnitsky. Performance decay in a single server exponential queueing model with long range dependence. *Operations Research*, 45(2), 1997, 235–243.

Wang, K.–H., C.–H. Wang, and S. X. Bai. Cost analysis of the R-unloader queueing system. *Journal of the Operational Research Society*, 48(8), 1997, 810–817.

Duenyas, I., D. Gupta, and T. L. Olsen. Control of a single-server tandem queueing system with setups. *Operations Research*, 46(2), 1998, 218–230.

Rossetti, M. D., and G. Clark. Evaluating a queueing approximation for the machine interference problem with two types of stoppages via simulation optimization. *Computers & Industrial Engineering*, 34(3), 1998, 655–668.

Cheng, R. C. H., and J. P. C. Kleijnen. Improved design of queueing simulation experiments with highly heteroscedastic responses. *Operations Research*, 47(5), 1999, 762–777.

Inventory

Wilson, M. Supply-chain simulation for manufacturing logistics. *Logistics Focus*, 5(1), 1997, 2–3.

Ardalan, A. Analysis of local decision rules in a dual-kanban flow shop. *Decision Sciences*, 28(1), 1997, 195–211.

Rubinstein, R. Y. Optimization of computer simulation models with rare events. *European Journal of Operational Research*, 99(1), 1997, 89–112.

Akkan, C. Finite-capacity scheduling-based planning for revenue-based capacity management. *European Journal of Operational Research*, 100(1), 1997, 170–179.

Barnes-Schuster, D., and Y. Bassok. Direct shipping and the dynamic single-depot/multiretailer inventory system. *European Journal of Operational Research*, 101(3), 1997, 509–518.

Fritsche, S. R., and M. T. Dugan. A simulation-based investigation of errors in accounting-based surrogates for internal rate of return. *Journal of Business Finance & Accounting*, 24(6), 1997, 781–802.

Szendrovits, A. Z., and T. Szabados. On the least cost safety inventory for large transfer lines. *Omega*, 25(4), 1997, 483–487.

Guide, V. D. R., Jr., and R. Srivastava. Repairable inventory theory: Models and applications. *European Journal of Operational Research*, 102(1), 1997, 1–20.

Fu, M. C., and Healy, K. J. Techniques for optimization via simulation: An experimental study on an (s,S) inventory system. *IIE Transactions*, 29(3), 1997, 191–199.

Tzafestas, S., G. Kapsiotis, and E. Kyriannakis. Model-based predictive control for generalized production planning problems. *Computers in Industry*, 34(2), 1997, 201–210.

Mohan, R. P., and L. P. Ritzman. Planned lead times in multistage systems. *Decision Sciences*, 29(1), 1998, 163–191.

Kim, Y.–D., D.–H. Lee, J.–U. Kim, and H.–K. Roh. A simulation study on lot release control, mask scheduling, and batch scheduling in semiconductor wafer fabrication facilities. *Journal of Manufacturing Systems*, 17(2), 1998, 107–117.

Huang, M., D. Wang, and W. H. Dingwei. Simulation study of CONWIP for a cold rolling plant. *International Journal of Production Economics*, 54(3), 1998, 257–266.

Kellerer, H., V. Kotov, F. Rendl, and G. J. Woeginger. The stock size problem. *Operations Research*, 46(3 (Supplement)), 1998, S1–S12.

Moon, I., and S. Kang. Rationing policies for some inventory systems. *Journal of the Operational Research Society*, 49(5), 1998, 509–518.

Robb, D. J., and E. A. Silver. Inventory management with periodic ordering and minimum order quantities.

Journal of the Operational Research Society, 49(10), 1998, 1085–1094.

Xie, X. Stability analysis and optimization of an inventory system with bounded orders. *European Journal of Operational Research*, 110(1), 1998, 126–149.

Rutten, W. G. M. M., and J. W. M. Bertrand. Balancing stocks, flexible recipe costs and high service level requirements in a batch process industry: A study of a small scale model. *European Journal of Operational Research*, 110(3), 1998, 626–642.

Bashyam, S., and M. C. Fu. Optimization of (s,S) inventory systems with random lead times and a service level constraint. *Management Science*, 44(12 (Part 2)), 1998, S243–S256.

Strader, T. J., F.–R. Lin, and M. J. Shaw. The impact of information sharing on order fulfillment in divergent differentiation supply chains. *Journal of Global Information Management*, 7(1), 1999, 16–25.

Metters, R., and V. Vargas. A comparison of production scheduling policies on costs, service level, and schedule changes. *Production & Operations Management*, 8(1), 1999, 76–91.

Lopez-Garcia, L., and A. Posada-Bolivar. A simulator that uses tabu search to approach the optimal solution to stochastic inventory models. *Computers & Industrial Engineering*, 37(1,2), 1999, 215–218.

Enns, S. T. The effect of batch size selection on MRP performance. *Computers & Industrial Engineering*, 37(1,2), 1999, 15–19.

Pfohl, H.–C., O. Cullmann, and W. Stolzle. Inventory management with statistical process control: Simulation and evaluation. *Journal of Business Logistics*, 20(1), 1999, 100–120.

Allen, D. S. Aggregate dynamics of (S, S) inventory management. *International Journal of Production Economics*, 59(1-3), 1999, 231–242.

de Kok, T. G., and J. W. C. H. Visschers. Analysis of assembly systems with service level constraints. *International Journal of Production Economics*, 59(1-3), 1999, 313–326.

Diks, E. B., and A. G. de Kok. Computational results for the control of a divergent N-echelon inventory system. *International Journal of Production Economics*, 59 (1-3), 1999, 327–336.

Johansen, S. G. Lot sizing for varying degrees of demand uncertainty. *International Journal of Production Economics*, 59(1-3), 1999, 405–414.

Bartezzaghi, E., R. Verganti, and G. Zotteri. Measuring the impact of asymmetric demand distributions on inventories. *International Journal of Production Economics*, 60–61, 1999, 395–404.

Fisher, M. L., and C. D. Ittner. The impact of product variety on automobile assembly operations: Empirical evidence and simulation analysis. *Management Science*, 45(6), 1999, 771–786.

Kulturel, S., N. E. Ozemirel, C. Sepil, and Z. Bozkurt. Experimental investigation of shared storage assignment policies in automated storage/retrieval systems. *IIE Transactions*, 31(8), 1999, 739–749.

Alfredsson, P., and J. Verrijdt. Modeling emergency supply flexibility in a two-echelon inventory system. *Management Science*, 45(10), 1999, 1416–1431.

Taylor, L. J. A simulation study of work-in-process inventory drive systems and their effect on operational measurements. *British Journal of Management*, 11(1), 2000, 47–59.

Job Shop Scheduling

Sotskov, Y., N. Y. Sotskova, and F. Werner. Stability of an optimal schedule in a job shop. *Omega*, 25(4), 1997, 397–414.

Hodgson, T. J., D. Cormier, A. J. Weintraub, and A. Zozom, Jr. Note. Satisfying due dates in large job shops. *Management Science*, 44(10), 1998, 1442–1446.

Holthaus, O. Design of efficient job shop scheduling rules. *Computers & Industrial Engineering*, 33(1,2), 1997, 249–252.

Enns, S. T. Lead time selection and the behavior of work flow in job shops. *European Journal of Operational Research*, 109(1), 1998, 122–136.

Enns, S. T. Evaluating shop floor input control using rapid modeling. *International Journal of Production Economics*, 63(3), 2000, 229–241.

Manufacturing

Hurrion, R. D. An example of simulation optimisation using a neural network metamodel: Finding the optimum number of kanbans in a manufacturing system. *Journal of the Operational Research Society*, 48(11), 1997, 1105–1112.

Mollaghasemi, M., K. LeCroy, and M. Georgiopoulos. Application of neural networks and simulation modeling in manufacturing system design. *Interfaces*, 28(5), 1998, 100–114.

Fisher, M. L., and C. D. Ittner. The impact of product variety on automobile assembly operations: Empirical evidence and simulation analysis. *Management Science*, 45(6), 1999, 771–786.

Pfeil, G., R. Holcomb, C. T. Muir, and S. Taj. Visteon's sterling plant uses simulation-based decision support in training, operations, and planning. *Interfaces*, 30(1), 2000, 115–133.

AGVs

Mezgar, I., C. Egresits, and L. Monostori. Design and real-time reconfiguration of robust manufacturing systems by using design of experiments and artificial neural networks. *Computers in Industry*, 33(1), 1997, 61–70.

Hwang, H., and S. H. Kim. Development of dipatching rules for automated guided vehicle systems. *Journal of Manufacturing Systems*, 17(2), 1998, 137–143.

Bing, W. X. The application of analytic process of resource in an AGV scheduling. *Computers & Industrial Engineering*, 35(1,2), 1998, 169–172.

Ho, Y.–C. A dynamic-zone strategy for vehicle-collision prevention and load balancing in an AGV system with a single-loop guide path. *Computers in Industry*, 42(2,3), 2000, 159–176.

Supply Chain Management

Berry, D., and M. M. Naim. Quantifying the relative improvements of redesign strategies in a P.C. supply chain. *International Journal of Production Economics*, 46–47, 1996, 181–196.

Wilson, M. Supply-chain simulation for manufacturing logistics. *Logistics Focus*, 5(1), 1997, 2–3.

Closs, D. J., A. S. Roath, T. J. Goldsby, J. A. Eckert, and S. M. Swartz. An empirical comparison of anticipatory and response-based supply chain strategies. *International Journal of Logistics Management*, 9(2), 1998, 21–34.

Rogers, D. S. Simulation takes pain out of trial and error. *Transportation & Distribution*, 38(4), 1997, 84–86.

Bowman, B. IBM's new supply-chain simulator. *World Trade*, 11(2), 1998, 56.

Swaminathan, J. M., S. F. Smith, and N. M. Sadeh. Modeling supply chain dynamics: A multiagent approach. *Decision Sciences*, 29(3), 1998, 607–632.

Bhaskaran, S. Simulation analysis of a manufacturing supply chain. *Decision Sciences*, 29(3), 1998, 633–657.

Petrovic, D., R. Roy, and R. Petrovic. Modeling and simulation of a supply chain in an uncertain environment. *European Journal of Operational Research*, 109(2), 1998, 299–309.

Strader, T. J., F.–R. Lin, and M. J. Shaw. The impact of information sharing on order fulfillment in divergent differentiation supply chains. *Journal of Global Information Management*, 7(1), 1999, 16–25.

Larsen, E. R., J. D. W. Morecroft, and J. S. Thomsen. Complex behaviour in a production-distribution model. *European Journal of Operational Research*, 119(1), 1999, 61–74.

Petrovic, D., R. Roy, and R. Petrovic. Supply chain modeling using fuzzy sets. *International Journal of Production Economics*, 59(1-3), 1999, 443–453.

Mason-Jones, R., and D. R. Towill. Total cycle time compression and the agile supply chain. *International Journal of Production Economics*, 62(1,2), 1999, 61–73.

Lin, G., M. Ettl, S. Buckley, S. Bagchi, D. D. Yao, B. L. Naccarato, R. Allan, K. Kim, and L. Koenig. Extended-enterprise supply-chain management at IBM personal systems group and other divisions. *Interfaces*, 30(1), 2000, 7–25.

Rao, U., A. Scheller-Wolf, and S. Tayur. Development of a rapid-response supply chain at caterpillar. *Operations Research*, 48(2), 2000, 189–204.

Group Technology

Kannan, V. R., and S. Ghosh. A virtual cellular manufacturing approach to batch production. *Decision Sciences*, 27(3), 1996, 519–539.

Cyr, B., S. Lambert, G. Abdul-Nour, and R. Rochette. Manufacturing flexibility: SMT factors study. *Computers & Industrial Engineering*, 33(1,2), 1997, 361–364.

Castillo, A., H. Seifoddini, and J. Abell. The development of a cellular manufacturing system for automotive parts. *Computers & Industrial Engineering*, 33(1,2), 1997, 243–247.

Logendran, R., and D. Talkington. Analysis of cellular and functional manufacturing systems in the presence of machine breakdown. *International Journal of Production Economics*, 53(3), 1997, 239–256.

Hurley, S. F., and D. C. Whybark. Inventory and capacity trade-offs in a manufacturing cell. *International Journal of Production Economics*, 59(1–3), 1999, 203–212.

Arzi, Y., and L. Iaroslavitz. Neural network-based adaptive production control system for a flexible manufacturing cell under a random environment. *IIE Transactions*, 31(3), 1999, 217–230.

Kannan, V. R., and S. W. Palocsay. Cellular vs process layouts: An analytic investigation of the impact of learning on shop performance. *Omega*, 27(5), 1999, 583–592.

Monahan, G., and T. L. Smunt. Processes with nearly-sequential routings: A comparative analysis. *Journal of Operations Management*, 17(4), 1999, 449–466.

Shambu, G., and N. C. Suresh. Performance of hybrid cellular manufacturing systems: A computer simulation investigation. *European Journal of Operational Research*, 120(2), 2000, 436–458.

Maintenance

Hwang, H.–S. A performance evaluation model for FMS based on RAM and LCC using FACTOR/AIM. *Computers & Industrial Engineering*, 31(3,4), 1996, 593–598.

Sinuany-Stern, Z., I. David, and S. Biran. An efficient heuristic for a partially observable markov decision process of machine replacement. *Computers & Operations Research*, 24(2), 1997, 117–126.

Logendran, R., and D. Talkington. Analysis of cellular and functional manufacturing systems in the presence of machine breakdown. *International Journal of Production Economics*, 53(3), 1997, 239–256.

Bowles, G., J. S. Dagpunar, and H. Gow. Financial management of planned maintenance for housing associations. *Construction Management & Economics*, 15(4), 1997, 315–326.

Potti, K., and S. J. Mason. Using simulation to improve semiconductor manufacturing. *Semiconductor International*, 20(8), 1997, 289–292.

Guide, V. D. R., Jr., and R. Srivastava. Repairable inventory theory: Models and applications. *European Journal of Operational Research*, 102(1), 1997, 1–20.

Gharbi, A., J. Girard, R. Pellerin, and L. Villeneuve. Bombardier turned to simulation to validate the CF-18 maintenance program. *Interfaces*, 27(6), 1997, 22–34.

Colquhon, I., A. Menendez, and R. Dovico. Method yields safety factor for in-line inspection data. *Oil & Gas Journal*, 96(38), 1998, 83–86.

Asano, M., and H. Ohta. Single machine scheduling to meet due times under shutdown constraints. *International Journal of Production Economics*, 60–61, 1999, 537–547.

Ntuen, C., and E. H. Park. Simulation of crew size requirement in a maintained reliability system. *Computers & Industrial Engineering*, 37(1,2), 1999, 219–222.

Heidergott, B. Optimisation of a single-component maintenance system: A smoothed perturbation analysis approach. *European Journal of Operational Research*, 119(1), 1999, 181–190.

Quality

Sterman, J. D., N. P. Repenning, and F. Kofman. Unanticipated side effects of successful quality programs: Exploring a paradox of organizational improvement. *Management Science*, 43(4), 1997, 503–521.

Sepulveda, A., and J. A. Nachlas. A simulation approach to multivariate quality control. *Computers & Industrial Engineering*, 33(1,2), 1997, 113–116.

Schmidt, D. C., J. Haddock, S. Marchandon, G. Runger, et al. A methodology for formulating, formalizing, validating, and evaluating a real-time process control advisor. *IIE Transactions*, 30(3), 1998, 235–245.

Celano, G., and S. Fichera. Multiobjective economic design of an X control chart. *Computers & Industrial Engineering*, 37(1,2), 1999, 129–132.

Fu, M. C., and J.–Q. Hu. Efficient design and sensitivity analysis of control charts using Monte-Carlo simulation. *Management Science*, 45(3), 1999, 395–413.

Computer Systems

Dragoni, A. F. Distributed decision support systems under limited degrees of competence: A simulation study. *Decision Support Systems*, 20(1), 1997, 17–34.

Starostanko, O., A. S. Aguilar, and S. Lobato. Simulation facilities for RISC processors data flow and performance optimizations. *Computers & Industrial Engineering*, 33(1,2), 1997, 109–112.

Neuse, D. M. Why simulate? *Capacity Management Review*, 26(2), 1998, 1–7.

Peyravian, M., and A. D. Kshemkalyani. Decentralized network connection preemption algorithms. *Computer Networks & Isdn Systems*, 30(11), 1998, 1029–1043.

Olsen, D. H., and S. Ram. An empirical analysis of the object-oriented database concurrency control mechanism O2C2. *Journal of Database Management*, 10(2), 1999, 14–26.

Kovacs, G. L., S. Kopacsi, J. Nacsa, G. Haidegger, and P. Groumpos. Application of software reuse and object-oriented methodologies for the modelling and control of manufacturing systems. *Computers in Industry*, 39(3), 1999, 177–189.

Yang, W.–L. A distributed processing architecture for a remote simulation system in a multiuser environment. *Computers in Industry*, 40(1), 1999, 15–22.

Rus, I., J. Collofello, and P. Lakey. Software process simulation for reliability management. *Journal of Systems & Software*, 46(2,3), 1999, 173–182.

Vazquez-Abad, F. J., and L. G. Mason. Decentralized adaptive flow control of high-speed connectionless data networks. *Operations Research*, 47(6), 1999, 928–942.

Prakash, S., E. Deelman, and R. Bagrodia. Asynchronous parallel simulation of parallel programs. *IEEE Transactions on Software Engineering*, 26(5), 2000, 385–400.

Chen, Y., and M. Winslett. Automated tuning of parallel I/O systems: An approach to portable I/O performance for scientific applications. *IEEE Transactions on Software Engineering*, 26(4), 2000, 362–383.

De Silva, F. N., and R. W. Eglese. Integrating simulation modelling and GIS: Spatial decision support systems for evacuation planning. *Journal of the Operational Research Society*, 51(4), 2000, 423–430.

Artificial Intelligence

Kamps, J., and M. Masuch. Partial deductive closure: Logical simulation and management science. *Management Science*, 43(9), 1997, 1229–1245.

Moss, S., and B. Edmonds. A knowledge-based model of context-dependent attribute preferences for fast moving consumer goods. *Omega*, 25(2), 1997, 155–169.

LeBaron, B., W. B. Arthur, and R. Palmer. Time series properties of an artificial stock market. *Journal of Economic Dynamics & Control*, 23(9,10), 1999, 1487–1516.

Neural Networks

Cheng, R., T. Tozawa, M. Gen, H. Kato, and Y. Takayama. AE behaviors evaluation with BP neural network. *Computers & Industrial Engineering*, 31(3,4), 1996, 867–871.

Nath, R., B. Rajagopalan, and R. Ryker. Determining the saliency of input variables in neural network classifiers. *Computers & Operations Research*, 24(8), 1997, 767–773.

Ramirez-Beltran, N. D., and J. A. Montes. Neural networks for online parameter change detections in time series models. *Computers & Industrial Engineering*, 33(1,2), 1997, 337–340.

Deshpande, P. B., and S. S. Yerrapragada. Predict difficult-to-measure properties with neural analyzers. *Control Engineering*, 44(10), 1997, 55–56.

Vukadinovic, K., D. Teodorovic, and G. Pavkovic. A neural network approach to the vessel dispatching problem. *European Journal of Operational Research*, 102(3), 1997, 473–487.

Badiru, A. B., and D. B. Sieger. Neural network as a simulation metamodel in economic analysis of risky projects. *European Journal of Operational Research*, 105(1), 1998, 130–142.

Aiken, M. Competitive intelligence through neural networks. *Competitive Intelligence Review*, 10(1), 1999, 49–53.

Mollaghasemi, M., K. LeCroy, and M. Georgiopoulos. Application of neural networks and simulation modeling in manufacturing system design. *Interfaces*, 28(5), 1998, 100–123.

Schmidt, D. C., J. Haddock, S. Marchandon, G. Runger, et al. A methodology for formulating, formalizing, validating, and evaluating a real-time process control advisor. *IIE Transactions*, 30(3), 1998, 235–245.

Franses, P. H., and P. Van Homelen. On forecasting exchange rates using neural networks. *Applied Financial Economics*, 8(6), 1998, 589–596.

Hurrion, R. D., and Birgil, S. A comparison of factorial and random experimental design methods for the development of regression and neural network simulation metamodels. *Journal of the Operational Research Society*, 50(10), 1999, 1018–1033.

Arzi, Y., and L. Iaroslavitz. Neural network-based adaptive production control system for a flexible manufacturing cell under a random environment. *IIE Transactions*, 31(3), 1999, 217–230.

Hurrion, R. D. A sequential method for the development of visual interactive meta-simulation models using neural networks. *Journal of the Operational Research Society*, 51(6), 2000, 712–719.

Health

Astin, J., A. Stewart, and E. McIntosh. Economic evaluation of chronic conditions: Framework for a quasi-markov process. *Journal of the Operational Research Society*, 48(6), 1997, 623–628.

Gapenski, L. C., B. Langland-Orban, and G. Strack. The financial risk to hospitals inherent in DRG, per diem, and capitation reimbursement methodologies / practitioner response. *Journal of Healthcare Management*, 43(4), 1998, 323–338.

Van der Ploeg, C. P. B., C. Van Vliet, S. J. De Vlas, J. O. Ndinya-Achola, L. Fransen, G. J. Van Oortmarssen, and J. D. F. Haberma. STDSIM: A microsimulation model for decision support in STD control. *Interfaces*, 28(3), 1998, 84–100.

Bernstein, R. S., D. C. Sokal, S. T. Seitz, B. Auvert, et al. Simulating the control of a heterosexual HIV epidemic in a severely affected east African city. *Interfaces*, 28(3), 1998, 101–126.

Rossi, C., and G. Schinaia. The mover-stayer model for the HIV/AIDS epidemic in action. *Interfaces*, 28(3), 1998, 127–143.

Liu, L., and X. Liu. Block appointment systems for outpatient clinics with multiple doctors. *Journal of the Operational Research Society*, 49(12), 1998, 1254–1259.

Davies, R., and P. Roderick. Planning resources for renal services throughout UK using simulation. *European Journal of Operational Research*, 105(2), 1998, 285–295.

Ridge, J. C., S. K. Jones, M. S. Nielsen, and A. K. Shahani. Capacity planning for intensive care units. *European Journal of Operational Research*, 105(2), 1998, 346–355.

Gonzalez, C. J., M. Gonzalez, and N. M. Rios. Improving the quality of service in an emergency room using simulation-animation and total quality management. *Computers & Industrial Engineering*, 33(1,2), 1997, 97–100.

Mejia, A., R. Shirazi, R. Beech, and D. Balmer. Planning midwifery services to deliver continuity of care. *Journal of the Operational Research Society*, 49(1), 1998, 33–41.

Lehaney, B., S. A. Clarke, and R. J. Paul. A case of an intervention in an outpatients department. *Journal of the Operational Research Society*, 50(9), 1999, 877–891.

Wolfe, B. L. Poverty, children's health, and health care utilization. *Economic Policy Review*, 5(3), 1999, 9–21.

Eldabi, T., R. J. Paul, and S. J. E. Taylor. Computer simulation in healthcare decision making. *Computers & Industrial Engineering*, 37(1,2), 1999, 235–238.

Folster, S. Social insurance based on personal savings. *Economic Record*, 75(228), 1999, 5–18.

Mui, S.–L. Projecting coronary heart disease incidence and cost in Australia: Results from the incidence module of the cardiovascular disease policy model. *Australian & New Zealand Journal of Public Health*, 23(1), 1999, 11–19.

Buchanan, J. L., and M. S. Marquis. Who gains and loses with community rating for small business? *Inquiry*, 36(1), 1999, 30–43.

Ledlow, G. R., D. M. Bradshaw, and M. J. Perry. Animated simulation: A valuable decision support tool for practice improvement. *Journal of Healthcare Management*, 44(2), 1999, 91–102.

Eldabi, T., R. J. Paul, and S. J. E. Taylor. Simulating economic factors in adjuvant breast cancer treatment. *Journal of the Operational Research Society*, 51(4), 2000, 465–475.

Davies, R., S. Brailsford, P. Roderick, C. Canning, and D. Crabbe. Using simulation modeling for evaluating screening services for diabetic retinopathy. *Journal of the Operational Research Society*, 51(4), 2000, 476–484.

Mather, D. A simulation model of the spread of hepatitis C within a closed cohort. *Journal of the Operational Research Society*, 51(6), 2000, 656–665.

Environment

White, T. P., R. Toland, J. A. Jackson, Jr., and J. M. Kloeber, Jr. Simulation and optimization of a new waste remediation process. *Omega*, 24(6), 1996, 705–714.

Penkuhn, T., T. Spengler, H. Puchert, and O. Rentz. Environmental integrated production planning for the ammonia synthesis. *European Journal of Operational Research*, 97(2), 1997, 327–336.

Von Winterfeldt, D., and E. Schweitzer. An assessment of tritium supply alternatives in support of the US nuclear weapons stockpile. *Interfaces*, 28(1), 1998, 92–112.

Toland, R. J., J. M. Kloeber, Jr., and J. A. Jackson. A comparative analysis of hazardous waste remediation alternatives. *Interfaces*, 28(5), 1998, 70–85.

Butler, J., and D. L. Olson. Comparison of centroid and simulation approaches for selection sensitivity analysis, *Journal of Multicriteria Decision Analysis*, 8(3), 1999, 146–161.

Accounting

Viezer, T. W., and Young, M. S. Apples to oranges? The capitalization versus expensing debate and performance return comparisons. *Real Estate Finance*, 14(3), 1997, 64–74.

Horgan, J. M. Stabilising the sieve sample size using PPS. *Auditing—A Journal of Practice & Theory*, 16(2), 1997, 40–51.

Fritsche, S. R., and M. T. Dugan. A simulation-based investigation of errors in accounting-based surrogates for internal rate of return. *Journal of Business Finance & Accounting*, 24(6), 1997, 781–802.

Brabazon, T. Using simulation in business decisions. *Management Accounting—London*, 75(2), 1997, 36–38.

Chen, Y., and R. A. Leitch. The error detection of structural analytical procedures: A simulation study. *Auditing—A Journal of Practice & Theory*, 17(2), 1998, 36–70.

Hitzig, N. B. Detecting and estimating misstatement in two-step sequential sampling with probability proportional to size. *Auditing—A Journal of Practice & Theory*, 17(1), 1998, 54–68.

Weber, B. R. The valuation of contaminated land. *Journal of Real Estate Research*, 14(3), 1997, 379–398.

Hsu, J., X.–M. Wang, and C. Wu. The role of earnings information in corporate dividend decisions. *Management Science*, 44(12 (Part 2)), 1998, S173–S191.

Finance

Stambauch, F. Risk and value at risk. *European Management Journal*, 14(6), 1996, 612–621.

Fishman, V., P. Fitton, and Y. Galperin. Hybrid low-discrepancy sequences: Effective path reduction for yield curve scenario generation. *Journal of Fixed Income*, 7(1), 1997, 75–84.

Edison, H. J., J. E. Gagnon, and W. R. Melick. Understanding the empirical literature on purchasing power parity: The post-bretton woods era. *Journal of International Money & Finance*, 16(1), 1997, 1–17.

Brown, S. J., W. N. Goetzmann, R. G. Ibbotson, and S. A. Ross. Rejoinder: The J-shape of performance persistence given survivorship bias. *Review of Economics & Statistics*, 79(2), 1997, 167–170.

Fluck, Z., B. G. Malkiel, and R. E. Quandt. The predictability of stock returns: A cross-sectional simulation. *Review of Economics & Statistics*, 79(2), 1997, 176–183.

Novomestky, F. A dynamic, globally diversified, index neutral synthetic asset allocation strategy. *Management Science*, 43(7), 1997, 998–1016.

Boyle, P., M. Broadie, and P. Glasserman. Monte-Carlo methods for security pricing. *Journal of Economic Dynamics & Control*, 21(8,9), 1997, 1267–1321.

Bowles, G., J. S. Dagpunar, and H. Gow. Financial management of planned maintenance for housing associations. *Construction Management & Economics*, 15(4), 1997, 315–326.

Songer, A. D., J. Diekmann, and R. S. Pecsok. Risk analysis for revenue dependent infrastructure projects. *Construction Management & Economics*, 15(4), 1997, 377–382.

Artikis, T. P., A. P. Voudouri, and J. I. Moshakis. A stochastic present value model in selecting risk management processes. *Journal of Applied Business Research*, 13(4), 1997, 119–125.

Gupta, K. L., and R. Lensink. Financial repression and fiscal policy. *Journal of Policy Modeling*, 19(4), 1997, 351–373.

Li, K. Bayesian inference in a simultaneous equation model with limited dependent variables. *Journal of Econometrics*, 85(2), 1998, 387–400.

Franses, P. H., and P. Van Homelen. On forecasting exchange rates using neural networks. *Applied Financial Economics*, 8(6), 1998, 589–596.

Diakosavvas, D., and C. J. Green. Assessing the impact on food security of alternative compensatory financing schemes: A simulation approach with an application to India. *World Development*, 26(7), 1998, 1251–1265.

Young, M. R., and P. J. Lenk. Hierarchical bayes methods for multifactor model estimation and portfolio selection. *Management Science*, 44(11 (Part 2)), 1998, S111–S124.

Driscoll, J. C., and A. C. Kraay. Consistent covariance matrix estimation with spatially dependent panel data. *Review of Economics & Statistics*, 80(4), 1998, 549–560.

Esty, B. C. Improving techniques for valuing large-scale projects. *Journal of Project Finance*, 5(1), 1999, 9–25.

Bers, J. A., G. S. Lynn, and C. Spurling. A computer simulation model for emerging technology business planning and forecasting. *International Journal of Technology Management*, 18(1,2), 1999, 31–45.

Kemp, A. G., and L. Stephen. Risk:reward sharing contracts in the oil industry: The effects of bonus:penalty schemes. *Energy Policy*, 27(2), 1999, 111–120.

Lu, Y.–C., S. Wu, D.–S. Chen, Dar-Shin, and Y.–Y. Lin. BOT projects in Taiwan: Financial modeling risk, term structure of net cash flows, and project at risk analysis. *Journal of Project Finance*, 5(4), 2000, 53–63.

Shell, J. Using options after FAS 133. *Corporate Finance*, (185), 2000, 36–39.

Marketing

Moss, S., and B. Edmonds. A knowledge-based model of context-dependent attribute preferences for fast moving consumer goods. *Omega*, 25(2), 1997, 155–169.

Franses, P. H., and A. B. Koehler. A model selection strategy for time series with increasing seasonal variation. *International Journal of Forecasting*, 14(3), 1998, 405–414.

Chiang, J., S. Chib, and C. Narasimhan. Markov chain Monte-Carlo and models of consideration set and parameter heterogeneity. *Journal of Econometrics*, 89(1/2), 1999, 223–248.

Silva-Risso, J. M., R. E. Bucklin, and D. G. Morrison. A decision support system for planning manufacturers' sales promotion calendars. *Marketing Science*, 18(3), 1999, 274–300.

Chiang, J., S. Chib, and C. Narasimhan. Markov chain Monte-Carlo and models of consideration set and parameter heterogeneity. *Journal of Econometrics*, 89(1/2), 1999, 223–248.

Project Management

Diekmann, J., D. Featherman, R. Moody, K. Molenaar, and M. Rodriguez-Guy. Project cost risk analysis using influence diagrams. *Project Management Journal*, 27(4), 1996, 23–30.

Rho, J.–J., and H. B. Kim. SIMBASE: An economic justification tool using project-based simulation. *Computers & Industrial Engineering*, 31(3,4), 1996, 737–741.

Thomas, P. R., and S. Salhi. An investigation into the relationship of heuristic performance with network-resource characteristics. *Journal of the Operational Research Society*, 48(1), 1997, 34–43.

Golenko-Ginzburg, D., and A. Gonik. Stochastic network project scheduling with nonconsumable limited resources. *International Journal of Production Economics*, 48(1), 1997, 29–37.

Titarenko, B. P. Robust technology in risk management. *International Journal of Project Management*, 15(1), 1997, 11–14.

Golenko-Ginzburg, D., and A. Gonik. Online control model for network construction projects. *Journal of the Operational Research Society*, 48(2), 1997, 175–183.

Yang, K.–K., and C.–C. Sum. An evaluation of due date, resource allocation, project release, and activity scheduling rules in a multiproject environment. *European Journal of Operational Research*, 103(1), 1997, 139–154.

Amirkhalili, R. Risk and capital budgeting. *Aace Transactions*, 1997, 80–83.

Ponce de Leon, G., and J. R. Knoke. Practical applications of probabilistic scheduling. *Aace Transactions*, 1997, 307–310.

Robinson, S., and M. Pidd. Provider and customer expectations of successful simulation projects. *Journal of the Operational Research Society*, 49(3), 1998, 200–209.

Badiru, A. B., and D. B. Sieger. Neural network as a simulation metamodel in economic analysis of risky projects. *European Journal of Operational Research*, 105(1), 1998, 130–142.

Basu, A. Practical risk analysis in scheduling. *Aace Transactions*, 1998, R1–R4.

Dawson, R. J., and C. W. Dawson. Practical proposals for managing uncertainty and risk in project planning. *International Journal of Project Management*, 16(5), 1998, 299–310.

Yang, K.–K. A comparison of dispatching rules for executing a resource-constrained project with estimated activity durations. *Omega*, 26(6), 1998, 729–738.

Fernandez, A. A., R. L. Armacost, and J. J. Pet-Edwards. Understanding simulation solutions to resource constrained project scheduling problems with stochastic task durations. *Engineering Management Journal*, 10(4), 1998, 5–13.

Bhuiyan, N., and V. Thomson. The use of continuous approval methods in defence acquisition projects. *International Journal of Project Management*, 17(2), 1999, 121–129.

Levitt, R. E., J. Thomsen, T. R. Christiansen, J. C. Kunz, J. Yan, and C. Nass. Simulating project work processes and organizations: Toward a micro-contingency theory of organizational design. *Management Science*, 45(11), 1999, 1479–1495.

Decision Analysis

Bielza, C., P. Muller, and D. R. Insua. Decision analysis by augmented probability simulation. *Management Science*, 45(7), 1999, 995–1007.

Butler, J., and D. L. Olson. Comparison of centroid and simulation approaches for selection sensitivity analysis. *Journal of Multicriteria Decision Analysis*, 8(3), 1999, 146–161.

Fischer, G. W., J. Jia, and M. F. Luce. Attribute conflict and preference uncertainty: The randMAU model. *Management Science*, 46(5), 2000, 669–684.

Software Products

Vandersluis, C. Risk analysis software: A definite safe bet. *Computing Canada*, 23(13), 1997, 27.

Petropoulakis, L., and L. Giacomini. Development of a hybrid simulator for manufacturing processes. *Computers in Industry*, 36(1,2), 1998, 117–124.

Bowden, R. The spectrum of simulation software. *IIE Solutions*, 30(5), 1998, 44–54.

Duffy, J. Simulation tools are as varied as the LANs they model. *Network World*, 15(8), 1998, 48–49.

Narayanan, S., D. A. Bodner, U. Sreekanth, T. Govindaraj, et al. Research in object-oriented manufacturing simulations: An assessment of the state of the art. *IIE Transactions*, 30(9), 1998, 795–810.

Deitz, D. Design optimization. *Mechanical Engineering*, 120(10), 1998, 24.

Chin, K. Making the most of your plant. *Chemical Engineering*, 106(3), 1999, 139–142.

Vasilash, G. S. The process is the key to the future. *Automotive Manufacturing & Production*, 111(4), 1999, 38–39.

Cruz, R. L. Communications networks simulated. *IEEE Spectrum*, 36(8), 1999, 86.

Index

SITE LICENSE AGREEMENT AND LIMITED WARRANTY

READ THIS LICENSE CAREFULLY BEFORE USING THIS PACKAGE. BY USING THIS PACKAGE, YOU ARE AGREEING TO THE TERMS AND CONDITIONS OF THIS LICENSE. IF YOU DO NOT AGREE, DO NOT USE THE PACKAGE. PROMPTLY RETURN THE UNUSED PACKAGE AND ALL ACCOMPANYING ITEMS TO THE PLACE YOU OBTAINED. *THESE TERMS APPLY TO ALL LICENSED SOFTWARE ON THE DISK EXCEPT THAT THE TERMS FOR USE OF ANY SHAREWARE OR FREEWARE ON THE DISKETTES ARE AS SET FORTH IN THE ELECTRONIC LICENSE LOCATED ON THE DISK:*

1. GRANT OF LICENSE and OWNERSHIP: The enclosed computer programs and data ("Software") are licensed, not sold, to you by Prentice-Hall, Inc. ("We" or the "Company") in consideration of your purchase or adoption of the accompanying Company textbooks and/or other materials, and your agreement to these terms. We reserve any rights not granted to you. You own only the disk(s) but we and/or licensors own the Software itself. This license allows you to use and display the enclosed copy of the Software on individual computers in the computer lab designated for this course at a single campus or branch or geographic location of an educational institution, for academic use only, so long as you comply with the terms of this Agreement.

2. RESTRICTIONS: You may <u>not</u> transfer or distribute the Software or documentation to anyone else. You may <u>not</u> copy the documentation or the Software. You may <u>not</u> reverse engineer, disassemble, decompile, modify, adapt, translate, or create derivative works based on the Software or the Documentation. You may be held legally responsible for any copying or copyright infringement which is caused by your failure to abide by the terms of these restrictions.

3. TERMINATION: This license is effective until terminated. This license will terminate automatically without notice from the Company if you fail to comply with any provisions or limitations of this license. Upon termination, you shall destroy the Documentation and all copies of the Software. All provisions of this Agreement as to limitation and disclaimer of warranties, limitation of liability, remedies or damages, and our ownership rights shall survive termination.

4. LIMITED WARRANTY AND DISCLAIMER OF WARRANTY: Company warrants that for a period of 60 days from the date you purchase this Software (or purchase or adopt the accompanying textbook), the Software, when properly installed and used in accordance with the Documentation, will operate in substantial conformity with the description of the Software set forth in the Documentation, and that for a period of 30 days the disk(s) on which the Software is delivered shall be free from defects in materials and workmanship under normal use. The Company does <u>not</u> warrant that the Software will meet your requirements or that the operation of the Software will be uninterrupted or error-free. Your only remedy and the Company's only obligation under these limited warranties is, at the Company's option, return of the disk for a refund of any amounts paid for it by you or replacement of the disk. THIS LIMITED WARRANTY IS THE ONLY WARRANTY PROVIDED BY THE COMPANY AND ITS LICENSORS, AND THE COMPANY AND ITS LICENSORS DISCLAIM ALL OTHER WARRANTIES, EXPRESS OR IMPLIED, INCLUDING WITHOUT LIMITATION, THE IMPLIED WARRANTIES OF MERCHANTABILITY AND FITNESS FOR A PARTICULAR PURPOSE. THE COMPANY DOES NOT WARRANT, GUARANTEE OR MAKE ANY REPRESENTATION REGARDING THE ACCURACY, RELIABILITY, CURRENTNESS, USE, OR RESULTS OF USE, OF THE SOFTWARE.

5. LIMITATION OF REMEDIES AND DAMAGES: IN NO EVENT, SHALL THE COMPANY OR ITS EMPLOYEES, AGENTS, LICENSORS, OR CONTRACTORS BE LIABLE FOR ANY INCIDENTAL, INDIRECT, SPECIAL, OR CONSEQUENTIAL DAMAGES ARISING OUT OF OR IN CONNECTION WITH THIS LICENSE OR THE SOFTWARE, INCLUDING FOR LOSS OF USE, LOSS OF DATA, LOSS OF INCOME OR PROFIT, OR OTHER LOSSES, SUSTAINED AS A RESULT OF INJURY TO ANY PERSON, OR LOSS OF OR DAMAGE TO PROPERTY, OR CLAIMS OF THIRD PARTIES, EVEN IF THE COMPANY OR AN AUTHORIZED REPRESENTATIVE OF THE COMPANY HAS BEEN ADVISED OF THE POSSIBILITY OF SUCH DAMAGES. IN NO EVENT SHALL THE LIABILITY OF THE COMPANY FOR DAMAGES WITH RESPECT TO THE SOFTWARE EXCEED THE AMOUNTS ACTUALLY PAID BY YOU, IF ANY, FOR THE SOFTWARE OR THE ACCOMPANYING TEXTBOOK. SOME JURISDICTIONS DO NOT ALLOW THE LIMITATION OF LIABILITY IN CERTAIN CIRCUMSTANCES, THE ABOVE LIMITATIONS MAY NOT ALWAYS APPLY.

6. GENERAL: THIS AGREEMENT SHALL BE CONSTRUED IN ACCORDANCE WITH THE LAWS OF THE UNITED STATES OF AMERICA AND THE STATE OF NEW YORK, APPLICABLE TO CONTRACTS MADE IN NEW YORK, AND SHALL BENEFIT THE COMPANY, ITS AFFILIATES AND ASSIGNEES. This Agreement is the complete and exclusive statement of the agreement between you and the Company and supersedes all proposals, prior agreements, oral or written, and any other communications between you and the company or any of its representatives relating to the subject matter. If you are a U.S. Government user, this Software is licensed with "restricted rights" as set forth in subparagraphs (a)-(d) of the Commercial Computer-Restricted Rights clause at FAR 52.227-19 or in subparagraphs (c)(1)(ii) of the Rights in Technical Data and Computer Software clause at DFARS 252.227-7013, and similar clauses, as applicable.

Should you have any questions concerning this agreement or if you wish to contact the Company for any reason, please contact in writing:

Director, Media Production

Pearson Education

1 Lake Street

Upper Saddle River, NJ 07458